5th International Conference New Achievements in Materials and Environmental Science (NAMES'16)

Journal of Physics: Conference Series Volume 879

Nancy, France
7 – 9 November 2016

ISBN: 978-1-5108-4784-2
ISSN: 1742-6588

Printed from e-media with permission by:

Curran Associates, Inc.
57 Morehouse Lane
Red Hook, NY 12571

Some format issues inherent in the e-media version may also appear in this print version.

Printed by Curran Associates, Inc. (2017)

For permission requests, please contact the Institute of Physics
at the address below.

Institute of Physics
Dirac House, Temple Back
Bristol BS1 6BE UK

Phone: 44 1 17 929 7481
Fax: 44 1 17 920 0979

techtracking@iop.org

Additional copies of this publication are available from:

Curran Associates, Inc.
57 Morehouse Lane
Red Hook, NY 12571 USA
Phone: 845-758-0400
Fax: 845-758-2633
Email: curran@proceedings.com
Web: www.proceedings.com

Table of contents

Volume 879

5th International Conference New Achievements in Materials and Environmental Science (NAMES'16)

7–9 November 2016, Nancy, France

Accepted papers received: 21 July 2017
Published online: 29 August 2017

Preface

Keynotes

Functional materials

Materials for environment

Natuaral resources and geomaterials

Continuous media mechanics

NAMES'16

Organised by

The University of Lorraine
The National University of Science and Technology "MISIS"

Supported by
The Ministry of education and research of Perm Krai, Russsia

Chair

L.O. Filippov,
Université de Lorraine
National University of sciences and technology MISIS

International Organising and Scientific Comittees

M.Aillerie, Université de Lorraine
A. Celzard, Université de Lorraine
X Deveaux, Université de Lorraine
S.Farrokhpay, Université de Lorraine and University of Queensland
I.V. Filippova, Université de Lorraine
D. Fornasiero, Future Institiute of Ressources, University of South Australia
T. Lyubimova, Institut of continouis media mechanics, Russian Academy of sciences
S. Nikulin, National University of sciences and technology MISIS
V. Petrov, Institute of Geology of Ore Deposits, Petrography, Mineralogy and Geochemistry (IGEM) Russian Academy of Sciences
D. Roizard, Université de Lorraine

Conference organised with the GeoRessources Laboratory in collaboration with Institut Jean Lamour, LMOPS and LRGP Laboratories

Sponsored by

Grand Est
ALSACE CHAMPAGNE-ARDENNE LORRAINE

ARCUS

NAMES a multidisciplinary Conference series

The NAMES conferences traditionally aim at reviewing and discussing current developments in the field of Materials Science, Natural resources and environmental sciences by researchers from Russia and Lorraine Region of France. More specifically and as set up by the organizers of the Conferences (*Université de Lorraine* and *National University of Sciences and Technology "MISIS"*), the Seminar's goals are the following:

- development of scientific and academic contacts by giving a new impulse to bring together fundamental research and technology transfer;
- development and consolidation of scientific, technical and business collaboration between regions of Russia and Lorraine through direct contacts between the universities, institutes and companies involved.

The First Seminar took place on October 27-29, 2004 at the « Institut National Polytechnique de Lorraine » (on the premises of the « Ecole Européenne d'Ingénieurs en Génie des Matériaux », Nancy, France). The first Seminar demonstrated a high quality and diversity of oral presentations and posters responding to highest international standards. Indeed, the first Seminar gathered 30 oral presentations and 72 posters, presented by:

- 19 participants from 5 Universities and 3 Institutes of the Russian Academy of Science,
- Participants from 11 laboratories of 3 Universities from the Lorraine region,
- Three industrial companies, including European Aeronautic Defense and Space Company - EADS, and ANVAR ("Agence nationale de valorisation de la recherché").

The 75th Anniversary of Moscow Institute of Steel and Alloys (MISIS) had been the occasion to set up the Second Seminar in Moscow, Russia, on November 10-12, 2005. The efficiency of scientific partnership between the groups of researchers from both Russia and France and initiated during the First Seminar had been brilliantly demonstrated during this Second Seminar. As a consequence of the presentations high standards and the Seminar widest organisation, the Second Seminar doubtlessly met the main standards of an international conference.
Reviews of the state-of-the-art developments in Materials Science were given by leading scientists from Moscow and from the Lorraine region. The three days long seminar was structured around four main topics:

- Functional Materials;
- Coatings, Films and Surface Engineering,
- Nanomaterials and Nanotechnologies,
- Environment;

and three Round Tables:

- defining practical means of carrying out Franco-Russian collaborations in technology transfer and innovation;
- Materials Science;
- ARCUS: Lorraine-Russian collaboration in materials science and environment.

The event totalised 32 oral and 25 poster reports spread between the four abovementioned topics and set out by 110 participants.

NAMES 2007, the 3rd France-Russia Confrence on "New Achievements in Materials and Environmental Sciences" took place in Metz (France) on November 7-9, 2007. The conference emphasized fundamentals and development of the five main themes linked to the Lorraine –Russia ARCUS project with possible extension to other topics.

The five mains subjects included in Arcus Project are the following:

- Bulk-Surface-Interface material sciences,

- Nanomaterials and nanotechnologies,
- Environment and Natural resources,
- Plasma physics —ITER Project, and
- Vibrational dynamics.

The 3rd conference totalised 46 oral and 52 poster presentations during 3 days with 154 participants.

The First, the Second and Third NAMES Seminars were financially supported by the following organisations

- Ambassade de France à Moscou and Région Lorraine,
- Communauté Urbaine du Grand Nanc and Conseil Général de Meurthe et Moselle
- Institut National Polytechnique de Lorraine, Université de Metz, Université Henry Poincaré
- CNRS and ANVAR
- Federal Agency on Science and Innovations of the Ministry of Education and Science of Russian Federation and Moscow Committee on Science and Technologies;
- National university of sciences and technology "MISIS".

The 4th NAMES Conference was supported by the Ministry of Foreign Affairs of France and the Lorraine Region Council.

The Conferences set out new perspectives for further research and stimulated the opportunities for enhanced cooperation between scientists from Lorraine and Russia.
The participants of the Conferences reviewed the remarkable worldwide progress with numerous breakthroughs in areas of basic research and industrial applications, particularly in the fields of nanomaterials and nanotechnologies, surface engineering, biomaterials and multifunctional coatings, functionally graded materials, new materials for microelectronics and optics, nanostructured thin films and nanodispersion strengthening coatings, combustion synthesis, new micro- and nanosystems and device, natural resources environmental sciences, clean technology, and recently, natural fibrous materials etc.

Lev Filippov,
Coordinator of NAMES conferences

Peer review statement

All papers published in this volume of *Journal of Physics: Conference Series* have been peer reviewed through processes administered by the proceedings Editors. Reviews were conducted by expert referees to the professional and scientific standards expected of a proceedings journal published by IOP Publishing.

IOP Conf. Series: Journal of Physics: Conf. Series **879** (2017) 012001 doi:10.1088/1742-6596/879/1/012001

Investigation of domain walls in PPLN by confocal raman microscopy and PCA analysis

Vladimir Ya. Shur1, Pavel Zelenovskiy1, Patrice Bourson2
[1]Institute of Natural Sciences, Ural Federal University, 51 Lenin Ave., Ekaterinburg, Russie
[2]Laboratoire Matériaux Optiques, Photonique et Systèmes, Université de Lorraine,4 Bd. Arago, Metz, France

Corresponding author: vladimir.shur@urfu.ru

Abstract

Confocal Raman microscopy (CRM) is a powerful tool for investigation of ferroelectric domains. Mechanical stresses and electric fields existed in the vicinity of neutral and charged domain walls modify frequency, intensity and width of spectral lines [1], thus allowing to visualize micro- and nanodomain structures both at the surface and in the bulk of the crystal [2,3]. Stresses and fields are naturally coupled in ferroelectrics due to inverse piezoelectric effect and hardly can be separated in Raman spectra. PCA is a powerful statistical method for analysis of large data matrix providing a set of orthogonal variables, called principal components (PCs).

PCA is widely used for classification of experimental data, for example, in crystallization experiments, for detection of small amounts of components in solid mixtures etc. [4,5]. In Raman spectroscopy PCA was applied for analysis of phase transitions and provided critical pressure with good accuracy [6].

In the present work we for the first time applied Principal Component Analysis (PCA) method for analysis of Raman spectra measured in periodically poled lithium niobate (PPLN). We found that principal components demonstrate different sensitivity to mechanical stresses and electric fields in the vicinity of the domain walls. This allowed us to separately visualize spatial distribution of fields and electric fields at the surface and in the bulk of PPLN.

References

[1] P.S. Zelenovskiy, V.Ya. Shur, P. Bourson, et al., *Ferroelectrics* **398**, 34 (2010).
[2] P.S. Zelenovskiy, M.D. Fontana, V.Ya. Shur, et al., *Appl. Phys. A* **99**, 741 (2010).
[3] V.Ya. Shur, P.S. Zelenovskiy, *J. Appl. Phys.* **116**, 066802 (2014).
[4] J.A. Falcon, K.A. Berglund, *Cryst. Growth Des.* **4**, 457 (2004).
[5] A.G. Ryder, G.M. O'Connor, T.J. Glynn, *J. Raman Spectrosc.* **31**, 221 (2000).
[6] R.Y. Sato-Berru, E.V. Mejia-Uriarte et al., *Spectrochim. Acta A* **66**, 557 (2007).

Innovations in the flotation of fine and coarse particles

D Fornasiero[1,2] and L O Filippov[2,3]

[1]Future Industries Institute, University of South Australia, Mawson Lakes SA 5095 Australia
[2]Université de Lorraine, Laboratoire GeoRessources laboratory UMR 7359 CNRS-CREGU, ENSG, 2 rue du Doyen Roubault, BP 40, F-54501, Vandœuvre-lès-Nancy, France
[3]National University of Science and Technology, MISIS, Moscow, Russia

Corresponding author: daniel.fornasiero@unisa.edu.au

Abstract. Research on the mechanisms of particle-bubble interaction has provided valuable information on how to improve the flotation of fine (<20 μm) and coarse particles (>100 μm) with novel flotation machines which provide higher collision and attachment efficiencies of fine particles with bubbles and lower detachment of the coarse particles. Also, new grinding methods and technologies have reduced energy consumption in mining and produced better mineral liberation and therefore flotation performance.

1. Introduction

Flotation is an efficient and economical technique to separate valuable minerals from gangue minerals in an ore in the particle size range 10-500 μm. As the rich ore bodies are depleted, miners have to process large volumes of rocks to make the process viable, which can be easily handled by flotation by simply increasing the size of flotation cells (~700 m^3). The separation by flotation is based on the difference in surface hydrophobicity of particles as they collide with gas bubbles (0.4-1 mm) in aqueous suspension in a flotation cell and only the hydrophobic particles attach to bubbles and rise to the top where they are recovered in the concentrate while the hydrophilic particles are sent to the tailings for further treatment. As most of the minerals are not naturally hydrophobic, short chain surfactants/collectors are added to selectively attach on and increase the surface hydrophobicity of the targeted minerals (usually the value minerals) while hydrophilic reagents, such as polymers, are added to depress the flotation of the unwanted minerals, and therefore further increase the mineral separation. The flotation process involves the interaction between particles and bubbles which can be described by the three consecutive and independent sub-processes of particle-bubble collision, attachment and stability. The flotation rate constant and therefore the recovery is the product of the probability or efficiency of these sub-processes, particle-bubble collision (E_c), attachment (E_a) and stability (E_s=1-E_d), multiplied by the particle-bubble collision frequency [1]. Although the flotation outcome can be controlled by chemical factors (such as reagents type and concentration, pulp pH and Eh) which mainly affect the particle-bubble attachment/adhesion (E_a), flotation also depends on the hydrodynamic conditions in flotation cells such as agitation/turbulence which affects the particle-bubble collision frequency and detachment efficiency (E_d). Of course, particle-bubble collision frequency and therefore flotation rate and recovery depend on particles and bubbles number (wt% solids and gas flow rate, respectively) and their size while the intensity of agitation/turbulence affects the velocity and trajectory of particles and bubbles. A high agitation/turbulence in the flotation cell is

IOP Conf. Series: Journal of Physics: Conf. Series **879** (2017) 012002 doi:10.1088/1742-6596/879/1/012002

essential for maximum particle-bubble collision frequency; however, a high solids percent and/or attractive interaction between fine particles will increase the pulp viscosity resulting in decreased turbulent energy dissipation rate, which in turn has a negative effect on particle-bubble collision frequency. This was clearly demonstrated by Schubert [2] who showed that turbulence damping (decrease in the turbulent energy dissipation rate) occurs with increased volume percent of particles, especially with very fine particles (<1 μm) and their attraction. Addition of dispersants was found to restore the level of turbulence and improve the rheological properties of the suspension. A recent study has shown that a high feed solids percent of molybdenite particles decreased the flotation recovery of molybdenite but the addition of a dispersant restored molybdenite recovery [3]. It is likely that the high concentration of molybdenite particles and attraction between them have produced turbulence damping in the flotation cell which results in lower particle-bubble collision frequencies, and therefore lower flotation recoveries.

Although a high agitation/turbulence in the flotation cell increases particle-bubble collision frequency and therefore flotation rate and recovery, too much turbulence is detrimental as the larger particles detach from the bubbles when the detachment force which can be simply represented by the centrifugal ($F_g=m.\gamma$) and gravity forces ($F_c=m.g$) becomes larger than the force of particle-bubble adhesion (proportional to the particle hydrophobicity or contact angle). As the centrifugal and gravity forces are proportional to the mass of the particle (density.π.diameter^3/6), the detachment force will increase with particle diameter and particle density, and correspondingly the flotation rate and recovery will decrease. Increasing the collector hydrocarbon chain length may increase the particle-bubble adhesion and therefore reduce particle detachment.

Conventional flotation machines were designed mostly for particles smaller than 150 μm, under which size most of the minerals in these particles are supposed to be liberated, and are most efficient for particles with sizes approximately between 10 and 150 μm [4]. Indeed, figure 1 shows typical recovery versus particle size curves for copper minerals along the bank of flotation cells. These curves present a recovery maximum at around 100 μm and a decrease in recovery for particles of smaller and larger sizes attributed to low particle-bubble collision efficiency and high detachment efficiency, respectively. The shape of these curves is common for all mechanical flotation machines independently of their size and the types of minerals floated [4] although the particle size of maximum recovery may slightly increase for lighter minerals (e.g., quartz) or decrease for denser minerals (e.g., galena), as shown in figure 2, as a result of an increased particle-bubble collision/attachment of the lighter minerals and detachment of the denser minerals. We note in the particular example in figure 1 that although the recovery of the finer particles is low in the first flotation cell it increases in the next flotation cells, which is characteristic of the slow flotation of fine particles [5] because of their low collision efficiency with bubbles (because of their low mass they follow the fluid streamlines around the bubble).

2. Diagnostic for the problem of low flotation

The analytical techniques of Scanning electron microscopy (SEM) and Quantitative evaluation of minerals by SEM (QEMSCAN) are used to inform on the degree of liberation of value minerals in particles which usually increases with decreasing particle size. These minerals are usually classified as liberated (90-100% liberation), middling (30-90%) or locked (0-30%). Larger particles are more likely to be composite particles where value minerals are locked. They have a relatively small proportion of their surface/volume made of value minerals covered with collectors and as a result their adhesion to bubbles is low and the detachment efficiency of these particles is high resulting in low recoveries. Figure 3 shows an example of two QEMSCAN images of particles in the size fraction +150-300 μm which clearly show that first most of the copper minerals in these particles are locked with gangue minerals and secondly the proportion of copper minerals in the particles is at least 30% in the concentrate and much less in the tailing. After collision, the air bubble spreads on the surface of a hydrophobic particle until an equilibrium contact angle is reached but for a composite particle the final contact angle on the hydrophobic domain is less than the equilibrium contact angle because the

IOP Conf. Series: Journal of Physics: Conf. Series **879** (2017) 012002 doi:10.1088/1742-6596/879/1/012002

perimeter of the bubble is pinned down at the interface between the hydrophobic and hydrophilic domains, which results in a weaker particle-bubble adhesion.

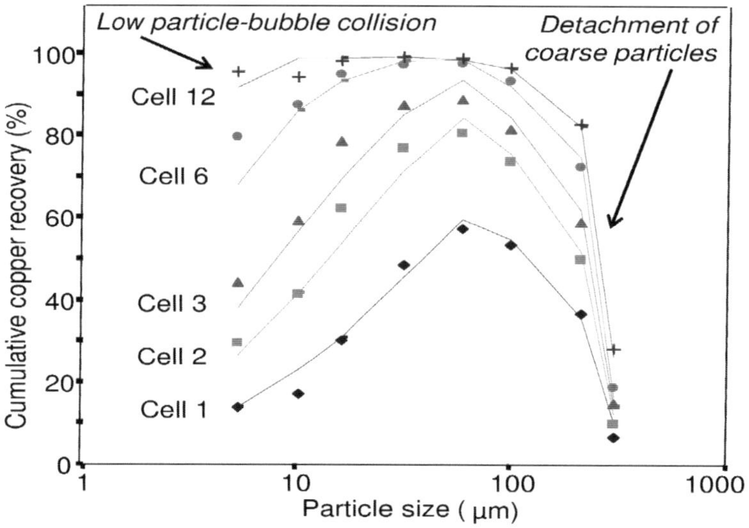

Figure 1. Cu flotation recovery as a function of particle size in a copper ore [6].

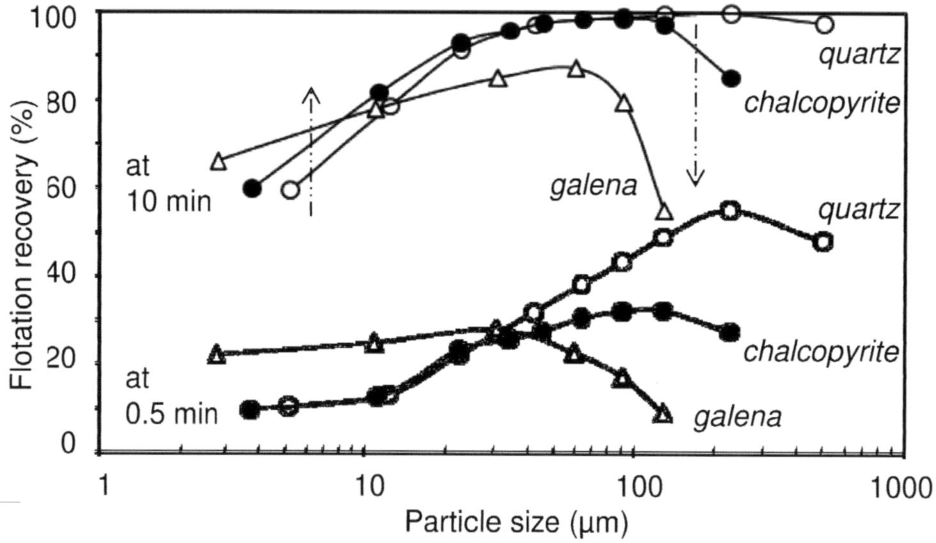

Figure 2. Flotation recovery of quartz, chalcopyrite and galena (density of 2.6, 4.3 and 7.6 g/cm^3, respectively) as a function of particle size at 0.5 and 10 min. The vertical arrows show the change in recovery with increasing mineral density.

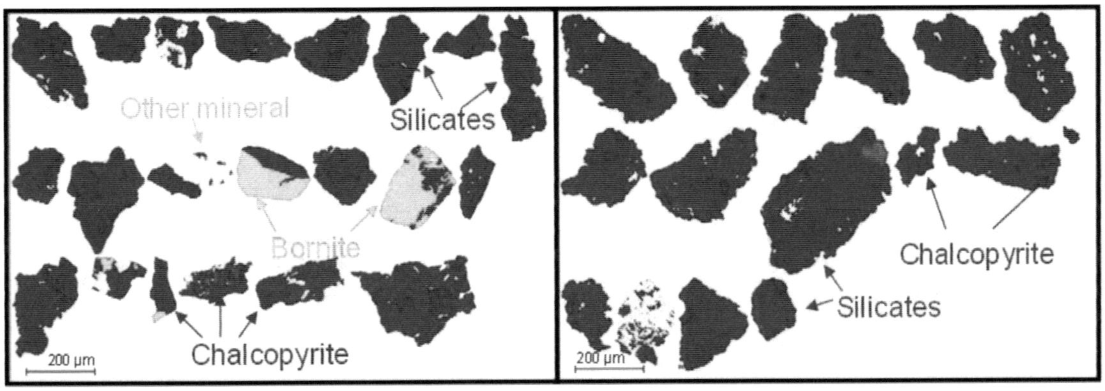

Figure 3. QEMSCAM images of particles in (left) the concentrate and (right) the tailing in the size fraction +150-300 μm in a copper ore [7].

Surface analysis of the particles can provide valuable information on the reasons why particles do not report to the concentrate, apart from the fact that they are composite particles. X-ray photoelectron spectroscopy (XPS), Time-of-flight secondary ion mass spectrometry (ToFSIMS) and Scanning Auger microscopy (SAM) are surface sensitive analytical techniques which can detect species on the top uppermost surface to a depth of 2 to 10 nm, and identify their chemistry and measure their concentration [8]. These surface species can be 1) the collector which can be in a too low concentration to provide enough hydrophobicity for a successful particle-bubble attachment; 2) fine gangue hydrophilic particles covering the value minerals, or slime coating, which need to be removed by chemical or mechanical methods (desliming) such as with charged polymeric and inorganic dispersants (e.g., carboxymethyl cellulose, carbonate, phosphate) or attritioning/high intensity conditioning. Figure 4 shows an example of the principle of ToFSIMS and its use to identify sphalerite and pyrite particles in a zinc ore after conditioning, and measure the species on their surface. The ToF-SIMS total ion yield image shows all the particles present in the sample while the other images are only representative of a particular element, Zn and S (sphalerite), Fe and S (pyrite), Cu (copper activation of sphalerite) or IBX (collector isobutyl xanthate) on these particles. The Fe image shows clearly the three pyrite particles surrounded by the sphalerite particles while the other images show that more copper and IBX collector are adsorbed on the sphalerite than on the pyrite particles, which produces a good separation of these two minerals during flotation [8].

3. Solutions to the problem of low flotation

The causes for the low flotation of fine and coarse particles are different and therefore these particles require different chemical and hydrodynamic conditions during their conditioning and flotation [9]. For this, the feed is split up using size classification such as with cyclones so that each size fraction received the most appropriate technology. Fines have a much larger surface area than the coarse particles and therefore need more collectors. They also require higher agitation/turbulence to increase particle-bubble collision frequency and attachment efficiency while for coarser particles low agitation/turbulence prevents their detachment from bubbles. The use of fine bubbles (<100 μm, produced by electroflotation, hydrodynamic cavitation or gas supersaturation) has been shown to increase the collision efficiency of fine particles with these bubbles and therefore increase the flotation rate and recovery (due in part to a higher number of fine bubbles for the same air volume and their lower rise velocity) [10]. Aggregation of particles is also a solution to improve the low recovery of fine particles because particle-bubble collision efficiency increases with particle size [10]. However, particles aggregation is not particularly selective and gangue particles may be trapped in the aggregate. For coarse particles, existing flotation cells can be modified to restrict the recirculation of the so-formed particle-bubble aggregates in the high turbulence regions of the rotor, and therefore decrease particle detachment (froth crowders) [9].

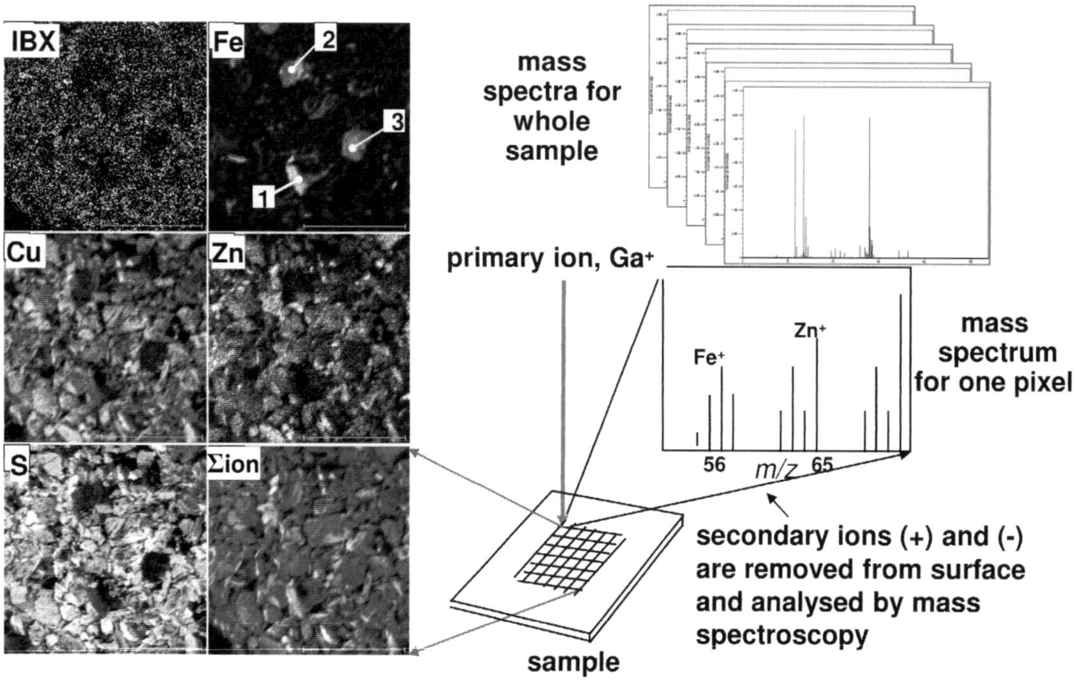

Figure 4. ToF-SIMS IBX, Fe, Cu, Zn, S and total ion yield (+ve SIMS) images (from top to bottom) of sphalerite and pyrite particles in the feed (the level of brightness is indicative of the abundance of that element). The scale for all images is 100 μm. The labelling in the Fe image represents three different pyrite particles [8].

New flotation cells have been developed to specifically address the problem of the low flotation of fine and coarse particles. For example, the HydroFloat™ uses an aerated dense fluidized-bed of particles where gas bubbles pass through the bed of particles and rise to the top with particles attached if the adhesion of the particle to the bubble and the buoyancy of the bubble are larger than the gravitation force [11]. With this set-up, shown in figure 5, there is maximum contact/collision between the gas bubbles and the particles, and no agitation/turbulence in this flotation cell and therefore particle detachment is reduced, which is particularly beneficial to the coarser particles. The flotation cells of type reactor-separator are more appropriate for the flotation of fine particles (figure 5). Air is injected at high pressure into the pulp in the confined space of the reactor, which insures optimum conditions to increase particle-bubble collision efficiency and frequency [12]. Moreover, the high pressure in the reactor (Reynold number of 10^4-10^5) and its release in the separator (Reynold number of 10^2-10^3) favour the nucleation of nano-bubbles only on the hydrophobic particles, which will further increase particle-bubble attachment. Results in Table 1 show that the reactor-separator produces a higher P_2O_5 recovery and grade compared to an Agitair mechanical flotation cell in the flotation of a phosphate ore.

Table 1. Difference in flotation performance between an Agitair and a Reactor-separator flotation cell.

Flotation cell	P_2O_5 recovery (%)	P_2O_5 grade (%)
Agitair	44.4	28.4
Reactor-separator	59.1	32.9

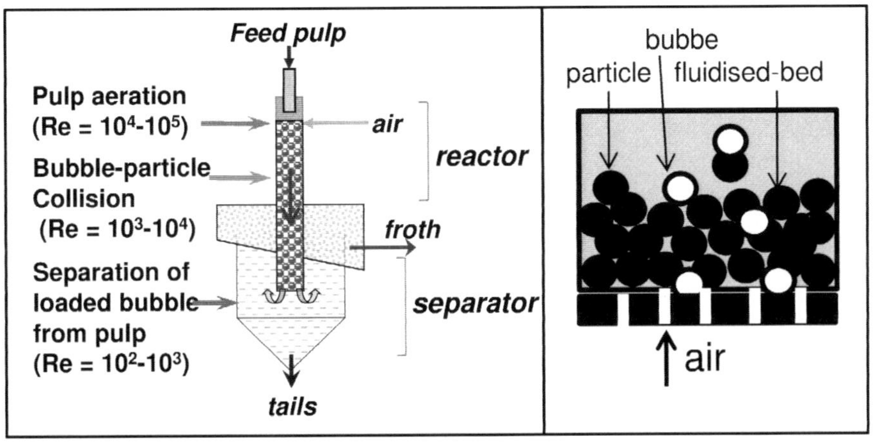

Figure 5. Schematic representation of flotation cells of type reactor-separator (left) and fluidised-bed (right). See text for explanation.

As the rich ore bodies are depleted, particles have to be ground finer to liberate the values minerals disseminated in these particles, which leads to huge increase in energy cost (grinding accounts for around 40% of the total energy used in mineral processing operations and 4% of the world's total electrical energy consumption [13]). Selecting the coarsest possible grind size reduces the overall grinding energy use. Floating coarse particles (such as with the HydroFloat^{TM}) allows to reject earlier those with no or little value minerals in them before further comminution in downstream processes of only those particles with economic interest. Reducing energy consumption can be achieved using new grinding methods such as with the High Pressure Grinding Rolls (HPGR) [9]. Also, high voltage pulses breakage technology, such as with SelFrag, provides a better liberation of the minerals of interest in solids [14]. With this process, a more selective fragmentation occurs along grains boundary and at the interface of minerals composing the solid, which necessitates less energy during finer grinding and provides a more selective mineral separation in flotation.

4. References

[1] Pyke B, Fornasiero D and Ralston J 2003 *J. Colloids Interface Sci.* **265** 141–51
[2] Schubert H 1999 *Int. J. Miner. Process.* **56** 257–76
[3] Zanin M, Ametov I, Grano S, Zhou L and Skinner W 2009 *Int. J. Miner. Process.* **93** 256–66
[4] Batterham R and Moodie J P *Centenary of Flotation Symp.* 2005 Brisbane pp 3–5
[5] Pease J D, Curry D C and Young M F *Centenary of Flotation Symp.* 2005 Brisbane pp 905–12
[6] Ralston J, Fornasiero D, Grano S, Duan J and Ackroid T 2007 *Int. J. Miner. Process.* **84** 89-98
[7] AMIRA P260, University of South Australia
[8] Boulton A, Fornasiero D and Ralston J 2003 *Int. J. Miner. Process.* **70** 205–19
[9] Tadosa E, Runge K and Duffy K A *Flotation'13* 2013 Cape Town
[10] Ahmed N and Jamesson G J 1985 *Int. J. Miner. Process* **14** 195–215
Sivamohan, R 1990 *Int. J. Miner. Process.* **28** 247-88
[11] Kohmuench J N, Luttrell G and Mankosa M 2001 *Miner. Metall. Process.* **18** 61–7
[12] Samygin V D, Panin V V, Filippov L O and Stenin N Yu 2010 *Metallurgist* **54** 395–400
Jameson G J 2010 *Miner. Eng.* **23** 835–41
[13] http://eex.gov.au/2013/11/crushing-energy-costs-in-the-mining-sector
[14] Wang E, Fengnian S and Manlapig E 2012 *Miner. Eng.* **34** 48–54, and references within

Acknowledgments
We gratefully acknowledge the financial support provided by the University of South Australia, the Australian Mineral Industries Research Association (P260), the Australian Research Council, the Université de Lorraine, the Région Lorraine (FEDER) and EMERALD.

Application of Z-scan technique for the study of nonlinear absorption in chemically reduced $LiNbO_3$ crystals

SM Kostritskii[1], M Aillerie[2,3], E Kokonyan[4,5], and OG Sevostyanov[6]

[1]RPC Optolink, Sosnovaya al. 6, bd. 5, Zelenograd, 124489, Moscow, Russia
[2]Université de Lorraine, LMOPS, EA 4423, 57070 Metz, France
[3]Centrale Supélec, LMOPS, 57070 Metz, France
[4]Armenian State Pedagogical University, Yerevan, Tigran Mets Ave., 17, Armenia
[5]Institute for Physical Researches, National Academy of Sciences of Armenia, Ashtarak-2, Armenia
[6]Phys. Dept., Kemerovo State University, Krasnaya str. 6, 650043, Kemerovo, Russia

E-mail: skostritskii@optolinkl.ru

Abstract. The nonlinear absorption (NLA) was studied by open-aperture Z-scan experiments in the chemically reduced nominally pure $LiNbO_3$ crystals at cw-illumination with the red (644 nm) and green (514.5 nm) laser beams. The magnitude of the measured NLA is considerably different from the reported Z-scan results obtained in as-grown $LiNbO_3$. The positive sign of NLA obtained with the red light has been related to the generation of the small bound polarons absorbing in red and near-IR ranges. Application of green light results in the light-induced transparency, i.e. the Z-scan traces show negative sign of NLA. Intensity dependence of Z-scan traces allows for conclusion that the photo-induced dissociation of small Nb_{Li}^{4+}:Nb_{Nb}^{4+} bipolarons and sequent generation of small polarons gives the dominating contribution to the nonlinear optical absorption in reduced crystals with a large bipolarons concentration.

1. Introduction

Recently, a new interest to $LiNbO_3$ crystals was stimulated by the study of small polarons and bipolarons [1,2]. Besides of fundamental aspects, polarons are very perspective entities for practical purposes: e.g., the polarons can be used in the holography as intrinsic photorefractive centers in undoped and doped $LiNbO_3$ crystals, including chemically reduced nominally pure ones [2-4]. There are many hints that short-living small polarons strongly influence the photorefractive sensitivity. Furthermore, it has been shown that small bound polarons may act as intermediate shallow traps in a multi-step recording scheme for phase holograms featuring non-destructive read-out [3,5]. To optimize the crystal properties for photorefractive holography, it is necessary to study the efficiency of single polarons excitation. We propose to use the open-aperture Z-scan method, as it has been established [6,7], that the photorefractive sensitivity is proportional to the light-induced absorption. Indeed, both effects present the same microscopic origin, i.e. the photo-generation of small single $Nb_{Li}^{3\bullet}$ polarons as result of light-induced dissociation of small Nb_{Li}^{4+}:Nb_{Nb}^{4+} bipolarons. Note, that the *Z*-scan technique is a well-established method for quantitative measurement of the nonlinear absorption (NLA) and nonlinear refraction (NLR) [7,8].

[1] To whom any correspondence should be addressed.

2. Experimental techniques

A lithium niobate ($LiNbO_3$) crystal grown from a melt of 48.5 mol % Li_2O has been used for preparation of the reduced samples studied, which are listed in Table 1. These six samples were annealed in pure hydrogen atmosphere (partial pressure of 0.1 atm) at different temperatures ranged from 500 to 700 ^{0}C for various times in range from 0.5 to 11 hours. One additional sample of the as-grown (unreduced) crystal was used as a reference specimen for the comparative study of the chemical reduction effect on NLA. All the samples were Z-cut 2-mm-thick plates with the front and back Z-surfaces polished at an optical-grade.

The typical optical absorption spectrum is shown in figure 1. As can be seen from our data, the strength of the bipolaron absorption band depends on the reduction temperature in a strongly nonlinear way. This fact makes it difficult to control the reduction process and may lead to arbitrarily different results for slightly altered experimental conditions [2,4]. Since the dependence of bipolaron concentration on the reduction conditions in nominally pure $LiNbO_3$ is not generally known, the most convenient characterization is given by the properties of the bipolaron band itself. According to the reference [2], we will refer to the linear absorption coefficient α_0 at $\lambda = 500$ nm, i.e., at a maximum of the bipolaron absorption band, as the degree of reduction.

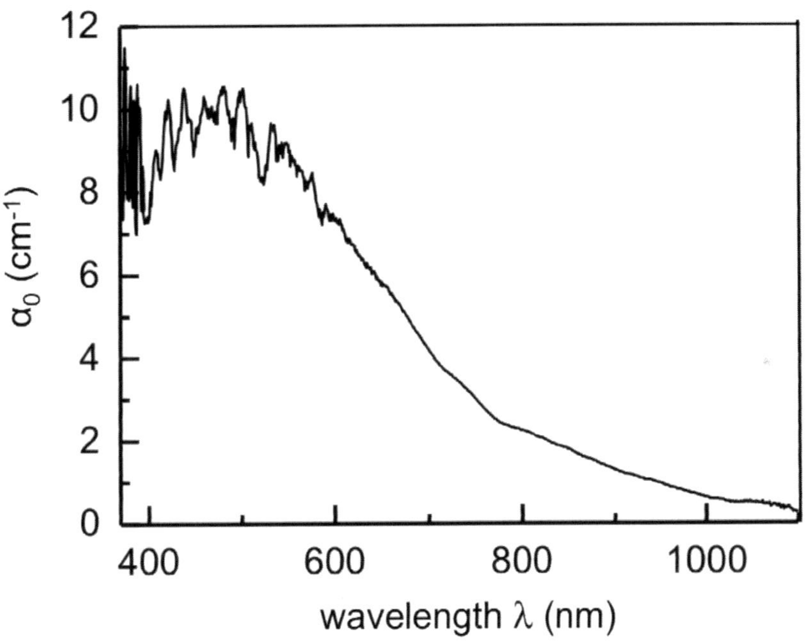

Figure 1. Spectral dependence of the linear absorption coefficient α_0 for the moderately reduced nominally pure $LiNbO_3$ crystal (sample #2). Sample notation is given in accordance with Table 1.

Experimental information on NLA was obtained from open aperture *Z*-scan measurements [7]. The samples were scanned in the vicinity of the lens focus ($z = 0$) starting from a position ahead of focus ($-z$) and moving toward a position behind the focus ($+z$). The scanning range was about ±25 mm in all the experiments. The data were collected at every 0.2 mm on the lens axis, by translating the sample under test through the focal point using a computer controlled stepper motor. An argon-ion laser operating at 514.5 nm was focused with a 200 mm focal length lens into the z-cut $LiNbO_3$ sample. This adjustable argon-ion laser generates a Gaussian light beam with an output power, corresponding to a light intensity I_0 at the focal point of the setup (i.e. at z = 0) in the 0.05 to 3.5

kW/cm^2 range. Besides, we use a diode laser at λ = 644 nm in a second series of NLA measurements with the same open-aperture Z-scan set-up. Calibrated neutral density filters were used to vary I_0 from 0.04 to 1.2 kW/cm^2. The thickness of a sample (2 mm) was chosen in such a way that it was smaller than the Rayleigh range z_R of the focusing lens. The Rayleigh range was evaluated by $z_R = \pi\omega_0^2/\lambda$; as the beam waist $\omega_0 = 4f\lambda/\pi d$ at the focal point (z = 0), where f is the focal length of the lens and d is the beam diameter at the aperture, i.e. z_R = 5.9 and 4.8 mm at λ = 644 and 514.5 nm, respectively.

Note, that the Z-scan response with a fully open aperture is insensitive to the nonlinear refraction and is only a function of the nonlinear absorption [6-9]. Such Z-scan traces with no aperture are expected to be symmetric with respect to the focus (z = 0) and the coefficient of nonlinear absorption can be easily calculated from a transmittance curve, when the low-intensity Z-scan background is subtracted to reduce the effects of linear sample inhomogeneity [7].

3. Experimental results

In figure 2 the normalized transmission is shown for an open aperture Z-scan as a function of the distance along the lens axis for 514.5-nm and 644-nm laser beams in the deeply reduced crystal having high values for the linear absorption coefficient α_0 at the both wavelengths. The transmission without an aperture was measured in the far field as the sample was moved through the focal point.

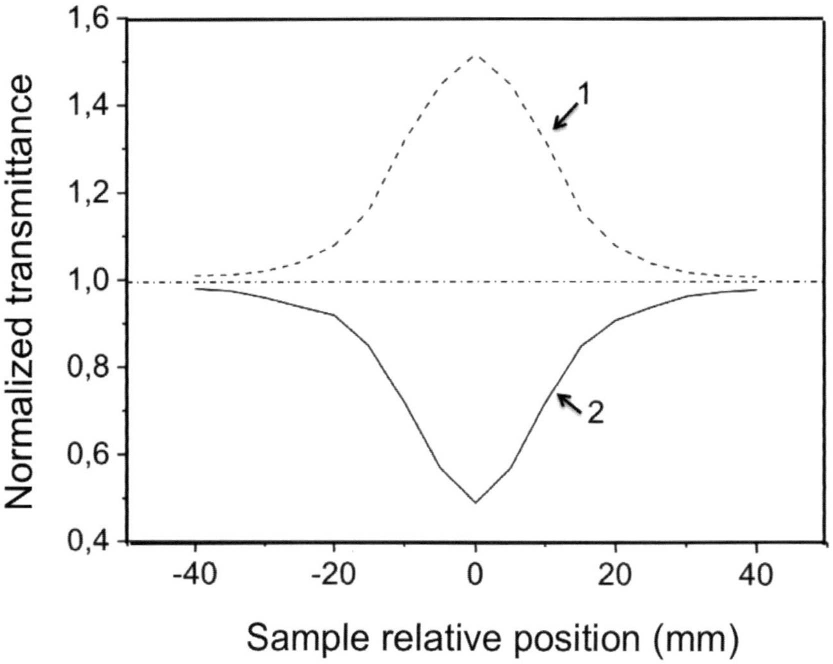

Figure 2. Open-aperture Z-scan traces measured in the sample #6 (Table 1) with 514.5-nm (curve 1) and 644-nm (curve 2) laser beams, respectively. The light intensities of 1.2 and 1.9 kW/cm^2 were used at 644 nm and 514.5 nm, respectively. Both transmittances $T(z)$ are normalized to the corresponding transmittances T_L in the regions of $|z| > 40$ mm. Thus, the normalized transmittance is $NT = (T(z)/T_L)\times100\%$.

The curve (2) of figure 2 depicts the open aperture response, which accounts for a decrease in the transmittance T due to nonlinear absorption of the 644-nm laser beam. Such a shape of the optical response is typical for open-aperture Z-scan measurements in the as-grown nominally pure and doped $LiNbO_3$ crystals [7-10]. In this case called Reverse Saturable Absorption (RSA) [11], the sample experiences the strongest I at focus ($I = I_0$ at $z = 0$); therefore it absorbs the most energy and allows least transmittance. Away from the focus in the directions of both increasing and decreasing z, transmittance increases evenly because the irradiance decreases symmetrically about the focus. However, the magnitude of this decrease of T has much larger values in the reduced crystals in comparison with an as-grown nominally pure Z-cut sample of same thickness, when the light intensity I_0 is fixed to any constant level within the range available with the 644-nm laser used. The open-aperture response, curve (1) in figure 2, obtained for the deeply reduced crystals ($\alpha_0 > 20$ cm^{-1}, i.e. in the samples #4-6, Table 1) with the 514.5-nm laser beam demonstrates the increase of T with I, i.e. T(z) has a maximum value at $z = 0$, that is typical for Saturable Absorption (SA) [11], when the nonlinear absorption has negative sign. It can be related to the bleaching of the defect centers absorbing at this laser wavelength in the deeply reduced crystals. In contrast to the reduced $LiNbO_3$, no alteration of nonlinear absorption sign has been observed in as-grown $LiNbO_3$ crystals at the change of light wavelength from 644 to 514.5 nm.

NLA can be determined directly from a $T(z)$ curve measured by the open aperture Z-scan. For a small nonlinear absorption the normalized change in transmittance ($\Delta T = T(z) - 1$) becomes [6,11]:

$$\Delta T(z) = - q_0/2\sqrt{2}, \qquad (1)$$

where

$$q_0 = \Delta\alpha L_{\text{eff}}/(1 + z^2/z^2_{\text{R}}) \qquad (2)$$

here, $L_{\text{eff}} = [1 - exp(-\alpha_0 L)]/\alpha_0$ is the effective length, α_0 is the linear absorption coefficient, L is the thickness of the sample, $\Delta\alpha$ is the light-induced change in absorption (i.e., the total absorption coefficient is $\alpha = \alpha_0 + \Delta\alpha$), z is the actual sample position relative to the focus and $z_{\text{R}} = (\pi w_0^2/\lambda)$ is the confocal parameter given for the beam waist radius w_0 at the focus and the laser wavelength λ. The change in optical absorption coefficient $\Delta\alpha$ refers to the ordinary polarization, since the optical axis of the crystal under test was always set parallel to the propagation direction of the laser beam.

The open aperture Z-scan data was used to evaluate the nonlinear absorption coefficient β (according to the common definition of two-photon absorption caused by third-order optical nonlinearity [6-9]: $\beta = \Delta\alpha/I_0$) in the as-grown $LiNbO_3$ crystals at excitation with the both laser wavelengths used in our experimental study. The experimental data on $T(z)$ yields the following values: $\beta = 0.45$ cm/kW at 514.5 nm and $\beta = 0.08$ cm/kW at 644 nm. Note, that such a spectral dependence is general feature of NLA [6]. These β values are much smaller than the previously reported values ($\beta = 2.7$ cm/kW was evaluated [10] at $\lambda = 532$ nm and $\beta = 5$ cm/kW was evaluated [9] at $\lambda = 514.5$ nm) obtained with cw-laser beams for the undoped as-grown lithium niobate crystals. This difference can be attributed to the significant influence of the photorefractive effect on NLA measurements, as the photorefractive effect depends strongly on many uncontrollable crystal parameters and specific excitation conditions [7,12].

Table 1. Samples notation, linear optical absorption coefficient α_0 values at $\lambda = 514.5$ nm, maximum light-induced absorption coefficient change $\Delta\alpha$ observed at saturation of NLA and modified nonlinear absorption coefficient β' for the reduced $LiNbO_3$ samples. β' was evaluated from the Z-scan data on $\Delta\alpha$ values measured with 514.5-nm laser beam at $I_0 = 59.5$ W/cm^2, as $\Delta\alpha = \beta'\sqrt{I_0}$. The signs of $\Delta\alpha$ and β' were defined in accordance with the expression for the total absorption coefficient $\alpha = \alpha_0 + \Delta\alpha$.

Sample #:	1	2	3	4	5	6
α_0, cm^{-1}	4.6	10.6	19.1	25.3	27.6	37.4
$\Delta\alpha$, cm^{-1}	-2.25	-6.8	-12.5	-16.8	-18.3	-24.3
β', W$^{-0.5}$	-0.09	-0.23	-0.44	-0.56	-0.61	-0.88

In the reduced crystals the light-induced absorption changes $\Delta\alpha$ have been found to be proportional to the square root of I_0 within the quite low-intensity range, Tables 1 and 2. These $\Delta\alpha(I_0)$ dependences for the both 644-nm- and 514.5-nm-excitations demonstrate the tendency to saturation even at the moderate intensities. However, there is a spectral dependence of the both modified NLA coefficient β' (obtained by fitting of experimental dependences by $\Delta\alpha(I_0) = \beta'\sqrt{I_0}$ [13]) and a maximum value of NLA, i.e. a value of $\Delta\alpha$ observed at maximum I_0.

4. Discussion

To explain the anomalous dependence of $\Delta\alpha$ on I_0 we took into account that the dominating mechanism of nonlinear absorption in reduced crystals at low and moderate intensities I_0 is the photo-dissociation of bipolarons B into single polarons P, which can be described by a second-order reaction $B \leftrightarrow P + P$ as proposed in reference [13]. Note, that this finding is in contrast to the other mechanisms of NLA in $LiNbO_3$ reported in references [6-11], as all these mechanisms can be described by a first-order reaction. The latter term is used in accordance with the definition given previously [13].

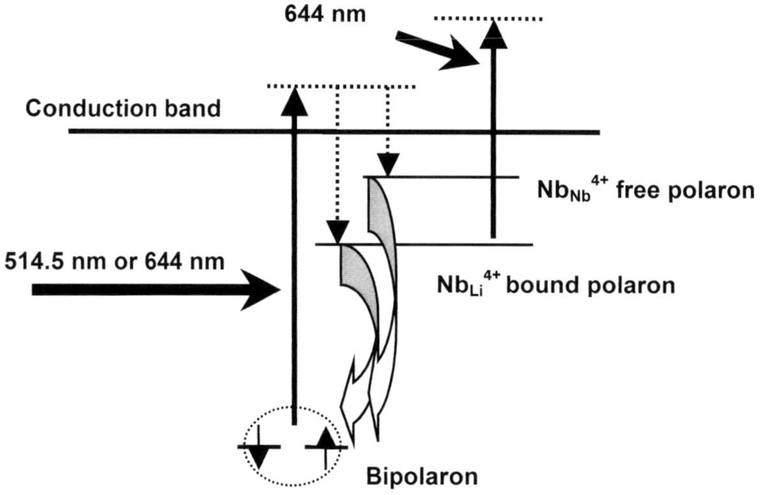

Figure 3. Schematic description for the mechanism of excitation (generation) and recombination (depopulation) of small polarons in reduced $LiNbO_3$. Thick arrows denote excitations by pump light, dashed arrows denote relaxations. Bent thick arrows stand for the transformation (annihilation) of single polaronic states without electronic transitions. Note that all the energy levels of polarons and bipolarons are within forbidden band and it are shown schematically; they do not represent defect energies.

At room temperature almost all of these polarons are paired forming the bipolarons in equilibrium. Bipolarons can be broken into single polarons again by illumination within visible range (see figure 3). At room temperature single polarons are short-living ($\leq$ ms) because of their small activation energy (0.3 eV) for annihilation [1,2]. Such a photo-dissociation of bipolarons changes the optical absorption spectrum of the crystals. The broad absorption band near 500 nm has been attributed to the presence of bipolarons and a narrower band at 760 nm (a short-wavelength tail is extended down to 600 nm) has been assigned to the formation of single polarons under illumination with proper wavelength [2,13,14]. Therefore, the existence of such a dissociation-annihilation process in continuous wave (cw) illuminated crystals is expected to manifest itself in the optical absorption change dependent on the wavelength of pump laser beam [2,13].

Table 2. Samples notation, linear optical absorption coefficient α_0 values at λ = 644 nm, maximum light-induced absorption coefficient change $\Delta\alpha$ observed at I_0 = 1200 W/cm^2, and modified nonlinear absorption coefficient β' for the reduced $LiNbO_3$ samples. β' was evaluated from $\Delta\alpha$ values measured with 644-nm laser beam at I_0 = 38 W/cm^2.

Sample #:	1	2	3	4	5	6
α_0, cm^{-1}	2.6	5.9	10.7	14.1	15.4	20.8
$\Delta\alpha$, cm^{-1}	0.43	1.15	2.01	2.56	2.85	3.63
β', W$^{-0.5}$	0.013	0.031	0.056	0.077	0.084	0.112

Thus, the photon-induced dissociation of small Nb_{Li}^{4+}:Nb_{Nb}^{4+} bipolarons, absorbing in the green spectral range, results in a light-induced transparency, i.e., SA effect, when $\Delta\alpha$ has negative sign (equation (3)), at excitation with 514.5-nm laser beam. The generation of the small bound Nb_{Li}^{4+} polarons, absorbing significantly in the red and near IR ranges, yields a positive light-induced absorption $\Delta\alpha$ related to the RSA at excitation with the 644-nm laser beam. According to the data on spectral shape of the bipolaron absorption band [2,4], the red light is still sufficient for one-photon dissociation of bipolarons due to a high light intensity within focused laser beam despite of a small linear absorption α_0 at this wavelength.

Combined solution of equations, describing the dissociation-annihilation process and the optical absorption, the following dependence of the light-induced absorption changes $|\Delta\alpha|$ on I_0 can be expressed by [13]:

$$|\Delta\alpha| = N\times s(gI_0/r)^{1/2} = \beta'\sqrt{I_0} \tag{3}$$

with $s = s_P - 0.5s_B$, where s_P denotes the absorption cross section of polarons and s_B that of bipolaron, g and r are generation and recombination rates, respectively, of small bound polarons in the photo-induced dissociation process of bipolarons.

Using the maximum transmission change ΔT measured at $z = 0$ in differently reduced crystals as input data for equations (1,2,3), the modified nonlinear absorption coefficient β' values were evaluated and presented in Tables 1 and 2. These β' values evaluated with the aid of the Z-scan technique have the same order of magnitude that its value reported previously (β' = 0.27 W$^{-0.5}$) [13]. At the same time, there is an alteration of the β' sign and a significant increase of $|\beta'|$ magnitude, when the pump wavelength changes from 644 to 514.5 nm for any deeply reduced crystal studied, as shown by the values reported in Table 2. Such a drastic spectral dependence of β' is in accordance with the data on the light-induced changes of the polaron/bipolaron absorption in the reduced lithium niobate crystals.

5. Summary

Chemically reduced nominally pure $LiNbO_3$ crystals were prepared and characterized for nonlinear absorption (NLA) using the open-aperture Z-scan technique at the wavelength of 514.5 and 644 nm. Z-scan spectra reveal the sign alteration of NLA across this wavelength range with a positive sign of NLA at 644 nm accompanied by a negative one for NLA at 514.5 nm. These anomalous nonlinearities in reduced $LiNbO_3$ induced by low- and moderate-intensity illumination are established to be mainly due to a two-photon absorption process associated with the photodissociation of bipolarons, which are generated in large quantity by the chemical reduction annealing. It has been found that the NLA of the deeply reduced crystals is much larger than that of the as-grown (unreduced) lithium niobate.

IOP Conf. Series: Journal of Physics: Conf. Series **879** (2017) 012003 doi:10.1088/1742-6596/879/1/012003

6. References

[1] Schirmer O F, Imlau M, Merschjann C and Schoke B 2009 *J. Phys.: Condens. Matter* **21** 123201
[2] Merschjann C, Schoke B and Imlau M 2007 *Phys. Rev. B* **76** 085114
[3] Guenther H, Macfarlane R, Furukawa Y, Kitamura K and Neurgaonkar R 1998 *Appl. Opt.* **37** 7611
[4] Kostritskii S M, Aillerie M, Margueron S, Bourson P 2013 *J. Phys.: Conf. Ser.* **416** 012033
[5] Hesselink L, Orlov S, Liu A, Akella A, Lande D, and Neurgaonkar R 1998 *Science* **282** 1089
[6] Christodoulides D N, Khoo I C, Salamo G J, Stegeman G I, and Van Stryland E W 2010 *Advances in Optics and Photonics* **2** 60
[7] Kostritskii S M, Aillerie M, Kokanyan E 2013 *Proc. SPIE* **9065** 906508
[8] Pálfalvi L, Hebling J, Almási G, Péter Á, Polgár K, Lengyel K and Szipocs R 2004 *J. Appl. Phys.* **95** 902
[9] Henari F Z, Cazzini K, Akkari F E, and Blau W J 1995 *Appl. Phys. Lett.* **78** 1373
[10] Chen Y, Liu S W, Wang D, Chen T, and Xiao M 2007 *Appl. Opt.* **46** 7693
[11] Wei T H, Huang T H, Lin H D, and Lin S H 1995 *Appl. Phys. Lett.* **67** 2266
[12] Kostritskii S M, Sevostyanov O G 1997 *Appl. Phys. B* **65** 527
[13] Jermann F, Simon M, Bower R, Krätzig E, and Schirmer O F 1995 *Ferroelectrics* **165** 319
[14] Badorreck H, Nolte S, Freytag F, Bäune P, Dieckmann V, and Imlau M 2015 *Opt. Mater. Express* **5** 2729

The clamped and unclamped effective electro-optic coefficients of zirconium-doped congruent lithium niobate crystals

Mustapha Abarkan[1,2,3,*], Anush Danielyan[2,3,4], Ninel Kokanyan[2,3], Michel Aillerie[2,3], Sergey Kostritskii[5], Edvard Kokanyan[4,6]

[1]LSI, Faculté Polydisciplinaire de Taza, Université Sidi Mohammed Ben Abdallah de Fès, BP 1223 Taza-Gare, Morrocco
[2]Université de Lorraine, LMOPS, EA 4423, 57070 Metz, France
[3]CentraleSupélec, LMOPS, 57070 Metz, France.
[4]Institute for Physical Researches, National Academy of Sciences of Armenia, Ashtarak-2, Armenia.
[5]RPC Optolink, Sosnovaya al. 6, bd.5, Zelenograd, 124489, Moscow, Russia.
[6]Armenian State Pedagogical University, Yerevan, Tigran Mets Ave., 17, Armenia.

E-mail: mustapha.abarkan@usmba.ac.ma, aillerie@metz.supelec.fr

Abstract: All coefficients of the unclamped and clamped electro-optic tensor of zirconium-doped congruent $LiNbO_3$ single crystals are determined as function of the dopant concentration at room temperature. With a distribution coefficient of zirconium closer to one in the considered range, the dopant concentration is in the range up to 2.5 mol% of ZrO_2. The electro-optic coefficients are measured by direct techniques based on interferometric and Sénarmont optical arrangements at the wavelength of 633 nm. It is found that all the unclamped and clamped effective electro-optic coefficients are relatively constant, except for the sample grown with 2 mol% of zirconium. The electro-optic behavior of LN:Zr as function of the dopant concentration was confirmed by dielectric characterizations. The concentration equal 2 mol% of ZrO_2 can be considered to a threshold concentration for various physical and optical properties.

1. Introduction

Zr-doped $LiNbO_3$ (LN) is a promising substrate material for nonlinear integrated optics as it combines excellent electro-optic (EO), acousto-optic, and nonlinear optical properties [1, 2]. Indeed, even based on pure composition, Lithium niobate (LN) crystals have good optical quality and uniformity, an excellent electro-optic, piezoelectric, acoustic and nonlinear optical properties. Generally, LN crystals with the molar ratio R= [Li]/[Nb]= 0.946 corresponding to the congruent composition, are used in a lot of optical and acoustic devices due to their physical properties but also due to their high optical quality and growth facilities. However, these devises suffer from serious photorefractive effect, as the main drawback of congruent LN is its relatively low optical damage resistance compared to other competing oxide crystals such as $LiTaO_3$ (LT) or BaB_2O_4 (BBO) [2, 3]. The photorefractive effect in LN is mainly linked to the presence in congruent LN crystals of intrinsic defects due to the presence of Nb ions on antisites, i.e. on Li sites (noted Nb_{Li}) [4, 5]. Different approaches have been developed to enhance or to reduce the photorefractive sensitivity in LN as function of the type of desired applications. In the first case, when an increase of the photorefractive properties is expected as involved for holographic applications, LN can be doped with metallic ions as chromium or iron ions. It was shown that the large dependence of the EO coefficients on the Cr concentration is the signature of a significant change in the structure of defect centers in the LN host lattice [6-8]. It was also pointed out by Chah et al. that the diminution of the EO properties caused by the addition of even a small amount of Fe to LN is due to the indirect iron doping effect, which modifies the crystal composition, mainly by changing the intrinsic defects content [9]. In the second case, when a decrease of the photorefractive properties, i.e. an increase of the damage threshold is expected as involved for laser applications (for example EO Q-switch applications), various approaches are currently considered. One of them consists of decreasing the number of intrinsic defects by growing LN crystals with the ratio R closer to the one corresponding to the stoichiometric composition [10, 11]. The second approach consists in doping congruent LN crystals by appropriate ions on lithium site, removing the antisites Nb_{Li} to their natural sites. It is currently well-known that doping LN crystals by divalent ions such as Mg^{2+} [12, 13] and Zn^{2+} [14, 15] can effectively suppress the photorefractive effect. Especially, a significant reduction of optical damage was observed in doped samples with a threshold of about 5.5 mol% MgO [16, 17], 7 mol% ZnO [18, 19], respectively. Same observations were also established for LN doped crystals by trivalent ions such as Sc^{3+} and In^{3+}, with the threshold about 1.5 mol% for both ions [20, 21]. Some years ago, the group of Kokanyan et al proposed a new set of non-photorefractive ions [22, 23]. It was shown that a reduction of the photorefractive sensitivity in LN crystals doped with tetravalent ions such as Hf^{4+} and Zr^{4+} could be achieved with more than 2 times lower impurity ions concentration than in the case of divalent ions. Moreover, this advantage is also associated to a distribution coefficient of these impurities at that concentration closer to one interesting during the growth process [24-28].

Among other optical and physical properties and closely linked in crystals of the LN family to the optical damage effect, the electro-optic properties are also mainly influenced by photorefractive-resistant impurities or dopants [19]. In previous contributions, we have shown that electro-optic properties of LN crystals doped with Zn or Mg ions present a huge oscillation with the concentration [13, 18]. By opposition, it seems that no significant dependence of the electro-optic properties on the doping concentration was observed in LN crystals doped with tetravalent ions as in hafnium doped LN crystals [29] or similarly, in zirconium doped LN crystals. In these crystals, the constant-stress electro-optic coefficients r_{222} and r_c of zirconium-doped LN crystals appears only to be slightly influence on Zr [27, 28].

In the present work, we focalized on the electro-optic properties of zirconium-doped congruent LN crystals, hereafter noted LN:Zr. We provide a determination of the values for all unclamped (noted r^T) and clamped (noted r^S) effective electro-optic coefficients $n_o^3r_{113}$, $n_e^3r_{133}$, $n_o^3r_{222}$, and $n_e^3r_c$ of a series of congruent LN:Zr crystals, with concentration from 0 up to 2.5 mol% ZrO_2. Samples were grown from the same charge, thus we emphasize that they have the same amount of unintended impurity trace. Our experiments lead to conclude that an effective and relatively abrupt kink in the electro-optical properties exists for crystals around 2 mol% ZrO_2 that is in full accordance with the threshold observed in optical damage. This behavior exists for all the unclamped and clamped coefficients of the third column of the electro-optic tensor, while the acoustic contribution remains constant in the studied concentration range.

2. Experimental

2.1 Sample preparation

Based on the Czochralski technique, a growth set-up using a single platinum crucible with rf heating element in air atmosphere was used to grow a set of Zr-doped lithium niobate crystals. In order to obtain directly during the growth process single-domain crystals, a dc electric current with a density of about 12 A/m^2 passed through the crystal-melt system. The starting materials used for sintering the lithium niobate charges of congruent composition were high purity Nb_2O_5 and $LiCO_3$ compounds from Johnson-Mattey and Merck. ZrO_2 was introduced into the melt with concentration equals to 0.625, 0.75, 0.875, 1.00, 1.25, 1.50, 2.00 and 2.50 mol%, respectively. In this range, the distribution coefficient of zirconium is closer to one and we considered the same concentration in melts than in crystals. Finally, samples were shaped in a parallelepiped shape with about x, y, z = 5, 10, 5 mm dimensions and were optically polished on all the surfaces.

The record of the UV-Visible-NIR optical transmission spectra reveals the optical qualification of the samples. Transmission spectra were recorded at room temperature with an un-polarized light using a Perkin Elmer Lamda 900 spectrometer. Samples were placed with their polished faces at normal incidence, with the c (z) axis parallel to the k vector of the incident light. The transmission spectra recorded in all crystals present same behaviors with a flat transparency response in the whole visible range and an absorption edge appearing in the UV. The fundamental absorption edges of the crystals were measured at the absorption coefficient of 20 cm^{-1}. Additionally to the good transparency without absorption parasitic peaks observed in all samples, these spectra bring interesting information on the role and influence of the dopant in lithium niobate crystals [30]. This point will be discussed further. However, we have plotted the interesting part for the present work of the absorption coefficient spectra and the absorption edges for the various samples, presented in figure 1. We observe a clear displacement of the absorption edge to UV when the concentration increases in the crystals up to 2 mol% of zirconium. Whereas for crystal doped with higher concentration, the displacement of the absorption edge is to the visible wavelengths. The zirconium concentration of 2 mol% in congruent LN crystal, which corresponds to the minimum observed in the absorption edge position, can be considered as a threshold concentration.

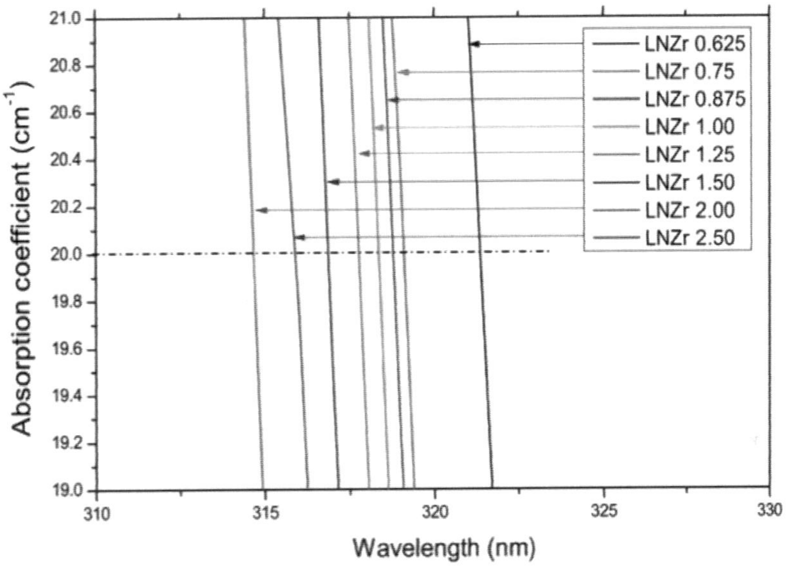

Figure 1. Absorption edge versus Zr concentration in congruent LN doped crystals evaluated from absorption spectra.

2.2 Electro-optic measurements

The measurements are performed using both a Mach-Zehnder type interferometric and the one-beam Sénarmont type ellipsometric techniques both for light propagation along the x-axis. With the first technique, with an applied electric field parallel to the z-axis of the crystal, we have investigated the electro-optic coefficients r_{113} and r_{133}. With the second technique, with an applied electric field parallel to the y-axis of the crystal we have investigated the electro-optic coefficient r_{222} and with an applied electric field parallel to the z-axis of the crystal, the combined coefficient r_c. All measurements were performed at room temperature and using a He-Ne laser (633 nm).

For the determination of the unclamped electro-optic coefficients r_{eff}^{T}, we used a method called the "Modulation Depth Method" (MDM) [31]. This method consists in the measurement of the amplitude of the modulated beam induced by an ac-voltage applied on the sample under test. In the present study, we performed measurements with an ac-voltage of up to peak-to-peak amplitude equal to 250 V at 1 kHz. For the determination of the clamped (i.e. constant strain) electro-optic coefficientsr_{eff}^{S}, we used a method called the "Time Response Method" (TRM), which consists in the measurement of the time response of the EO crystal to a fast step voltage [32]. This method also alloys the determination of the unclamped EO coefficients when the long time range contribution is considered in the EO response. We performed measurements with a pulse voltage of amplitude 300 V having a rise time equal to 5 ns and a repetition rate equal to 10 µs. It is to be of note that the unclamped coefficients can be determined by both methods MDM and TRM allowing the validation of the experimental results. In the same manner, we can note that the combined EO coefficients can be directly measured by the Sénarmont arrangement, offering a high accuracy and verified by its determination from the direct EO coefficients as measured by the interferometric setup.

Under the above-mentioned conditions, within both MDM and TRM methods, the absolute value of the effective EO coefficient r_{eff} can be determined within an equation having the form [30, 32]

$$n^3 r_{eff}(\nu) = A \frac{\lambda\ D}{\pi\ (I_{max} - I_{min})\ L} \frac{\Delta i(\nu)}{\Delta V(\nu)} \quad (1)$$

where A = 2 or 1 for interferometric or Sénarmont measurements, respectively. In this equation, D is the inter-electrode distance, L the length of the crystal along the beam-propagation direction, λ is the vacuum wavelength, n the refractive index seen by the eigen-wave propagating in the crystal, I_{Max} and I_{min} the maximum and minimum of the transmitted intensity and ΔI_E is the peak-to-peak amplitude of the sinusoidal (in MDM method) ΔV is the corresponding amplitude of the modulated voltage applied to the crystal.

For TRM technique we measure the instantaneous variation of the transmitted beam intensity Δi(t) induced by the applied voltage ΔV(t) which is given by:

$$\Delta i(t) = \frac{\pi\ L\ (I_{max} - I_{min})}{A\ \lambda\ D}\ n^3 r_{eff}(t) \otimes \Delta V(t) \quad (2)$$

where ⊗ is the convolution operator and $r_{eff}(t)$ is the instantaneous value of the EO coefficient.

In this technique, we display and measure with the help of an oscilloscope, the time signals Δi(t) and ΔV(t).

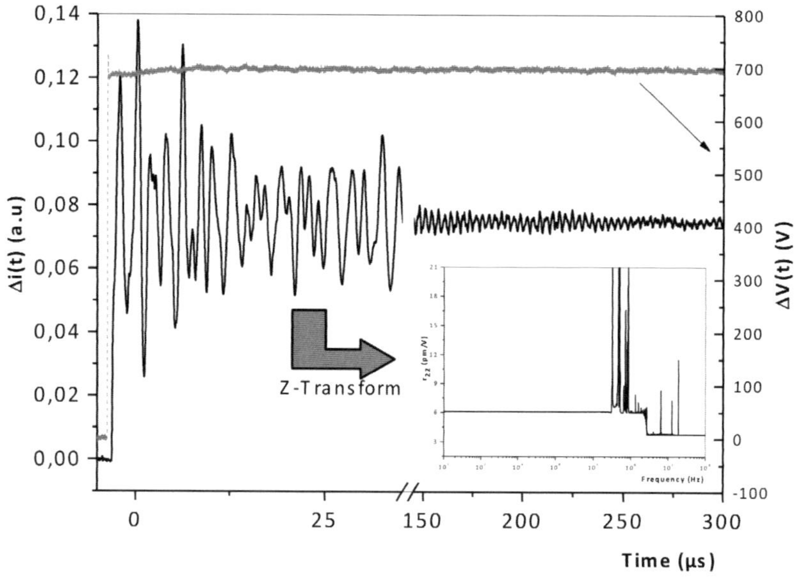

Figure 2. Time and frequency responses of the r_{222} EO coefficient. Response Δi(t) to one step voltage ΔV(t) and the frequency dependence obtained by the Z-transform (the insert) in the case of the EO coefficient r_{222} in 0.8%-Zr-doped LN single crystal. Measurements were performed at the wavelength of 633nm.

As example of obtained results, figure 2 shows the recording of both the applied voltage and the optical signal measured in the 0.8 mol%-Zr doped LN sample at different time within the TRM method in the case of r_{222} EO coefficient. In the long-time range, the optical signal oscillates with periods corresponding to the main piezo-electric frequency resonances. For time shorter than 350 ns the oscillations do not exist since the acoustic waves need more time to propagate across the crystal. The value of the intensity corresponding to the plateau appearing for time below 350 ns is much smaller than this obtained for time larger than 200 μs, which indicates a large acoustic contribution. After the Z-transform of the two time dependent signals, Δi(t) and ΔV(t), the frequency dispersion of the EO coefficients is calculated from the ratio of Δi(ν) and ΔV(ν) according to Eq. 1, as it is the case with the MDM method [33]. We report in the insert of figure 2 the frequency dependence of the EO coefficient r_{222} in 0.8%-Zr-doped LN single crystal. We can observed that the answer is flat and quite at the same level on both sides of the piezoelectric resonances.

The measurements of all studied EO coefficients were performed for all crystals under investigation. All crystals in the r_{222} opto-geometric configuration, present the same behavior with frequency as that of the 0.8%-Zr-doped LN single crystal reported in figure 2. By cons, in the r_{113} and r_{333} configuration, the contribution of the piezo-optic effect was observed smaller that the accuracy of the methods for all the crystals under studies. Thus, we can experimentally consider that when the electric field is applied onto samples in the z direction, $r_{xx3} = r^S_{xx3} = r^T_{xx3}$, with xx =11 or 33.
The dependence of all clamped and unclamped EO coefficients, on the molar Zirconium concentration is shown in figure 3 and presented in Table 1 and Table 2. The coefficients obtained experimentally are given within accuracy better than 10% with the different methods used in this study.
In these results, the effective EO coefficient $n_e^3 r_c$ defined by

$$n_e^3 r_c = n_e^3 r_{333} - n_o^3 r_{113} \qquad (3)$$

is obtained by direct measurements using the Sénarmont set-up and, also calculated by equation 3, allowing the check of the methods and the results.

The curves in figure 3 highlight the quasi-constant values, within the associated error-bars for all the electro-optic coefficients with dopant concentration except for the sample doped with 2 mol% of zirconium where a kink exists, which corresponds to a lower value of the coefficients presenting a change greater than the amplitude of the experimental uncertainty. This concentration of 2 mol% zirconium in LN crystal can be considered as the threshold concentration and could suggest the existence of a doping-induced lattice reorganization.
In ferroelectric inorganic materials, it was established that the EO properties are linked to the linear dielectric properties and their temperature or frequency dependence reproduced the behavior of the dielectric permittivity ε [34]. This relationship between EO coefficients and permittivity also exist in their dependence on crystal composition or dopant concentration [35]. It is for this reason that we have experimentally measured the dielectric permittivity for all samples in the studied zirconium doped series. These measurements were done in a large frequency range up to 1 GHz by means of the impedance analyzers HP4151 and HP4191A. We can note that the behavior of the dielectric permittivity is the same observed in the case of the effective EO coefficients with frequency for all samples for as well.

However, contrary to r_c coefficient, where the difference between r_c^T et r_c^S is smaller than the accuracy of the methods, the difference $\Delta\varepsilon_{33}$ between the dielectric permittivities ε_{33}^T and ε_{33}^S gives a delta epsilon equal to 5. This variation has to be found in the electromechanical process mainly originate from the elastic stiffness contribution.

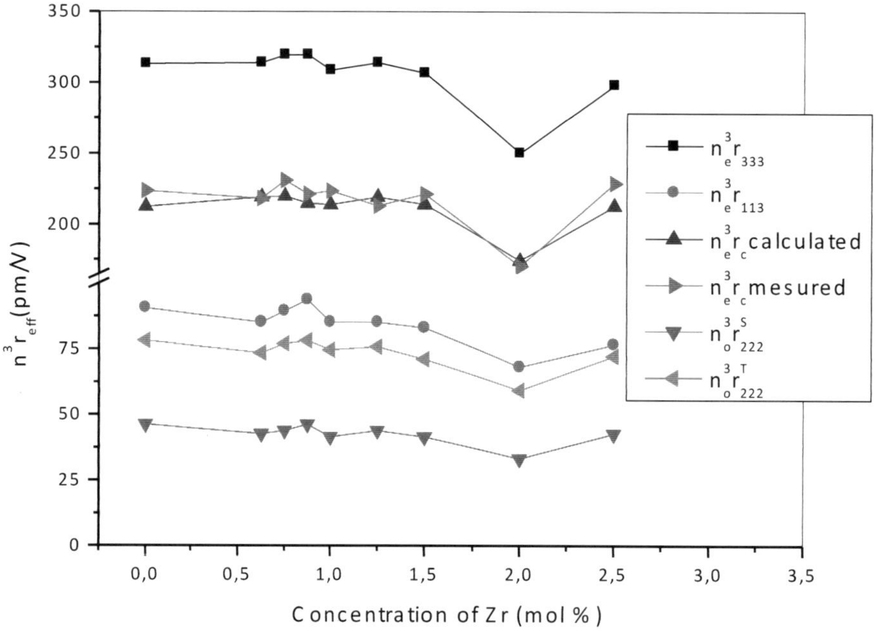

Figure 3. High and low frequency effective EO coefficients n^3r_{eff} versus Zr concentration in congruent LN.

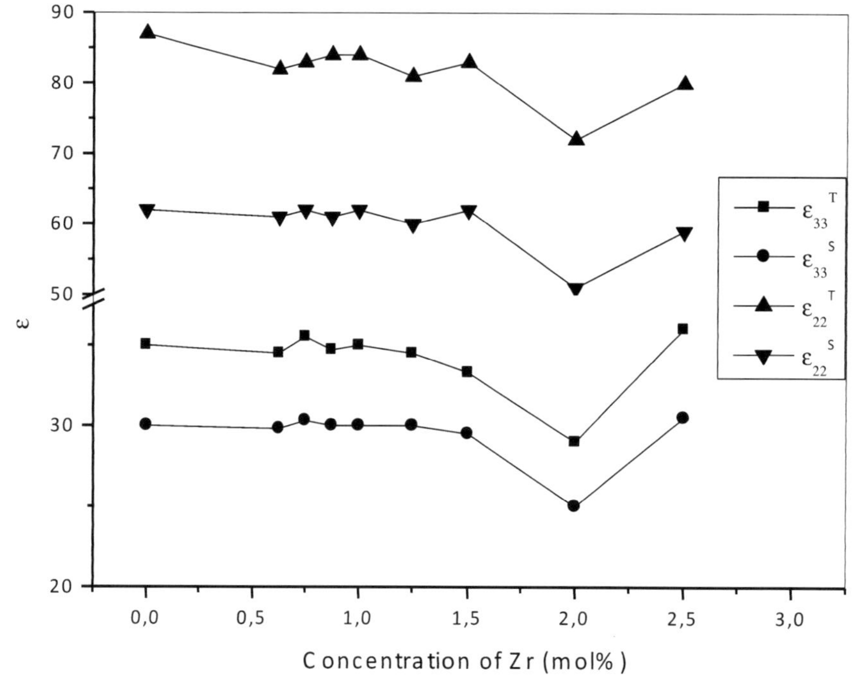

Figure 4. Dielectric constants versus Zr concentration in congruent LN.

Results for the unclamped and clamped dielectric constants obtained for all crystals are plotted in figure 4 and reported in Table 1 and Table 2.

Table 1. Absolute values of the r_{222} EO coefficient and related parameters of LN:Zr crystals as function of zirconium concentration: The EO coefficient r_{222} is obtained when the beam propagates along the c-axis of the sample and an external electric field was applied along the a-axis (or equivalently the b-axis). The EO coefficients at constant stress (r^T) were obtained by both MDM and TRM methods and at constant strain (r^S) by the TRM method, at 633 nm and at room temperature. The dielectric permittivities $\varepsilon_{22}{}^T$ and $\varepsilon_{22}{}^S$ were measured at room temperature.

	Sénarmont setup			Dielectric constant	
	TRM method		MDM method		
	$r_{222}{}^T$ (pm/V)	$r_{222}{}^S$ (pm/V)	$r_{222}{}^T$ (pm/V)	ε_{22}^T	ε_{22}^S
LN-Cg	6,6±0.4	3,9±0.3	6,4±0.4	87±5	62±3
0,625%ZrO_2	6,1±0.4	3,5±0.3	6,1±0.4	82±5	61±3
0,75%ZrO_2	6,5±0.4	3,7±0.3	6,2±0.4	83±5	62±3
0,875%ZrO_2	6,6±0.4	3,9±0.3	6,4±0.4	84±5	61±3
1%ZrO_2	6,2±0.4	3,5±0.3	6±0.4	84±5	62±3
1,25%ZrO_2	6,4±0.4	3,7±0.3	6,3±0.4	81±5	60±3
1,5%ZrO_2	6±0.4	3,5±0.3	6.1±0.4	83±5	62±3
2%ZrO_2	**5±0.4**	**2.9±0.3**	**5.2±0.4**	**72±4**	**51±3**
2.5%ZrO_2	6,1±0.4	3,6±0.3	6 ±0.4	80±5	59±3

Table 2. Absolute values of the third column of the electro-optic tensor r_{113} and r_{333} EO coefficients of LN:Zr crystals as function of zirconium concentration measured by interferometric setup at 633 nm. Configuration obtained when a light propagation along the axis (b axis) and an external electric field was applied along the c-axis. The combined coefficient r_c is measured by Sénarmont setup is compared with the calculated values given b y Eq. 3. The dielectric permittivities $\varepsilon_{33}{}^T$ and $\varepsilon_{33}{}^S$ were measured at room temperature.

	Interferometric setup			Sénarmont setup		Dielectric constant	
	TRM method			TRM method	MDM method		
	$r_{113}^{T,S}$	$r_{333}^{T,S}$	$r_c^{T,S}$ (calc.)	$r_c^{T,S}$ (meas.)	r_c^T (meas.)	ε_{33}^T	ε_{33}^S
LN-Cg	8.5±0.6	29.4±2	20±2	21±1	20.7±1	35±2	30±2
0,625%ZrO_2	8±0.6	29.5±2	20.6±2	20.5±1	20.3±1	34.5±2	29.8±2
0,75%ZrO_2	8.4±0.6	30±2	20.7±2	21.7±1	21.3±1	35.5±2	30.3±2
0,875%ZrO_2	8.8±0.6	30±2	20.2±2	20.8±1	20.8±1	34.7±2	30±2
1%ZrO_2	8±0.6	29±2	20.1±2	21±1	21.2±1	35±2	30±2
1,25%ZrO_2	8±0.6	29.5±2	20.6±2	20±1	20.3±1	34.5±2	30±2
1,5%ZrO_2	7.8±0.6	28.8±2	20.1±2	20.8±1	20.6±1	33.3±2	29.5±2
2%ZrO_2	**6.4±0.4**	**23.5±2**	**16.4±2**	**16±1**	**16.2±1**	**29±2**	**25±2**
2.5%ZrO_2	7.2±0.5	28±2	20±2	21.5±1	21.2 ±1	36±2	30.5±2

We have demonstrated in our experiments that both the unclamped and clamped effective EO coefficients and the dielectric permittivity present a kink at a threshold zirconium concentration of around 2 mol%. We can note that this concentration corresponds to the threshold concentration determined in absorption measurements. However, Kovacs et al. have shown that the absorption edge is affected by the [Li]/[Nb] ratio [31]. Therefore, below the threshold concentration, the reduction in the absorption edge can be attributed to the increase of the [Li]/[Nb] ratio due to the substitution by Zr ions of Nb ions on lithium site, i.e. on antisite Nb_{Li}. The increase of the absorption edge for higher concentration, above the threshold concentration, is linked to the disorder induces by zirconium incorporation on both Li and Nb sites and to the necessary charge compensation process. Thus, the threshold concentration can be explained by the existence of doping-induced lattice reorganization. The threshold observed in the EO and optical properties is in coincidence with the photorefractive concentration threshold observed by straightforward photorefractive measurements in Ref. [36].

Conclusion

We have determined the unclamped and clamped electro-optic (EO) coefficients of zirconium doped lithium niobate crystals, for two opto-geometric configurations of samples. Two optical arrangements, Mach-Zehnder and Sénarmont were used for the determination of all coefficients of the electro-optic tensor of the LN:Zr crystal series from pure congruent to zirconium doped with 2.5 mol% of ZrO_2 in the crystal considering a distribution coefficient of zirconium closer to one. The first configuration, with an electric field parallel to the crystallographic c axis involves r_{113} and r_{333} direct coefficients, determined with the interferometric setup, and r_c combined coefficient measured with the Sénarmont arrangement. The second configuration, with an electric field parallel to the crystallographic a or b axis allows the determination of the r_{222} coefficient with the Sénarmont setup.

Experimental results clearly highlight a specific zirconium concentration that can be considered as a threshold concentration where the EO properties decrease of about 20% in all opto-geometric configurations. This threshold appears in crystal doped with 2 mol% of zirconium.

The EO results were analyzed by the consideration of the dielectric properties of the crystals, which are also characterized in this work. The experiments show that the low and high frequencies dielectric coefficients, present also a kink at the concentration 2 mol%. The presented analysis corroborates the existence of a doping-induced reorganization of the lattice in the doped LN crystal appearing at a zirconium concentration of 2 mol%. This concentration value corresponds to a threshold for numerous physical properties.

The present study confirms the interest for electro-optic applications of all zirconium-doped LN crystals and, when considering also the photorefractive sensitivity, for laser applications with the crystal doped at the concentration threshold value equal to 2 mol% of zirconium.

References

[1] Räuber A, 1978 in E. Kaldis E (Ed.), *Current Topics in Material Sciences, North-Holland Publishing, Amsterdam*, 481.
[2] Krätzig E, Schirmer O F, 1988 in P. Gunter, J.-P. Huignard (Eds.), *Photorefractive Materials and Their Applications I, Springer, Berlin – Heidelberg*, 131.
[3] Volk T, Wöhlecke M 1998 *Ferroelectr. Rev.* B **195.**
[4] Volk T, Wöhlecke M 2008 *Springer Series in Material Science* **115.**
[5] Kostritskii S M, Aillerie M 2012 *J. Appl. Phys.* **111** 013519
[6] Chah K, Aillerie M, Fontana M D, Malovichko G, Kokanyan E 1996 *Ferroelectrics* **186** 13-16.
[7] Chah K, Aillerie M, Fontana M D, Malovichko G, Betzler K, Kokanyan E 1997 *Opt. Comm.* **136** 231-234.
[8] Chah K, Fontana M D, Aillerie M, Bourson P, Malovichko G *Appl. Phys. B: Optics and Lasers* **67** 65-71.
[9] Chah K, Aillerie M, Fontana M D, Malovichko G 2000 *Opt. Comm.* **176** 261-265.
[10] Malovichko G I, Grachev V G, Kokanyan E P, Schirmer O P, Betzler K, Gather B, Jermann F, Klauer S, Schlarb U, Wohlecke M1993 *Appl. Phys. A. Solids and Surfaces* **56** 103.
[11] Abdi F, Aillerie M, Bourson P, Fontana M D, Polgar K 1986 *J. Appl. Phys.* **84(4)** 2251.
[12] Abdi F, Aillerie M, Bourson P, Fontana M D *J. of Appl. Phys.* 2009 **106(3)** 033519.
[13] Polgár K, Kovács L, Földvári I, Cravero I 1986 *Solid State Comm.* **59(6)** 375.
[14] Aillerie M, Bourson P, Mostefa M, Abdi F, Fontana M D 2013 *J. Physics* **416** 012001.
[15] Volk T, Pryalkin V, Rubinina N 1990 *Opt. Lett.* **15** 996.
[16] Aillerie M, Bourson P, Mostefa M, Abdi F, Fontana M D 2013 *J. Phys.* **416** 012002.
[17] Schlarb U, Betzler K 1994 *Phys. Rev. B* **50** 751.

[18] Iyi N, Kitamura K, Yajima Y, Kimura S, Furukawa Y, Sato M 1995 *J. Solid State Chem.* **118(1)** 148.
[19] Volk T, Maximov B, Sulyanov S, Rubinina N, Abdi F, Aillerie M, Bourson P, Fontana M D 1999 *OSA Trends in Optics and Photonics* **27** 145.
[20] Abdi F, Aillerie M, Fontana M D, Bourson P, Volk T, Maximov B, Sulyanov S, Rubinia N, Wöhlecke M 1999 *Appl.Phys. B* **68** 795.
[21] Kong Y, Wen J, Wang H 1995 *Appl. Phys. Lett.* **66** 280.
[22] Yamamoto J.K.,. Kitamura K, Iyi N., Kimura S., Furukawa Y., Sato M. 1992 *Appl. Phys. Lett.* **61**, 2156.
[23] Petrosyan A M, Hovsepyan R K, Kokanyan E P, Feigelson R S 2000 *Proc. SPIE* **4060** 106
[24] Kokanyan E P, Razzari L, Cristiani I, Degiorgio V., Gruber J., 2004 *Appl. Phys. Lett.* **84** 1880
[25] Li S, Liu S, Kong Y, Deng D, Gao G, Li Y, Gao H, Zhang L, Hang Z, Chen S, Xu J 2006 *J. Phys. Condens. Matter* **18** 3527.
[26] Minzioni P, Cristiani I, Yu J, Parravicini J, Kokanyan E P, Degiorgio V 2007 *Opt. Express* **15** 14171.
[27] Aillerie M, Abarkan M, Kostritskii S M, Kokanyan E P 2014 *Opt. Mat.* **36(7)** 1238.
[28] Abarkan M, Aillerie M, Kokanyan N, Teyssandier C, Kokanyan EP 2014 *Opt. Mat. Express* **4(1)** 179.
[29] Argiolas N., Bazzan M., Ciampolillo M.V., Pozzobon P., Sada C., Saoner L., Zaltron A.M., Bacci L., Minzioni P., Nava G., Parravicini J., Yan W., Cristiani I.,.Degiorgio V. 2010 *J. Appl. Phys.* **108**, 093508.
[30] Abarkan M, Aillerie M, Salvestrini J P, Fontana M D, Kokanyan E P 2008 *Appl. Phys B* **92** 603.
[31] Kovacs L, Ruschhaupt G, Polgar K, Corradi G, Wohlecke M 1997*Appl. Phys. Lett.* **70(21)** 2801.
[32] Aillerie M, Theofanous N, Fontana M D 2000 *Appl. Phys. B.* **70** 317.
[33] Abarkan M, Salvestrini J P, Fontana M D, Aillerie M 2003 *Appl. Phys. B* **76** 765.
[34] Salvestrini J P, Fontana M D, Wyncke B, Brehat F 1997 *Nonlinear Optics* **17** 271.
[35] Kaminow I P 1974 An Introduction to Electro-Optic Devices *Academic Press.*
[36] Nava G, Minzioni P, Yan W, Parravicini J, Grando D, Musso E, Cristiani I, Argiolas N, Bazzan M, Ciampolillo M V, Zaltron A, Sada C, Degiorgio V 2011 *Opt. Mater. Express* **1(2)** 270.

The r_{22} electro-optic coefficients in indium-doped congruent lithium–niobate crystals

Mustapha Abarkan[1,2,*], Anush Danielyan[2,3,4], Sasha Sewastianow[2,3], Michel Aillerie[2,3], Nicéphore Théofanous[5], Sergey Kostritskii[6], Edvard Kokanyan[4,7]

[1]Université Sidi Mohammed Ben Abdallah - Fès, LSI, Taza-Gare, Morrocco.
[2]Université de Lorraine, LMOPS, EA 4423, 57070 Metz, France.
[3]Centrale Supélec, LMOPS, 57070 Metz, France.
[4]Institute for Physical Researches, National Academy of Sciences of Armenia, Ashtarak-2, Armenia.
[5]University of Athens, Dept. Informatics and Telecommunications, 15784, Athens, Greece.
[6]RPC Optolink, Sosnovaya al. 6, bd.5, Zelenograd, 124489, Moscow, Russia.
[7]Armenian State Pedagogical University, Yerevan, Tigran Mets Ave., 17, Armenia.

E-mail: mustapha.abarkan@usmba.ac.ma, aillerie@metz.supelec.fr

Abstract. The high- and low- frequency electro-optic coefficients r_{222} of In-doped lithium niobate (LN) and the corresponding dielectric permittivity as well, have been experimentally determined and compared with the results obtained in undoped congruent LN crystals. Compared to pure congruent lithium niobate, a low acoustic contribution of the electro-optic and dielectric properties are originally found in indium (In)-doped congruent lithium niobate (LN:In) crystals in the low indium concentration range [0.12-1.7 mol%]. All reported results confirm that the LN:In is a very promising candidate for several non-linear devices as Pockels cells for laser Q-switching.

1. Introduction

It is now well established that the electro-optic effect is one of the most convenient properties that is could be exploited to obtain fast Q switching. Although, $LiNbO_3$ (LN), $LiTaO_3$ (LT), $KTiOPO_4$ (KTP) and BaB_2O_4 (BBO) are widely used as Pockels cells for laser Q-switching [1, 2]. $LiNbO_3$ offers good transmission and high extinction ratio with modest half-wave voltage in transverse configuration r_{22} and it is not hygroscopic [3]. When it is grown in its usual composition, i.e. congruent composition, it suffers of a relatively low optical damage threshold being a major drawback of a LN Pockels cell [4]. It can be noted that this low optical damage threshold is related to high photorefractive sensitivity interesting for holographic applications. Thus, for the realization of integrated devices based on this property, LN crystals in congruent composition or doped with metallic ions are good candidate [5-9]. On contrary, the low optical damage threshold generally observed in LN crystals limits the use of LN in high power laser systems [10]. Nevertheless, doping by ions of transitions elements or rare earths makes possible to modify the physical properties and seems to be a way to mitigate this problem. The optical damage threshold is considerably increased in stoichiometric LN and in congruent LN doped with specific ions, as it depends on the amount of intrinsic defects present in the lattice. Some

IOP Conf. Series: Journal of Physics: Conf. Series **879** (2017) 012005 doi:10.1088/1742-6596/879/1/012005

researches point out that divalent ion such as Mg^{2+} [11-16] and Zn^{2+} [17-19] and more recently tetravalent ions as Hf [20-26] and Zr [27-32] can improve the optical damage resistance of LN crystals for specific concentrations. It was found a threshold concentration around [Hf] = 2 mol% in the melt to lead a significant increase of the photorefractive resistance and it has been found that the light-induced birefringence changes of $LiNbO_3$ crystal doped with 4 mol% of HfO_2 were comparable to that of 6 mol% MgO doped crystals [23]. It was also shown the existence of a threshold concentration around 2.0 mol% in zirconium doped LN crystals. Moreover, hafnium and zirconium ions have the advantage to possess a distribution coefficient near one at the threshold concentration. Therefore, high quality crystals can be grown easier than the usual LN:Mg crystal with 6% of MgO having a distribution coefficient closer to 1.2 [26]. This fact suggests that Hf and Zr might represent an excellent alternative for obtaining higher optical damage resistance crystals with high optical quality [27-32]. It was proved that the EO coefficients reveal a small dependence on the Hf and Zr content introduced in the LN lattice with just a kink at 2.0 mol% attributed to the strain contribution related to the introduction of Hf and Zr ions, respectively [23, 27, 28]. Otherwise, if compared to undoped congruent crystal, the hafnium and zirconium doped lithium niobate crystals present the advantage to have smaller optical damage, especially for concentration equal to 2.0 mol%, and therefore should be more suitable for EO and NLO applications [26, 29]. In these divalent and tetravalent-doped LN crystals, electro-optic behaviors were studied and the obtained results largely confirm the threshold in dopant concentrations mentioned above and as observed experimentally by other techniques. Nevertheless, only few studies mentioned the doping of LN with trivalent ions such as Sc^{3+} and In^{3+} in the aim of increasing the damage threshold for EO and NLO applications [33, 34].

Concerning this last point, we remember that piezoelectric ringing in $LiNbO_3$ crystal in r_{22} configuration is big, as evidenced by comparison to the r_c configuration. It has been proved that in $LiNbO_3$ Pockels cells the applied voltage induces a mechanical compression of the crystal via the piezoelectric effect, which remains for some interval of time after the voltage has been removed, creating an additional birefringence modulation via the elasto-optic effect. A polarization sensitive element inside the cavity transforms this modulation into an additional time-dependent cavity loss. The acoustic oscillations following the initial relaxation of the compression were considered irrelevant for Q switching due to their long periods, on the order of 2 μs [1]. Abarkan et al. [35, 36] showed that the resonances amplitude have diminuend when the pulse duration is short (less than 100 ns).

In the present work we report, as function of indium concentration, experimental results and analysis obtained in the characterization of the EO coefficients r_{222}^T and r_{222}^S mainly involved in Q-switch laser applications. The coefficient r_{222} is obtained when a light-beam is propagating along the optical axis (c-axis) of the crystal and, therefore with an optical polarization of the transmitted beam in the isotropic plane, which renders it unaffected by the temperature dependence of the birefringence within this configuration. The EO results would be then discussed. To complete the study, we have measured the frequency dependence of the corresponding dielectric permittivity ε_{22} and finally established the figure of merit allowing the comparison between materials, used as modulator, dedicated for Q–switching applications.

Five samples were grown by the Czochralski growing method from a congruent melt with various concentrations of In_2O_3 (0.12%, 0.25%, 0.5%, and 1.7%) corresponding to 0.24, 0.5, 0.9, and 2.1 at% in the crystal [33]. These crystals were considered for comparison and discussion, with a congruent lithium niobate and a Zr doped lithium niobate crystals grown from a melt with a concentration of 2mol% of ZrO_2 [27, 28]. It is to be of note that the Zr doped sample was chosen as its concentration of dopant, 2 mol% of ZrO_2 corresponds to the threshold concentration where the optical damage is the smaller in Zr doped lithium niobate series [37]. All samples were prepared, cut and polished at optical grade.

2. Electro-optic measurements

According to [38, 39], we remind that the linear or Pockels EO coefficients in dielectric materials may be described as

$$r^T = r^a + r^S. \tag{1}$$

In this relation, r^T is the EO coefficient for the unclamped sample characterized under low-frequency electric field or DC electric field, r^a is the acoustic or piezo-optic contribution related to the crystal deformation via the piezo-electric and the elasto-optic effects, and r^S is the corresponding EO coefficient for the clamped sample measured at high frequency electric field.
For EO measurements, optical interferometric and ellipsometric arrangements can be used. In the current contribution, we used the ellipsometric one with an experimental setup based on the Sénarmont setup [38]. The transfer function is presented in figure 1.

The point M_1 located at 50% transmission $(I_{max}-I_{min})/2$ point corresponds to the so-called linear working point. When a sinusoidal electric field is applied on the sample under tests, this point is associated with the "Modulation Depth Method" (MDM) [38] and can be used to determine the EO coefficient as a function of frequency from DC to 1 MHz. This maximum frequency limit is defined by the specifications of power supply and signal acquisition electronic apparatus used for these experiments. In this method, the peak-to-peak amplitude i_{pp} of the modulated signal is measured at the point M_1, and the EO coefficient is directly obtained from the following equation [38]:

$$r_{eff}(\nu) = \frac{2\lambda d}{\pi n_{eff}^3 I_0 L} \frac{i_{pp}(\nu)}{V_{pp}(\nu)}, \tag{2}$$

Here, $I_0 = I_{max} - I_{min}$ represents the total intensity shift of the transfer function, L is the length of the crystal along the beam-propagation direction, d is the sample thickness along the applied electric field direction, n_{eff} is the effective refractive index, λ is the laser wavelength and V_{pp} is the peak-to-peak value of the applied ac field at the frequency ν.

The M_1 point can be also associated to a method called "Time Response Method" (TRM) [39]. The shapes of signals involved in this method are represented in figure 1. In this method, time signals $\Delta i(t)$ and $\Delta V(t)$ are displayed and measured on an oscilloscope screen. To obtain the various contributions of the EO coefficient this method is associated with performing power supply and oscilloscope i.e. apparatus allowing very short rising-time. At the working point M_1, the instantaneous variation of the transmitted beam intensity $\Delta i(t)$, induced by the applied voltage $\Delta V(t)$ on the crystal, is given by:

$$\Delta i(t) = \frac{\pi n_{eff}^3 L I_0}{2\lambda D} r_{eff}(t) \otimes \Delta V(t), \tag{3}$$

where $\otimes$ is the convolution operator and $r_{eff}(t)$ is the instantaneous value of the EO coefficient.

The Z-transformation of these signals allows the determination of the frequency dependence of the response $\Delta i(\nu)$ with the applied voltage $\Delta V(\nu)$. Thus, the EO coefficients can be derived within the use of the same equation 2 as established in the MDM method. This allows the full determination of the frequency dependence of the EO coefficients, possibly above their acoustic resonance if special attention is taken in the choice of apparatus. The optical response at short time leads to the clamped (high frequency) coefficient r^S, while the optical response at longer time provides the unclamped (low frequency) coefficient r^T.

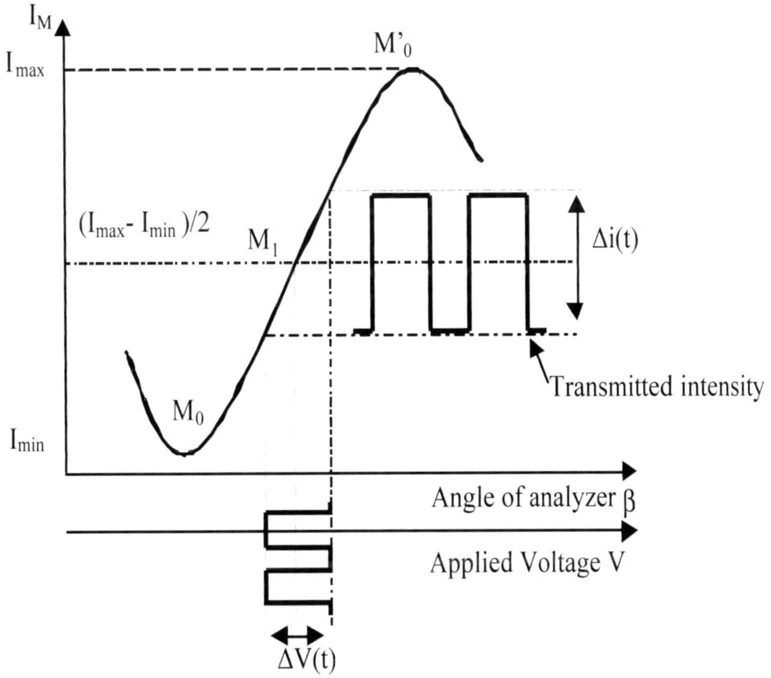

Figure 1. Optical transmission function of Sénarmont setup versus the analyzer β and applied voltage V. The point M_0 is the minimum transmission point for which the output optical signal has a frequency twice the frequency of the applied electric field. M_1 is the 50% transmission point yielding the linear replica of the ac voltage.

In practice, we have shown that this technique allows obtaining the frequency dispersion of the EO coefficient from DC up to at least 500 MHz, mainly limited by the rising time of the voltage pulse [38]. It is to be mentioned that the values of the coefficients obtained by these techniques are absolute values.

For the specific experiments that are concerned by this work, electro-optic measurements were carried out by both MDM and TRM methods presented above. Within experiments devoted to the determination of the EO coefficient r_{222} the beam propagates along the c-axis of the sample and an external electric field was applied along the a-axis (or equivalently the b-axis).

The measurements were carried out at room temperature using a He-Ne laser (λ = 633 nm). In the MDM method, we used an ac voltage of 220 V peak-to-peak at 1kHz, whereas in the TRM method, a pulse of amplitude equal to 700 V was applied onto the sample. Within these conditions, the sample dimensions and the performing optical and electric arrangements used for the experiment, the final uncertainty is in the order of 6% on the EO coefficients. Moreover, in the limit of the accuracy of the methods, no additional physical effects, as the distortion of the laser beam due to photorefractive effect or the presence of the quadratic electro-optic effect that could disturb the measurements were detected. Figure 2 shows the recording of the optical signal measured in the 1.7 mol%-In doped LN and LN in congruent composition samples at different time within the TRM method.

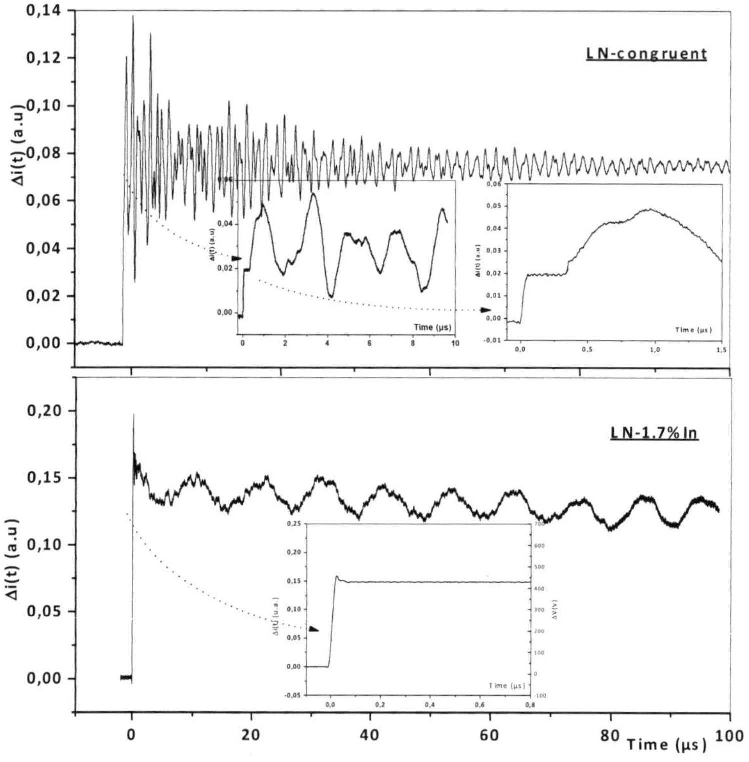

Figure 2. Responses Δi(t) to a step voltage ΔV(t) at different time scales in the case of the EO coefficient r_{222} in 1.7%-In doped LN single crystal and in LN in congruent composition. Measurements were performed at the wavelength of 633nm.

In the long-time range, the optical signal oscillates with periods corresponding to the main piezo-electric frequency resonances. In the short time, the inset of figure 2 illustrates the EO response for a small time scales. For this time the oscillations do not exist since the acoustic waves need more time to propagate across the crystal. Contrary to LN in congruent composition, no difference in the level of the signal in the two time ranges is observed with In doped LN sample pointing out the negligible amplitude of the acoustic contribution.

The frequency dispersion of the EO coefficient r_{222} is calculated according to equation 2. As shown in figure 3, the frequency dependence of the EO coefficient r_{222} of the crystal is flat on both sides of the piezo-resonances, giving, for this crystal doped with 1.7 mol% of indium the values of the EO coefficient at high frequency $r_{222}^S = 4.4 \pm 0.4$ pm/V which is identical to the value at low frequency $r_{222}^T = 4.6 \pm 0.4$ pm/V which are also in a good agreement with those measured with the MDM method (see table 1).

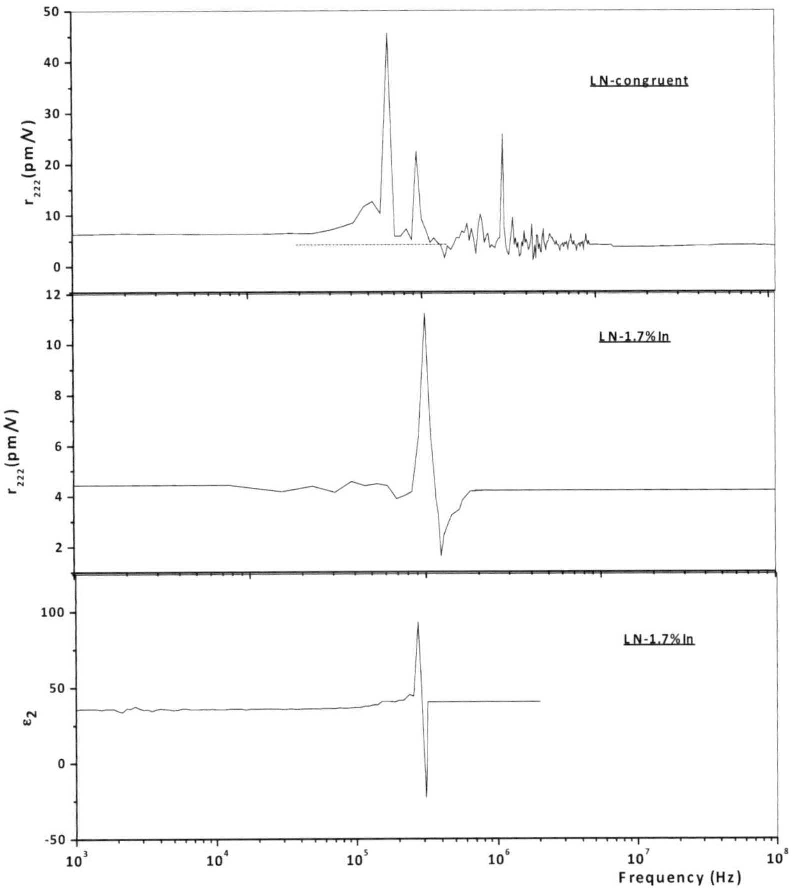

Figure 3. The frequency dispersions of the EO coefficient r_{222}in the 1.7 mol% In-doped LN crystal and Congruent LN. Comparison of the frequency dispersions of the dielectric permittivity ε_2 and of the EO coefficient r_{222} in the 1.7 mol% In-doped LN crystal.

The measurements of the coefficient r_{222} were performed for each crystal under investigation. All crystals present the same behavior as function of frequency. The dependence of the clamped and unclamped EO coefficients, r_{222}^T and r_{222}^S on the molar Indium concentration is shown in figure 4 and presented in table 1.

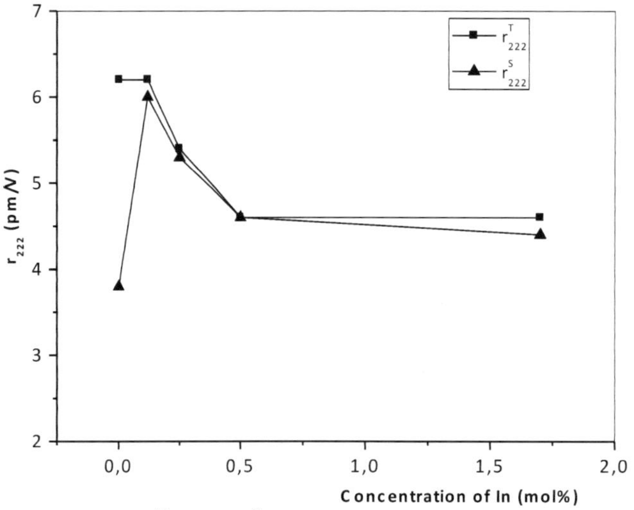

Figure 4. EO coefficient r^T_{222} and r^S_{222} versus In concentration in congruent LN.

We can see in figure 4 that both clamped and unclamped EO coefficients r_{222} decrease versus indium concentration and present a kink probably at around 1.0 mol% of In in the crystal. The values of the low- and high- frequency r_{222} coefficients in In-doped LN in the case of 1.7 mol% are equal to $r_{222}^T = 4.6 \pm 0.4$ pm/V and $r_{222}^S = 4.4 \pm 0.4$ pm/V respectively. It corresponds to a decrease of about 25% of the value obtained in the 0.1 % In doped sample. Moreover, this result also prevents that the high frequency electro-optic coefficient value in all In doped samples remains higher than in the case of pure congruent LN pointing out a lower acoustic contribution attributed to indium incorporation.

We note that contrary to Hf-doped LN crystals [13-19], the value of r_{222} coefficient versus dopant concentration in LN:In crystals has to be emphasized since this coefficient presents a non-monotonous dependence as already observed for crystals doped with other ions such as Zr, Zn or Mg [27, 28, 37, 40-42].

3. Dielectric measurements

EO properties are linked to the linear dielectric properties in ferroelectric inorganic materials. The frequency dependence [43] of an EO coefficient reproduces the behavior of the corresponding dielectric permittivity ε and a link between ε and r still exists in their dependence on the doping composition.

Using a low voltage equal to 1 V, inducing an electric field in sample under test equal to 1 kV/m, the measurements of the dielectric permittivity ε_{22} as function of frequency for all crystal-samples were done by means of two impedance analyzers HP4151 and HP4191A in frequency ranges from 1 Hz to 13 MHz and from 1 MHz to 1GHz, respectively. The frequency dispersion of dielectric permittivity ε_{22} of the 1.7 mol%-In doped LN crystal measured along the c-axis is shown in figure. 5. All samples present the same general response with the frequency being such that the In-concentration dependence of the clamped and unclamped permittivities ε_{22}^T and ε_{22}^S present a same kink-behavior as the EO coefficient r_{222} shown in figure 4.

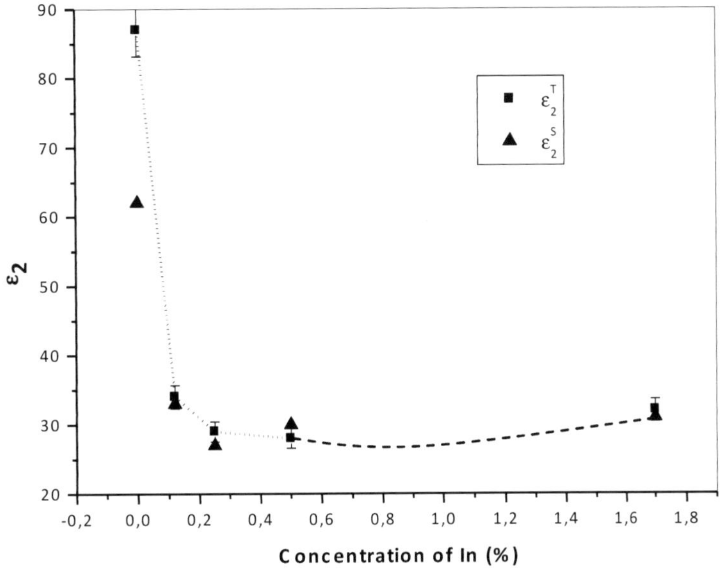

Figure 5. Dielectric constant ε_{22}^T and ε_{22}^S versus In concentration in congruent LN.

All data are reported in table 1. As expected, the behavior of ε_{22} with frequency is very similar to this of r_{222}. In figure 3, we can see that ε_{22} remains constant and displays only a very small jump between both sides of the piezoelectric resonances. We remind that this jump corresponds to the small electromechanical contribution $\Delta\varepsilon$ to the static permittivity ε^T. This will be commented and discussed below.

Table 1. Absolute values of the r_{222} EO coefficient and related parameters of LN:Zr crystals as function of zirconium concentration: The EO coefficients at constant stress (r^T) were obtained by both MDM and TRM methods and at constant strain (r^S) by the TRM method, at 633 nm and at room temperature. The dielectric permittivities ε_{22}^T and ε_{22}^S were measured at room temperature. The figure of merit $F = n^7(r_{222}^S)^2/\varepsilon_{222}^S$ was calculated from experimental values.

	MDM method	TRM method		Acoustic contribution	Dielectric constants		Figure of merit
	r_{222}^T (pm/V)	r_{222}^T (pm/V)	r_{222}^S (pm/V)	$r_{222}^T - r_{222}^S$ (pm/V)	ε_{22}^T	ε_{22}^S	$n^7(r_{222}^S)^2/\varepsilon_{22}^S$ (pm/V)^2
LN-Cg	6,4±0.4	6,6±0.4	3,9±0.3	**2.7±0.4**	87±5	62±3	**79**
2%ZrO_2	5.2±0.4	5±0.4	2.9±0.3	**2.1±0.4**	72±4	51±3	**157**
1.7%In	4.8±0.3	4.6±0.3	4.4±0.3	**0.2 ±0.4**	32±1	31±1	**200**
0.5%In	4.6±0.3	4.6±0.3	4.5±0.3	**0.1 ±0.4**	32±1	30±1	**216**
0.25%In	5.5±0.4	5.4±0.4	5.3±0.4	**0.1 ±0.4**	29±1	27±1	**333**
0.12%In	6.4±0.4	6.2±0.4	6 ±0.4	**0.2 ±0.4**	34±2	33±2	**349**

4. Discussion

In inorganic crystals the contribution arising from the optical phonons is responsible for a large value of the EO coefficient at high frequency r^S [43]. The acoustic contribution can be thus derived from the experimental values of the EO coefficients measured below and above the piezo-electric resonances. Compared to congruent composition where we found $r_{222}{}^{a} = r_{222}{}^{T} - r_{222}{}^{S} = 2.7 \pm 0.4$ pm/V, we can observe a small dependence of the acoustic contribution in the case of the In-doped LN crystals, which is found $r_{222}{}^{a} = r_{222}{}^{T} - r_{222}{}^{S} = 0.2 \pm 0.4$ pm/V.

The dependence of all physical properties on indium concentration in lithium niobate is still not available. Nevertheless, we are able to estimate the acousto-optic contribution to the electro-optic effect, r^{a} from the elasto-optic (Pockels) p^{E}_{ijkl} at constant electric field, and from the piezoelectric d_{kij} tensors and the electromechanical contribution to the static permittivity, i.e. the difference $\Delta\varepsilon_{ij}$ between the low- and the high- frequency values (ε^{T} and ε^{S}, respectively) of dielectric permittivity recorded on both sides of the acoustic resonances [43]

$$r^{a}_{ij,k} = \sum_{lm} p^{E}_{ij,lm} d_{lm,k}, \tag{4}$$

$$\Delta\varepsilon_{ij} = \varepsilon^{T}_{ij} - \varepsilon^{S}_{ij} = \sum_{kl} d_{ij,k} e_{kl} = \sum_{kl} d_{ij,k} C^{E}_{ijkl} d_{l,ij}, \tag{5}$$

where e is the piezoelectric stress tensor and C^{E} is the tensor of the elastic constants at constant electric field or elastic stiffness.

It is to be noted that the coefficients $p^{E}_{ij,lm}$ can be also expressed with coefficients C^{E}_{ijkl} and e_{kl} using the thermodynamic relations. According to the point group 3m of LN crystal, [44] and by application of Neumann's principle to the p, d and C^{E} tensors followed by the use of the reduced-subscript notation, we obtain the piezo-optic contribution to r_{222} and to ε_{22} from Eqs. 4 and 5 as

$$r^{a}_{222} = -\left(p^{E}_{11} - p^{E}_{12}\right) d_{22} + p^{E}_{14} d_{15}, \tag{6}$$

$$\Delta\varepsilon_{22} = 2d^{2}_{22}\left(C^{E}_{11} - C^{E}_{12}\right) - 4d_{22}d_{15}C^{E}_{14} + d^{2}_{15}C^{E}_{44}, \tag{7}$$

Using the values of piezo-electric and elasto-optic coefficients, available in literature [45, 47] for the pure congruent composition only, we found (Eq. 6) $r^{a}_{222} = 2.7$ pmV^{-1} in a good agreement with the experimental value, within the experimental error (10%). This piezo-optic contribution is relatively large in LN since it constitutes nearly 40% of the total value $r^{T}_{222} \sim 6.6$ pmV^{-1}. Within Eq.7, also evaluated for pure congruent lithium niobate, we found $\Delta\varepsilon_{22} = 35$, which is close to the step directly detected in the experiments between both sides of the piezo-resonances.

We have demonstrated in our experiments that both the unclamped and clamped EO coefficient r_{222} and the dielectric permittivity ε_{22} decrease and present probably a kink around In-doped LN 1 mol%. From the above consideration, the small step between low- and high-frequencies in both the EO coefficient r_{222} and the dielectric permittivity ε_{22} in LN:In is mainly due to the quasi-absence of electromechanical contribution in LN:In, which constitutes the large contribution in the congruent material. We also note that the term included C_{44} is the largest in the congruent crystal even if it is too small to explain itself the piezo-optic contribution to ε_{22}. Thus, the important changes in the electro-optic coefficient and dielectric constant can be attributed to large changes in piezo-electric and elasto-optic coefficients in doped crystals and generally to the strain effects along a (or b) axis.

Furthermore, indium is an important optical damage resistance dopant in lithium niobate, which would improve the photoconductivity σ_{ph} of the crystal and we can note that indium impurity is the most ''efficient'' among optical damage resistant ions due to the lowest threshold concentration, which was believed about 1.5–2 mol % [34]. It is important to emphasize that incorporation of In impurity ions do not affect the optical absorption of lithium niobate compared to the spectrum of pure congruent lithium niobate grown from the same charge [34].

In addition, as concerns optoelectronic applications of lithium niobate crystals, we have calculated the figure of merit linked to the driving voltage and the power (switching speed) of an EO modulator [44] or of a Pockels cell [37]. This figure of merit qualify EO devices used as modulators or Q-switches and is defined as $F = n^7(r^S_{222})^2/\varepsilon_{22}$. The values of F obtained in the present study for the LN:In samples are listed in table 1. LN:In crystals exhibit higher values of the figure of merit F compared to crystals of pure congruent LN composition. Thus, LN:In is a good candidate for Pockels cell [23]. Moreover, in LN:In series, the figure of merit F is improved for crystal having a concentration of 1.7 mol% of indium. This concentration corresponds also to an increase of the optical damage resistance threshold [34]. In the other hand, comparing to the latter, the indium-doped crystal possesses several advantages, such as larger photoconductivity, lower absorption coefficient, and higher photorefractive sensitivity [34, 48]. So, the current results and analyzes point out the high potential of LN:In to be a promising materiel for EO modulation and Q-switching applications.

5. Conclusion

In a series of Indium doped lithium niobate crystals, the frequency dependence of the EO coefficient r_{222} and of the associated permittivity ε_{22} are experimentally measured and the piezo-optic contributions were calculated. The piezo-optic contribution to the EO coefficient was found much smaller in r_{222} compared to congruent composition. This is mainly related to the smaller deformation along the a (b) axis induces by the applied electric field as reflected by the difference between the low-and high-frequency values of ε_{22}.Both EO and dielectric coefficients reveal a dependence on the In content introduced in the LN lattice with a kink probably located at 1.0 mol% and attributed to the introduction of In ions and to the related strain contribution. Compared to undoped congruent crystal, the indium doped lithium niobate crystals present the advantage to have smaller optical damage, especially for concentration equal 1.7 mol%, and therefore is a promising material for EO modulation and Q-switching applications.

References

[1] Salvestrini J P, Abarkan M, Fontana M D 2004 *Opt. Mat.* **26** 449
[2] Volk T, Wöhlecke M 2008 Lithium Niobate: Defects, Photorefraction and Ferroelectric Switching, Springer-Verlag
[3] Räuber A, 1978 in E. Kaldis E (Ed.), Current Topics in Material Sciences, North-Holland Publishing, Amsterdam, 481.
[4] Krätzig E, Schirmer O F, 1988 in P. Gunter, J.-P. Huignard (Eds.), Photorefractive Materials and Their Applications I, Springer, Berlin – Heidelberg, 131.
[5] Kostritskii S M, Aillerie M 2012 *J. Appl. Phys.***111** 013519
[6] Chah K, Aillerie M, Fontana M D, Malovichko G, Kokanyan E 1996 *Ferroelectrics* **186** 13-16.
[7] Chah K, Aillerie M, Fontana M D, Malovichko G, Betzler K, Kokanyan E 1997 *Opt. Comm.* **136** 231-234.
[8] Chah K, Fontana M D, Aillerie M, Bourson P, Malovichko G *Appl. Phys. B: Optics and Lasers* **67** 65-71.
[9] Chah K, Aillerie M, Fontana M D, Malovichko G 2000 *Opt. Comm.* **176** 261-265.
[10] Eimerl D, Velsko S, Davis L, Wang F 1990 Progress in nonlinear optical materials for high power lasers *Prog. Cryst. Growth Charact. Mater.* **20(1–2)** 59
[11] Zhong G, Jin J, Wu Z, 1980 in Proceedings of the 11th International Quantum Electronics

Conference IEEE-IQEC'80 **63**
[12] Volk T R, Rubinina M N, Woehlecke M 1994 *J.Opt. Soc. Am. B* **11** 1681
[13] Furukawa Y, Sato M, Kitamura K, Yajima Y, Minakata M 1992 *J. Appl. Phys.* **72(8)** 3250
[14] Bryan DA, Gerson R, Tomaschke HE 1984 *Appl. Phys. Lett.* **44(9)** 847
[15] Aillerie M, Bourson P, Mostefa M, Abdi F, Fontana M D 2013 *J. Phys.* **416** 012002
[16] Abdi F, Aillerie M, Bourson P, Fontana M D 2009 *J. Appl. Phys.* **106** 033519
[17] Volk T, Maximov B, Sulyanov S, Rubinina N, Abdi F, Aillerie M, Bourson P, Fontana M D 1999 *OSA Trends in Optics and Photonics* **27** 145
[18] Volk T, Pryalkin V, Rubinina N 1990 *Opt. Lett.* **15** 996
[19] Aillerie M, Bourson P, Mostefa M, Abdi F, Fontana M D 2013 *J. Physics* **416** 012001
[20] Kokanyan E P, Razzari L, Cristiani I, Degiorgio V, Gruber J B 2004 *Appl. Phys. Lett.* **84** 1880
[21] Razzari L, Minzioni P, Cristiani I, Degiorgio V, Kokanyan E P 2005 *Appl. Phys. Lett.* **86** 131914
[22] Li S, Liu S, Kong Y, Deng D, Gao G, Li Y, Gao H, Zhang L, Hang Z, Chen S, Xu J 2006 *J. Phys. Condens. Matter* **18** 3527
[23] Abarkan M, Aillerie M, Salvestrini J P, Fontana M D, Kokanyan E P 2008 *Appl. Phys. B* **92** 6038
[24] Minzioni P, Cristiani I, Degiorgio V, Kokanyan E P 2007 *J. Appl. Phys.* **101** 116105
[25] Minzioni P, Cristiani I, Yu J, Parravicini J, Kokanyan E P, Degiorgio V 2007 *Opt. Express* **15** 14171.
[26] Kokanyan E P 2006 *Ferroelectrics* **341** 119
[27] Abarkan M, Aillerie M, Kokanyan N, Teyssandier C, Kokanyan E P 2014 *Opt. Mat. Express* **4(1)** 179
[28] Aillerie M, Abarkan M, Kostritskii S M, Kokanyan E P 2014 *Opt. Mat.* **36(7)** 1238
[29] Petrosyan A M, Hovsepyan R K, Kokanyan E P, Feigelson R S 2000 *Proc. SPIE* **4060** 106
[30] Kong Y, Liu S, Zhao Y, Liu H, Chen S, Xu J 2007 *Appl. Phys. Lett.* **91(8)** 081908
[31] Liu H, Liang Q, Zhu M, Li W, Liu S, Zhang L, Chen S, Kong Y, Xu J 2011 *Opt. Express* **19(3)** 1734
[32] Argiolas N, Bazzan M, Ciampolillo M V, Pozzobon P, Sada C, Saoner L, Zaltron A M, Bacci L, Minzioni P, Nava G , Parravicini J, Yan W, Cristiani I, Degiorgio V 2010 *J. Appl. Phys.***108(9)** 093508
[33] Volk T R, Rubinina N M 1992 *Ferroelectrics Lett.* **14**
[34] Qiao H, Xu J, Wu Q, Yu X, Sun Q, Zhang X, Zhang G, Volk T R 2003 *Opt. Mat.* **23** 269
[35] Abarkan M, Salvestrini J P, Fontana M D, Aillerie M 2003 *Appl. Phys. B* **76** 765
[36] Salvestrini J P, Abarkan M, Zaccaro J, Ibanez A 2003 *J. Opt. Soc. Am. B* **20(8)** 1661
[37] Nava G, Minzioni P, Yan W, Parravicini J, Grando D, Musso E, Cristiani I, Argiolas N, Bazzan M, Ciampolillo M V, Zaltron A, Sada C, Degiorgio V 2011 *Opt. Mater. Express* **1(2)** 270
[38] Aillerie M, Theofanous N, Fontana M D 2000 *Appl. Phys. B.* **70** 317
[39] Abarkan M, Salvestrini J P, Aillerie M, Fontana M D 2003 *Appl.Opt.* **42** 2346
[40] Abdi F, Aillerie M, Fontana M D, Bourson P, Volk T, Maximov B, Sulyanov S, Rubinia N, Wöhlecke M 1999 *Appl.Phys. B* **68** 795
[41] Grabmaier B C, Otto F 1986 *J. Cryst. Growth* **79** 682
[42] Salvestrini J P, Fontana M D, Wyncke B, Brehat F 1997 *Nonlinear Optics* **17** 271
[43] Kaminow I P 1974 An Introduction to Electro-Optic Devices *Academic Press*
[44] Nye J F 1957 Physical Properties of Crystals *Oxford Univ. Press*
[45] Jazbinsek M, Zgonik M 2002 *Appl. Phys. B* **74** 407
[46] Warner A W, Onoe M, Coquin G A 1966 *J. Acoust. Soc. Am.* **46(6)**, 1223
[47] Dixon R W, Cohen M G 1966 *Appl. Phys. Lett.* **8** 205
[48] Kong Y, Wen J, Wang H 1995 *Appl. Phys. Lett.* **66** 280

Thickness optimization of the ZnO based TCO layer in a CZTSSe solar cell. Evolution of its performance with thickness when external temperature changes.

Meriem Chadel[1,2,3,a)], Mohammed Moustafa Bouzaki[1,2,3], Asma Chadel[1], Michel Aillerie[2,3,b)], and Boumediene Benyoucef[1]

[1]University of Tlemcen, URMER, 13000 Tlemcen, Algeria.
[2]Université de Lorraine, LMOPS, EA 4423, 57070 Metz, France.
[3]Centrale Supélec, LMOPS, 57070 Metz, France.

E-mail: [a)] ch_meriem_ph@yahoo.fr [b)] aillerie@metz.supelec.fr

Abstract. The influence of the thickness of a Zinc Oxide (ZnO) transparent conductive oxide (TCO) layer on the performance of the CZTSSe solar cell is shown in detail. In a photovoltaic cell, the thickness of each layer largely influence the performance of the solar cell and optimization of each layer constitutes a complete work. Here, using the Solar Cell Capacitance Simulation (SCAPS) software, we present simulation results obtained in the analyze of the influence of the TCO layer thickness on the performance of a CZTSSe solar cell, starting from performance of a CZTSSe solar cell commercialized in 2014 with an initial efficiency equal to 12.6%. In simulation, the temperature was considered as a functioning parameter and the evolution of tthe performance of the cell for various thickness of the TCO layer when the external temperature changes is simulated and discussed. The best efficiency of the solar cell based in CZTSSe is obtained with a ZnO thickness equal to 50 nm and low temperature. Based on the considered marketed cell, we show a technological possible increase of the global efficiency achieving 13% by optimization of ZnO based TCO layer.

1. Introduction

Solar cells based on an absorber layer of $Cu_2ZnSn(S,Se)_4$ (CZTSSe) are considered as a promising photovoltaic device [1] for low cost [2] and high efficiency thin film solar cells. CZTSSe possesses an optimum optical band gap energy between 1.0 and 1.5 eV [3, 4], a high absorption coefficient higher than $10^4 cm^{-1}$ [5, 6]. It consists of earth abundant elements and also considered as an alternative to the expensive and toxic $CuIn_{1-x}Ga_x(S_y, Se_{1-y})_2$ (CIGSSe) and CdTe currently used as thin film solar cells [7, 8]. Watson, in 2013 obtained the best performances for CZTSSe based solar cells with power conversion efficiency equal to 12.6% [8, 9]. Several research groups reported the growth and fabrication of CZTSSe solar cells using various methods such as sputtering followed sulfo-selenization [10, 11], solution growth process [12, 13], nano-particle approach [14, 15], electrodeposition [16, 17], ink jet printing [18], pulsed laser deposition [19, 20], ball milling and hybrid ink followed by sulfo-selenization [21, 22].

In the reference cell [8], the absorber layer CZTSSe was deposit on a Mo/glass-substrate followed by a Cadmium Sulfide (CdS) buffer layer and a ZnO transparent conducting oxide layer. CdS material with the energy gap of 2.4 eV is used for the buffer layer in numerous photovoltaic cells [1-2, 7-9]. Experimental details of the cell fabrication are given in Refs. [1, 3, 8]. In the reference cell, the widths of the various layers are as follows: ZnO (200nm)/ CdS (50 nm)/ CZTSSe (2μm)/Mo/glass-substrate. Thicknesses of all layers play individually, but also collectively important role in the global efficiency of the cell. For this purpose, in the current contribution, we analyze by simulation the influence of the TCO thickness on the performance of the cell considering, for a clear analysis, constant the parameters of the other layers. This simulation is performed by using the Solar Cell Capacitance Simulation (SCAPS) simulator.

Within the same procedure we analyze the influence of the temperature on the performance of the CZTSSe solar cell considering the optimized structure found by simulation with the ZnO based transparent conductive oxide (TCO) layer, and by comparison with the reference cell, we discuss the influence of the thickness of this layer on the performance of the cell when the temperature change.

2. The reference solar cell

The structure of the reference CZTSSe solar cell is represented in Figure 1. It is designed with the CZTSSe absorber layer deposited on a Mo/glass-substrate. On this layer, with respect of the lattice parameters, a CdS buffer layer was grown and followed by a deposition of ZnO transparent conductive oxide layer. On the top surface of the TCO layer, metal flat bands assume the electrical contacts.

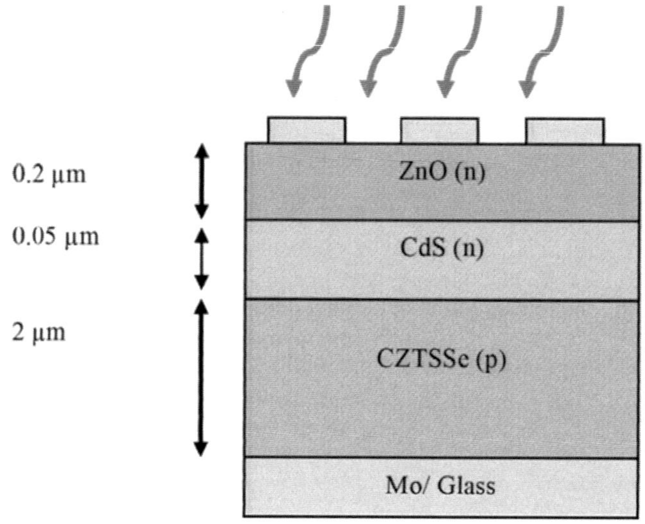

Figure 1. Structure of the simulated solar cell based in CZTSSe.

The band gap of the CZTSSe depends of the ratio S/Se in the material and can vary from 1 to 1.5 eV. In CZTSSe solar cell, the band gap (E_g) of the absorber layer is generally chosen equal to 1.13 eV [23, 24]. For each layer, all useful material parameters were introduced in the simulator code. In addition, Closely to the reference structure, in this simulation, we have introduced flat bands as front contacts, for which SCAPS software automatically calculates the metal work function depending upon whether it is n-type or p-type or intrinsic type layers [25-26]. For the flat band calculations, the layer adjacent to the contact was considered.

The necessary parameters of the materials for each layers in the reference solar cell are the thicknesses for each layer, as mentioned in figure 1 and, additionally to the above mentioned values for the energy gaps $E_{g\text{-}CZTSSe}$=1.13 eV for the CZTSSe layer, $E_{g\text{-}CdS}$=2.4 eV for the CdS layer, and $E_{g\text{-}ZnO}$=3.37 eV for the ZnO based TCO layer. We note that the necessary parameters, used in this contribution were taken from literature values [8, 27, 28].

The external parameters considered as fixed for this simulation are the solar spectrum, equal to AM.1.5, the series resistance (R_S) , R_S=0.36 Ω.cm^2 and the shunt resistance (R_{Sh}) R_{Sh}=1000 Ω.cm^2.

As far as we are concerned, namely the ZnO layer, we recall that this material is no toxic material, low cost, and presents a high stability [29, 30].

Before starting the performance analysis of our structure, we validate our model for the commercial cell [8] with the simulator SCAPS. The result of this first study is illustrated in figure 2 by the current density-voltage (J-V) characteristic for the reference CZTSSe solar cell.

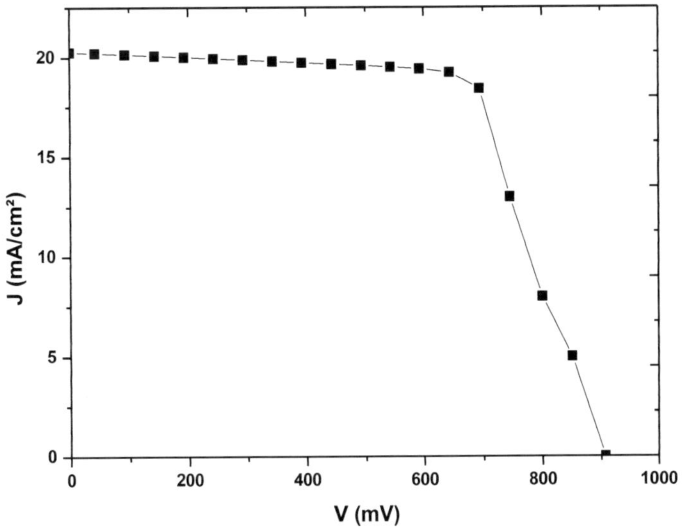

Figure 2. J-V Characteristics of the reference solar cell.

The efficiency of CdS/CZTSSe solar cell is 12.75%, which is not very different from that of a practical solar cell CZTSSe obtained by W. Wang et al and equal to 12.6% in ref [8]. This shows that our solar cell model is valid and accurate, with an error gap compare to practical values less than 1.2% and that it can be used to study the effects of the ZnO based TCO layer thickness on the performance of the cell.

3. Optimization of the TCO layer thickness

Figure 3 shows the evolution of the performance parameters of the cell, i.e. the short-circuit current density (J_{SC}), open-circuit voltage (V_{OC}), fill factor (FF), and efficiency (η) of the CZTSSe solar cell as a function of the thickness ZnO based TCO layer from 0.05 μm to 0.2 μm. These two considered values for the thickness of the TCO layer in our simulation are industrially possible for commercial cells.

The open circuit voltage is relatively constant with the thickness of the ZnO layer. On the contrary, J_{SC} decreases from the 20.68 mA/cm^2 to 20.27 mA/cm^2 when the thickness of the ZnO layer increased from 0.05 μm to 0.2 μm. In Ref. [31] Gordon shows that the increase of the TCO thickness directly induces an increase of the optical absorption in the layer of the incident light at the front side of the cell. This optical absorption is linked to the internal scattering inducing a longer light path-length proportional to the thickness. Thus this loss in the TCO layer due to absorption decreases the amount of photons taking part in the photo-electronic conversion process in the active layers of the cell, with a direct consequence of the decrease of the current and of course, its short circuit value J_{SC}.

By cons, we observe that the FF is only slightly improved starting at 69% for a layer with a thickness equal to 0.05 μm growing close 69.5% for that having a thickness equal to 0.2 μm. in fact, the main photovoltaic process takes place in the buffer and absorber layers and the form of the J-V characteristic is not significantly changed by a change of the thickness of the TCO layer.

Finally, we can see the huge influence of the thickness of the TCO layer on the efficiency of the cell, which is the highest, equal to 13% for the thinnest TCO layer, and after a continuous decrease with the thickness when the layer possesses the thickness of the reference cell, i.e. down to an efficiency equal to 12.75% for a thickness equal to 0.2 μm. This loss of efficiency is straightforwardly linked to the decrease of the number of photons crossing through the TCO layer with the thickness, i.e. the decrease of the current density participating in the photovoltaic process, as explained above. Thus, the transparent conductive layer thickness increases deteriorates the cell performances.

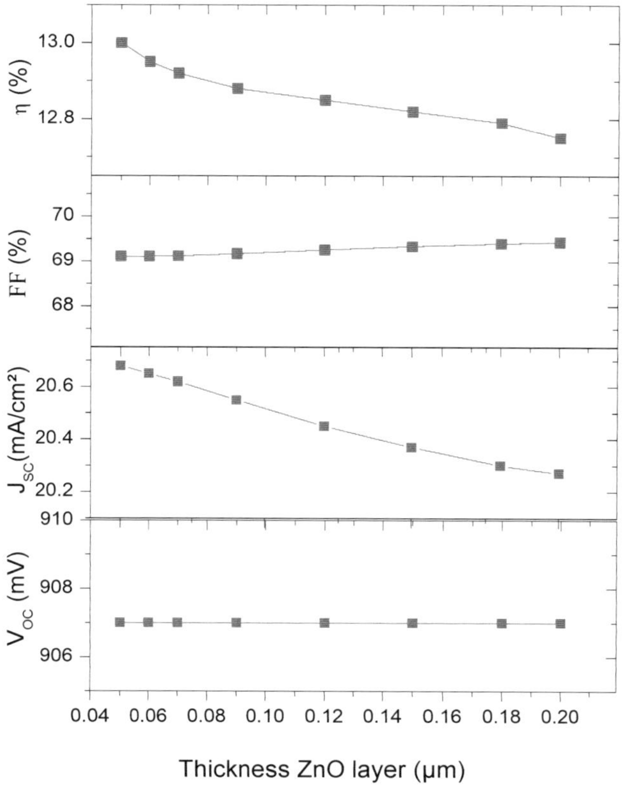

Figure 3. Effects of the thickness of the TCO layer on the performance of the CZTSSe solar cell.

Now, the relative questions to the above simulation results concern the knowledge of the performance of the optimization cell with a change of temperature and if the integration of a thinner TCO layer presents significant advantages in term of efficiency over the entire operating temperature range.

4. Influence of the temperature on the performance of the optimized solar cell

In this section, we consider the same performance parameters for the cell than those considered above except the thickness of the ZnO based TCO layer. In this paragraph, simulations were done with two different thicknesses of the TCO layer, one corresponding to that of the reference cell, i.e. a thickness equal to 0.2 μm and the second to the optimized one, i.e. a thickness equal to 0.05 μm. We have plotted in Figure 4, the evolution of the performance parameters of the CZTSSe solar cell in the temperature range from 5°C to 70°C, corresponding to the generally considered temperature range in practical functioning conditions for both structures.

It is clear from the curves in Figure 4 that J_{SC} slightly increases with increasing temperature with a higher current in all the temperature range for the cell having the thinner TCO layer compared to the reference cell as already mentioned and explained in the previous paragraph. According to the well-known Varshni relationship for temperature dependence of the bandgap energy, this slight increase of J_{SC} reveals small bandgap shrinkage with temperature of the CZTSSe cell. The decrease of Eg with temperature results of a higher absorption for incident light at visible and IR wavelengths inducing an increase of J_{SC}.

By cons, V_{OC} largely decreases when the temperature increase but no difference occurs in both reference and optimized cells confirming the same V_{OC} observed above for cells having various TCO thicknesses in the all temperature range.

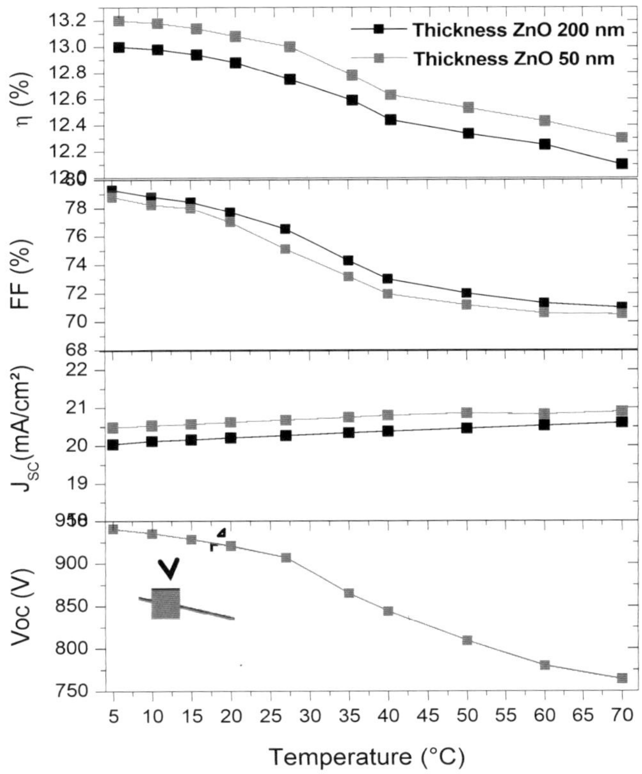

Figure 4. The performance of the CZTSSe solar cell as function of temperature for two thicknesses of the cell.

In Ref. [32], Cai et al. have modeled the dependence of V_{OC} with the temperature for CZTSSe cells. Thus, the relation they suggested between V_{OC} and the temperature can be written as [32]:

$$\frac{dV_{OC}}{dT} = -\frac{V_{go}-V_{OC}+\frac{\gamma kT}{q}}{T} \tag{1}$$

where V_{g0} is the equivalent voltage of the semiconductor band gap at absolute zero temperature (T) obtained by linear extrapolation; γ includes temperature related factors determining the reverse saturation current density; q is the electronic charge and is equal to 1.6×10^{-19} eV and k is the Boltzman constant.

Thus, Eq. 1 shows that V_{OC} is independent of the TCO thickness and is only linked to the thermal agitation of the charge carriers being important at the atomic level. Indeed, an increase of the electronic vibration leads to an increase of the recombination rate, and thus, a decrease in the voltage values of the solar cell with temperature.

We also show in Figure 4 that the evolution of FF and η with temperature present same behavior that Voc. It is due to the fact that FF is a function of V_{OC} and presents the same change with temperature and, thus, also decreases when temperature increase. In fact, FF depends of other functional parameters of the cell but results of simulation prove that, even if other factors as J_{SC} increase with temperature, the decrease of V_{OC} plays a more important role in FF, which finally decreases with temperature. In the same way, as already well known in solar cell performances, the efficiency of the solar cell decreases with the temperature.

In our simulation we observe that the gap in the efficiency between both reference and optimized cells is found equal to around 0.25%. This difference remains quasi-constant in the whole temperature range, with an absolute decrease for both cell of around 8% for a change of temperature equal to 65° temperature range (from 5°C to70°C).

5. Conclusion

In this contribution, we have studied the thickness of the ZnO based TCO layer as a parameter to be optimized in order to improve the performances of a CZTSSe solar cell. We have considered the range of the TCO layer thickness by the technologically feasible thickness in commercial cells. From the results of simulation obtained with SCAPS software on the basis of a commercial CZTSSe solar cell, we have shown that the performances of a CZTSSe solar cell can be improved by the decrease of the thickness of its TCO layer. Thus, in our study, the best performances of the solar cell based on CZTSSe, compared with those of a reference cell, are obtained with a cell having a thickness of 50 nm of the ZnO based TCO layer whatever is the considered temperature in the 5-70° range. This thickness remains a technological possible solution for commercial cells.

Finally, we found that the decrease of the efficiency is equal to 8%, passing from 13.2% to 12.2% for a change of temperature equal to 65° (from 5°C to70°C). It results at room temperature, for the optimized cell having a ZnO based TCO layer equal to 0.05 μm, an efficiency equal to 13%, a fill factor equal to 69.1%, a current density of short circuit equal to 20.68 mA/cm^2 and an open circuit of voltage equal to 907 mV.

References

[1] Varley J B, Lordi V, He X and Rockett A 2016 *J. of Appl. Phys.* **119** 025703
[2] Subbaiah Y V, Prathap P and Reddy K R 2016 *Appl. Surf. Scien.* **253(5)** 2409
[3] Halim M A, Islam M M, Luo X et al. 2016 *AIP Advances.* **6** 035216
[4] Li J, Shen H,Wang W, Chen J, Shang H, Li Y, Zhai Z. 2016 *Materials Letters.* **172** 90
[5] Das S, Chaudhuri S K, Bhattacharya R N, Mandal K C 2014 *Appl. Phys. Lett.* **104** 192106
[6] Benmir A, Aida M S. 2016 *Superlattices and Microstructures.* **91** 70
[7] Park J Y, Chalapathy R B, Lokhande A C, Hong C W, & Kim, J H. 2016 *J. All and Comp.*
[8] Wang W, Winkler M T, Gunawan O, Gokmen T, Todorov T, Zhu Y, Mitzi D B. 2013 *Adv. Energy Mater.* **4** 1301465
[9] Kim J H, Hiroi H, Todorov T K, Gunawan O, Kuwahara M, Gokmen T, Nair D, Hopstaken D, Shin B, Lee Y S, Wang W, Sugimoto H, Mitzi D B 2014 *Adv. Mater.* **26** 7427
[10] Gang M G, Shin S W, Hong C W, Gurav K V, Gwak J, Yun J H, Lee J Y, Kim J H,2016 *Green Chem.* **18** 700
[11] Hiroi H, Sakai N, Iwata Y, Kato T, Sugimoto H, 2015 *IEEE Photovolt. Special. Conf*
[12] Xin H, Vorpahl S M, Collord D, Braly I L, Uhl A R, Kruger B W, Ginger D S, Hill House H W,

2015 *Phys. Chem. Chem. Phys.* **17** 23859
[13] Schnabel T, Abzieher T, Friedlmeier T M, Ahlswede E, 2015 *IEEE J. Photovolt.* **5** 670
[14] Miskin C K, Yang W C, Hages C J, Carter N J, Joglekar C S, Scatch E A, Agrawal R,2015 *Prog. Photovolt. Res. Appl.* **23** 654
[15] Singh S, Brandon M, Liu P, Laffir F, Redington W, Ryan K M 2016 *Chem. Mater.* **28** 5055
[16] Seo S W, Jeon J O, Seo J W, Yu Y Y, Jeong J H, Lee D K, Kim H G, Ko M J, S H J, Jang H W, Kim J Y 2016 *ChemSusChem.* **9** 439
[17] Gurav K V, Shin S W, Patil U M, Suryawanshi M P, Pawar S M, Gang M G, Vanalakar S A, Yun J H, Kim J H, 2015 *J. Alloys Compd.* **631** 178
[18] Lin X, Kavalakkatt J, Lux Steiner M C, Ennaoui A, 2015 *Adv. Sci.* **2** 1500028
[19] Hea J, Suna L, Dinga N, Konga H, Zuo S, Chena S, Chena Y, Yanga P, 2012 , *J. Alloys Compd.* **529** 34
[20] Vanalkar S A, Mali S S, Agwane G L, Kamble A, Kim I Y, Patil P S, Kim J Y, Kim J H, 2016 *Sol. Energy Mater. Sol. Cells.* **157** 331
[21] Li C, Yao B, Li Y,.Xiao Z, Ding Z, Zhao H, Zhang L, Zhang Z 2015 *J. Alloys Compd.* **643** 152
[22] Ma R, Yang F, Li S, Zhang X, Cheng S, Liu Z 2016 *Appl. Surf. Sci.* **368** 8
[23] Todorov, T., Sugimoto, H., Gunawan, O., Gokmen, T., & Mitzi, D. B. 2014 *Absorbers. IEEE Journal of Photovoltaics.* **4(1)** 483
[24] Gershon, T., Gokmen, T., Gunawan, O., Haight, R., Guha, S., & Shin, B 2014 *MRS Communications.* **4(04)** 159
[25] Niemegeers A, Burgelman M, Decock K, Verschraegen J, Degrave S 2014 *<http://scaps.elis.ugent.be/>*
[26] Burgelman M, Nollet P, & Degrave S 2000 *Thin Solid Films.* **361** 527
[27] Asma, C., Boumdienne, B., & Meriem, C. 2016 *Inter. J. Nanoelec. Mat.* **9(2)**
[28] Bedia A, Bedia F Z, Aillerie M, Maloufi N, & Benyoucef B 2014 *Energy Procedia.* **50** 603
[29] Lundberg O, Bodegård M, Malmström J, & Stolt L. 2003 *Prog. Photovo. Rese. Appl.* **11(2)** 77
[30] Kim D.2009 *Trans. Electr. Electron. Mater.* **10** 169
[31] Gordon, R G.2000 *MRS bulletin.* **25(08)** 52
[32] Cai X, Zeng S, Li X, Zhang J, Lin S, Lin A, Zhang B. 2011 *Inter. Conf. Elect. Contr. Engin.* 1535

Growth of new borate crystals with fiber shape by the micro-pulling down technique

Farah Assi[1,2], Michel Ferriol[1,2], Michel Aillerie[1,2], Marianne Cochez[1,2]

[1]Université de Lorraine, LMOPS, EA 4423, 57070 Metz, France.
[2]CentraleSupélec, LMOPS, 57070 Metz, France.

E-mail: michel.ferriol@univ-lorraine.fr, aillerie@metz.supelec.fr, marianne.cochez@univ-lorraine.fr

Abstract. Borate-based materials are of high interest to generate an UV laser light from a crystal. Fiber-shaped crystals combining the advantages of fiber lasers and crystalline lasers, we have investigated the growing conditions required to obtain usable fibers of $Bi_2ZnB_2O_7$ (BZBO) and $LaBGeO_5$ (LBGO). For BZBO, the major drawback was the pronounced color of the fibers and for LBGO, it was the high viscosity of its melt leading to use a flux. This paper presents our state of the art to obtain good quality BZBO and LBGO crystal fibers.

1. Introduction

To generate an ultraviolet laser light from a crystal, the only efficient way is currently by frequency conversion of a visible of near-infrared source. For that purpose, several crystals can be used on condition that they possess high non-linear properties, wide window of transparency and high optical damage threshold. A lot of crystals of the borate family possess more or less these characteristics. Borates are considered as essential materials for the achievement of lasers able to replace excimer lasers and more generally for solid state lasers in the blue-UV range. Among crystals of the borate family, one can cite β-BaB_2O_4 (BBO), LiB_3O_5 (LBO), CsB_3O_5 (CBO), $CsLiB_6O_{10}$ (CLBO), $Ca_5(BO_3)_3F$ (CBF), and $YCa_4O(BO_3)_3$ (YCOB) widely used for efficient second harmonic generation (SHG) of Nd:YAG lasers. [1–5]. Despite recent significant progress in growth and characterization, these crystals also possess some drawbacks such as difficult crystal growth or hygroscopicity.

The Czochralski technique is generally the technique used for growing borate crystals. This method does not allow both quickly growing the samples and enhancing their crystal quality and optical properties. By else, it is however time consuming. On the contrary, the production of crystal fibers by specific growing method can present a solution to the drawbacks of the Czochralski technique. The most common methods used to grow crystalline fibers are laser heated pedestal growth (LHPG) and micro-pulling down (μ-PD) techniques [6–8]. They have shown to be versatile methods for the growth of high quality crystals of a large variety of materials. By else, in addition to their unique set of properties allowing various uses in optoelectronic devices, the final shape of crystal fibers is suitable for physical and optical testing, reducing time and cost of preparation. In addition, this specific shape of crystal fibers is directly usable for laser applications with an interlacing length/thickness ratio allowing to combine the advantages of fiber lasers and crystalline lasers, in particular the large heat exchange surface.

The goal of our research was to investigate fiber-shaped borate crystals for solid-state lasers and optical applications. We have investigated the growth of two new materials belonging to the borate family: $Bi_2ZnB_2O_7$, noted BZBO and $LaBGeO_5$ noted LBGO. Only few publications, from a few research groups, deal with these materials in literature: for BZBO [9-13] and for LBGO [14-18]. These choices are driven by applications. BZBO is considered as an interesting alternative of the well-known KDP (KH_2PO_4) because its doubling efficiency would be three or four times higher [9] and LBGO crystals are phase-matchable for Nd-laser and its SHG, while $LaBGeO_5$-Nd are promising materials for self frequency doubling laser applications.

In this paper, we report the recent progress achieved in the growth of BZBO and LBGO fiber crystals by the micro-pulling down method (μ-PD). Indeed, the viscosity being very high, it appeared to be a serious drawback for pulling LBGO by the μ-PD technique. In particular, we discuss here the contributions to the quality of the fibers obtained by the use of different fluxes that we have selected to overcome it.

2. Experimental

BZBO was synthesized from the starting materials Bi_2O_3 (Aldrich 99.9%), ZnO (Aldrich 99.99%) and H_3BO_3 (Acros 99.99%). In a first method [19] the samples were introduced in a furnace at 665°C during two weeks before reducing the temperature to 550°C with a cooling rate of $5°C.h^{-1}$ temperature at which the sample was maintained for 1 h before cooling down to room temperature by increasing the rate up to $20°C.h^{-1}$. It exists a second method [11] much simpler in which the samples were placed in the furnace and then, the following temperature program was applied: up to 630°C with a heating rate of $6°C.h^{-1}$, keeping at 630°C for 48 h and cooling down to room temperature with the same rate.

LBGO was prepared using a conventional solid-state reaction method. A mixture of 25% La_2O_3 (99,99% from Materion), 25%H_3BO_3 (99,99% from Acros) and 50% GeO_2 (99,999% from Neyco) was prepared and grinded in an agate mortar. Then, it was pressed into pellets, heated to 1050°C at $12°C.h^{-1}$, annealed in air for 82 h, then cooled down to room temperature at $24°C.h^{-1}$. The LBGO powders were checked by X-ray diffraction and DTA analysis.

In the μ-PD method, as used with the device installed in our laboratory, the material to be grown is molten in a small platinum crucible fitted with a pipe-shaped capillary at its bottom. The crucible is resistively heated but RF furnace is also used when melting temperatures above 1500 1C are required. When complete melting is obtained, a seed is placed into contact with the droplet of liquid at the end of the capillary. The fiber is obtained by pulling the seed in the downward direction. An after-heater allows controlling axial temperature gradients.

The pulled fibers were characterized by differential thermal analysis (DTA), scanning electron microscopy (SEM), EDX microprobe and Raman microspectroscopy.

3. Results

3.1. BZBO

BZBO has a non-centrosymmetric orthorhombic structure, belonging to the Pba2 space group. It possesses large nonlinear optical coefficients (d_{31}=0.91 $pm.V^{-1}$ [13]), a pretty large birefringence (0.085–0.106 [12]) and is transparent from 350 nm to more than 2500 nm [11]. Its growth is less complicated that the standard BBO because it has a congruent melting with no undesirable solid phase transition. It is non-hygroscopic and resistant to deliquescence [12].

Several fibers were easily pulled in air atmosphere with a platinum wire as seed and pulling rates ranging from 4.5 to 12 $mm.h^{-1}$. With a quasi-circular cross-section, their diameter was regular and between 300 and 550 μm (according to the pulling speed) for a fiber length up to 100 mm (figure 1). As-grown fibers were transparent and free of cracks and bubbles, but were uniformly colored from yellow to orange-red whichever was growth atmosphere (figures 1 and 2). X-ray diffraction showed that the pulled fibers were c-axis oriented with the lattice parameters: a=10.8233(4) Å, b=11.0203(6) Å and c=4.8864(2) Å. These values are in very good agreement with those given in the literature [19].

In all cases, SEM revealed that all along the fibers, small clear parts of different shapes were dispersed in a darker matrix and elongated in the growth direction (figure 3). EDX microprobe showed that oxygen was very slightly less concentrated in the clear parts than in the dark ones (less bright color in the clear zones). On the contrary, bismuth seemed slightly more concentrated in the clear zones because its color was brighter. However, the differences were too small to be quantitatively measured. That microstructure and the color of fibers were initially ascribed to a reaction between BZBO melt and platinum crucible. The growth of other fibers by LHPG, a crucibleless growth technique, invalidated this hypothesis because the obtained fibers were always colored and presented the same microstructure.

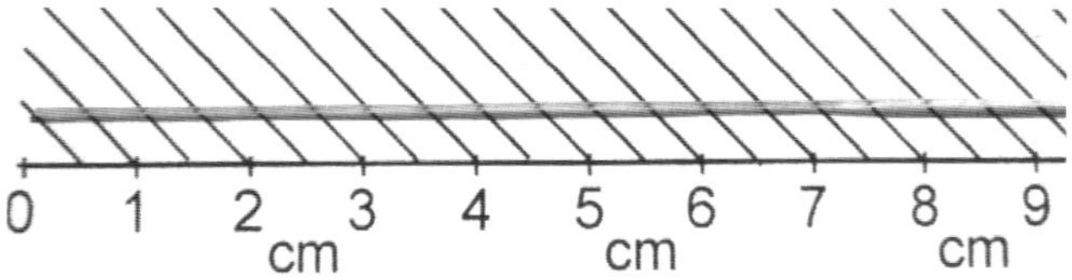

Figure 1. $Bi_2ZnB_2O_7$ fibers grown in air atmosphere

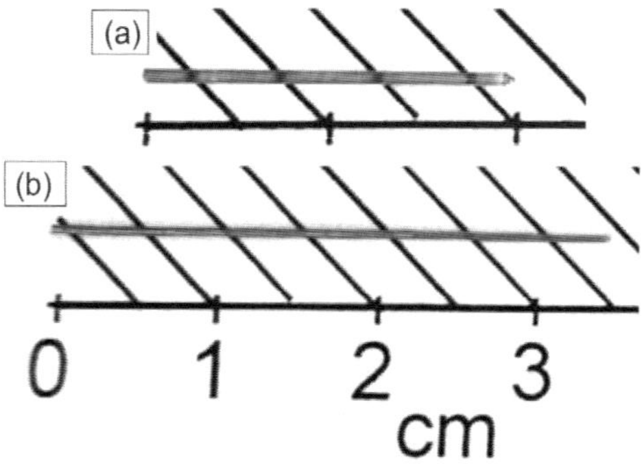

Figure 2. Examples of $Bi_2ZnB_2O_7$ fibers grown in oxygen and nitrogen

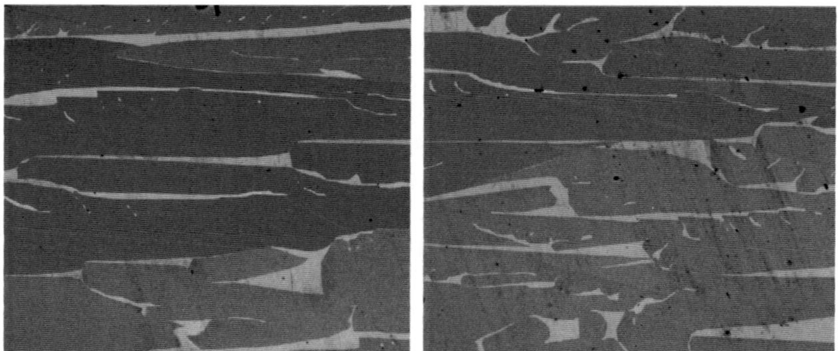

Figure 3. SEM images of different parts of the microstructure of a BZBO fiber grown in air

The elemental analysis of several crushed fibers was performed by inductively coupled plasma (ICP-AES) after complete mineralization. It revealed, during the growth process, a non-negligible loss of 13.5% boron oxide (molar composition) compared to the starting composition, probably evaporated mainly from the melt.

Correlated to DTA measurements, high temperature X-ray powder diffraction experiments performed on crushed fibers between 590 and 680°C (just before melting) showed that phase transformations occurred due to a composition shift in the Bi_2O_3-B_2O_3-ZnO system corresponding to the loss of boron oxide. All these observations suggested an incongruent behavior of the BZBO melting, possibly, of peritectic type. Then, low pulling rates are required to allow the completion of the invariant reaction and to obtain pure BZBO crystals. To validate that interpretation, a pulling rate of 0.15 mm.h^{-1} was used to grow a small transparent and colorless crystal with a 3 mm length. The very low pulling rate did not allow to obtain longer fibers. Thoroughly studied by Raman spectroscopy, the quality of the crystal was confirmed (figure 4). However, a deeper investigation of the composition is necessary on a longer fiber to assess the true stoichiometry of the crystal

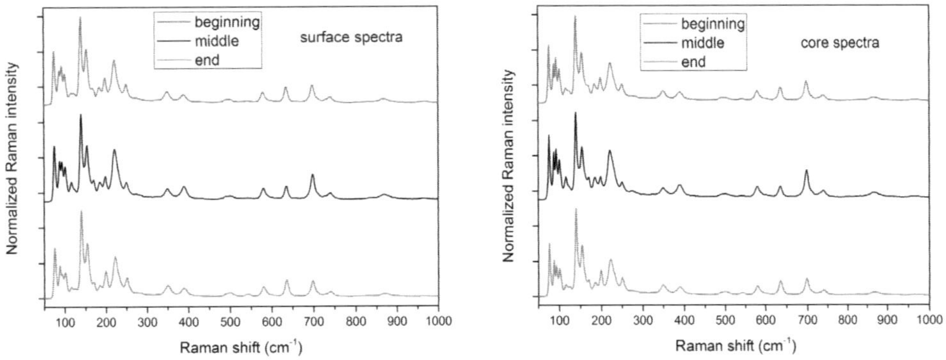

Figure 4. Typical Raman spectra of a BZBO crystal pulled at 0.15 mm.h^{-1}

3.2. LBGO

LBGO structure is similar to that of stillwellite belonging to the C_3^2 space group. The trigonal unit-cell parameters are a=7.020(5) Å, c=6.879(4) Å and Z = 3. The differential thermal analysis of LBGO showed that this crystal is congruently melting [15, 16]. By else, due to the high viscosity of the melt and violent evaporation of B_2O_3, at the melting temperature (T = 1150–1200°C), during pulling, the growth procedure is complicated.

The growth of LBGO crystal fibers directly from the melt was impossible due to its high viscosity which entailed the separation of the seed from growth interface after less than 1 mm of pulling of an opaque, white crystal. The, the use of a flux was necessary to reduce the viscosity and allow the growth of good quality crystals with usable sizes. Fluoride ions being known to break extended –O-B-O- chains and then reduce viscosity, LiF was considered as a possible flux after determination of the phase diagram LBGO-LiF (figure 5) because of the wide crystallization range of LBGO between 1200 and 760°C.

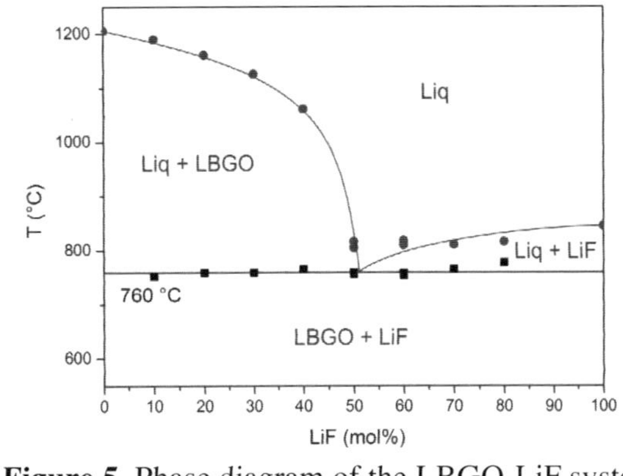

Figure 5. Phase diagram of the LBGO-LiF system

Numerous attempts to grow LBGO single crystal fibers using LiF flux were made. From the LBGO-30%LiF and LBGO-50%LiF mixtures, pulling was not successfull. The melt was still very viscous combined with a strong LiF evaporation. In order to check the evaporated LiF amount, a growth test was made from a LBGO-80%LiF mixture. With a 3 mm.h^{-1} pulling rate, we obtained an opaque and white polycrystalline fiber (Fig. 6(a)). The length of the as-grown fiber is limited due to the high volatility of LiF and an important migration of the melt on the crucible walls.

The fiber was analyzed by Raman spectroscopy and compared to the Raman spectrum of LBGO powder of literature [20]. Fig. 6(b) shows that it corresponds well to that of LBGO. This shows that a strong evaporation of LiF occurred since with the LiF starting amount used, the crystallizable compound is not LBGO according to the phase diagram (Fig. 5). It can be concluded that more than 70% of the initial LiF amount evaporated, mainly during the first times of the growth experiment where temperature can vary rather quickly during the adjustment period required to attain the steady state leading to regular fibers.

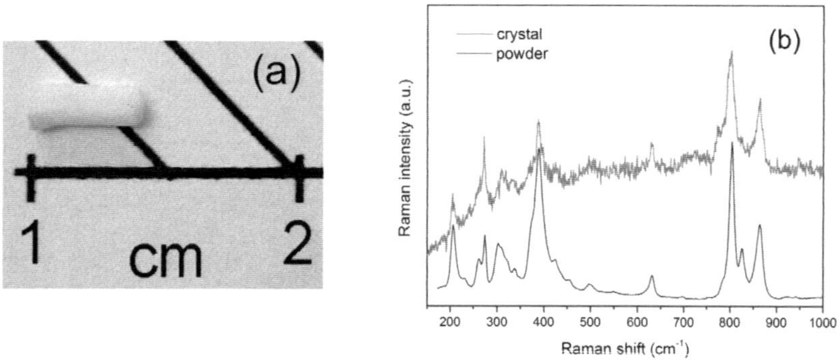

Figure 6. LBGO crystal (a) and Raman spectra of powder and crystal (b)

To limit LiF evaporation during the growth, B_2O_3 was added to the starting LBGO-LiF mixture as reported earlier [21]. Several ratios LiF/B_2O_3 were investigated to find the optimal mixture to grow good quality LBGO fiber crystals. Growth attempts from LBGO-30 mol% LiF-10 mol% B_2O_3 and LBGO-40 mol% LiF-10 mol% B_2O_3 were unsuccessful, the liquid still being too viscous. From LBGO-40 mol% LiF-20 mol% B_2O_3, the pulling was difficult and we obtained a small polycrystalline white fiber. From LBGO-50 mol% LiF-10 mol% B_2O_3 and LBGO-50 mol% LiF-7 mol% B_2O_3 mixtures, with a 3 mm.h^{-1} pulling rate, we obtained a 3 cm in length opaque white fiber. Its Raman spectrum confirmed that the fiber contained LBGO, but supplementary bands could be seen, attributed to a still unknown compound.

Fig. 7 shows the microstructure of a LBGO crystal fiber pulled from the 50mol% LiF-7 mol% B_2O_3, observed by SEM.

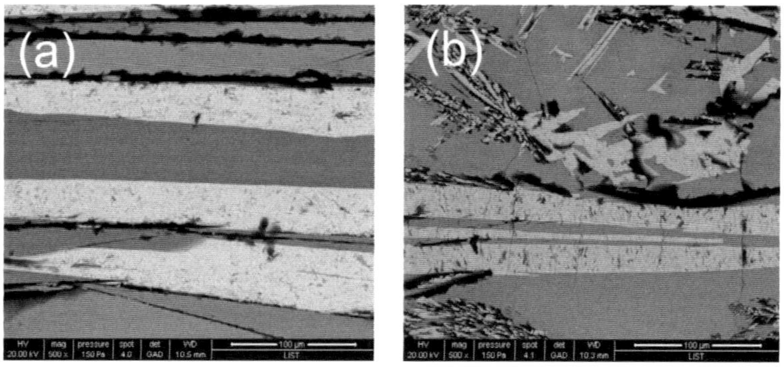

Figure 7. Microstructure of a fiber pulled from LiF-B_2O_3 flux

Three different areas can be seen, generally oriented parallel to the growth axis (Figure 7(a)) or more or less disoriented (Figure 7(b)). The darkest zones were found empty of matter except some amounts of silica due to polishing products and identified by EDX microprobe. The deficiency of matter was already found in our study of the growth of $Ca_5(BO_3)_3F$ crystal fibers by μ-PD [22]. The clearest zones were also shown to be more concentrated in lanthanum and boron than the grey ones, but poorer in germanium.

Typical Raman spectra of these two areas are given in Fig. 8. They show that they do not correspond to LBGO phase but to unknown compounds. The Raman spectrum of the clearest zone was reported earlier by Stone et al. [23] and attributed to a La-rich and B-depleted phase, in comparison to the LBGO phase. This confirms our observations.

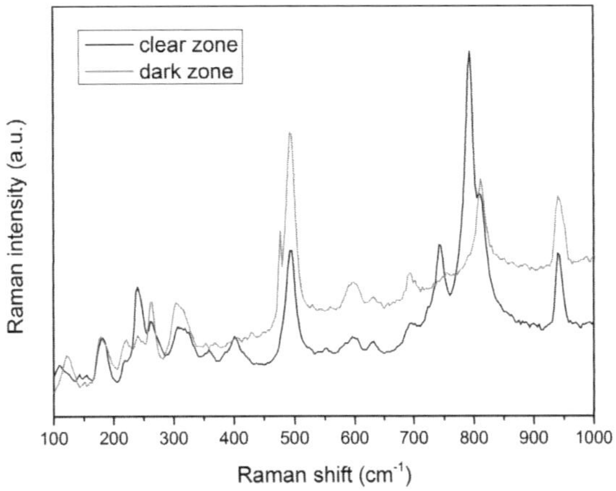

Figure 8. Raman spectra of clear and dark zones of a LBGO fiber pulled from a $LiF-B_2O_3$ flux

The observed microstructure can be interpreted taking the viscosity of the melt into account. Indeed, in a standard resistively heated micro-pulling down crucible, the melt and the meniscus are strongly stirred by thermo-capillary convection (Marangoni effect) whereas the flow through the capillary is a steady Poiseuille flow [24]. Then, one can assume that, in our experiments, the viscosity of the melt is still too high and strongly disturbs the thermo-capillary convection leading to chemical homogeneities in the melt which impact the flow through the capillary and lead to crystallization of different species inside the fiber. To solve this problem, the flux needs to be improved.

4. Conclusion

In this work, first of all, we have studied the conditions necessary to grow transparent and colorless BZBO crystal fibers. Due to the evaporation of some boron oxide from the melt, its composition is displaced into the ternary system B_2O_3-Bi_2O_3-ZnO undergoing an incongruent melting and crystallization behavior. Then, the growth of crystals needs very slow pulling rates. Using them, we succeeded in growing transparent and colorless BZBO fibers.

Concerning LBGO, the high viscosity of its melt requires the use of a flux to grow usable fibers. The phase diagram LBGO-LiF showed that LiF would be convenient. However, its evaporation combined by an insufficient decrease of viscosity led to combine boron oxide and lithium fluoride in the flux to obtain fibers with suitable length. The use of such a flux led to multiphase fibers containing unknown compounds and the flux needs to be improved. This work is in progress.

References

[1] D. Y. Tang,W. R. Zeng, Q. L. Zhao 1992 *J. Cryst. Growth* **123** 445
[2] C. Chen, Y.Wu, A. Jiang, B.Wu, G. You, R. Li, and S. Lin 1989 *J. Opt. Soc. Am. B* **6** 616
[3] Y. Wu, T. Sasaki, S. Nakai, A. Yokotani, H. Tang, and C. Chen 1993 *Appl. Phys. Lett.* **62** 2614
[4] Y. Mori, I. Kuroda, S. Nakajinma, T. Sasaki, and S.Nakai 1995 *Appl. Phys. Lett.* **67** 1818
[5] Q. Ye and B. H. T. Chai 1999 J. Cryst. Growth **197** 228
[6] R.S. Feigelson, Springer 1985 *Ser. Opt. Sci.* **47** 129

[7] R.S. Feigelson 1986 *J. Cryst. Growth* **79** 669
[8] D.H. Yoon, I. Yonenaga, T. Fukuda, N. Ohnishi 1994 *J. Cryst. Growth* **142** 339
[9] F. Li, S. Pan, X. Hou, J. Yao 2009 *Cryst. Growth Des.* **9** 4091
[10] N. Li, P. Fu, Y. Wu, J. Zhang 2009 *J. Cryst. Growth* **311** 3476
[11] F. Li, X. Hou, S. Pan, X. Wang 2009 *Chem. Mater.* **21** 2846
[12] F. Li, S. Pan, X. Hou, Z. Zhou 2010 *J. Cryst. Growth* **312** 2383
[13] F. Li, S. Pan 2011 *J. Crys. Growth* **318** 629
[14] A. A. Kaminskii, A. V. Bustashin, I. A. Malyanizin, B. V. Mill, V. S. Mironov, and S. P. Rozov 1991 *Phys. Status Solidi A* **125** 671
[15] S. Miyazawa, T. Kanamori, S. Ichikawa, and H. Nakae 2011 *CLEO/Europe and EQEC 2011 Conference Digest, OSA Technical Digest (CD)* **paper CE1_6**
[16] E. L. Belokonova, B. V.Mill, A. V. Bustashin, and A. A. Kaminskii 1991 *Izv. Akad. Nauk. SSSR, Ser. Neorg. Mater.* **27** 556
[17] V. N. Sigaev, S. Yu. Stevanovich, P. D. Sarkisov, and E. V. Lopatina 1995 *Mat. Sci. Eng. B-Solid* **32** 17
[18] S. Miyazawa, T. Kanamori, S. Ichikawa, and H. Nakae 2011 *Phys. Status Solidi A* **208** 1195
[19] A.H. Reshak, X. Chen, I.V. Kityk, S. Auluck, K. Iliopoulos, S. Couris, R. Khenata 2008 *Curr. Opin. Solid St. M.* **12** 26
[20] J. Barbier, N. Penin, L.M. Cranswick 2005 *Chem .Mater.* **17** 3130
[21] M.B. Smirmov, A.V. Menschikova, I. Kratochvilova-Hruba, Z. Zikmund 2004 *Phys. Status Solidi B* **241** 1017
[22] K.Xu, L. Gheorghe, P. Loiseau, G. Aka, J. Lejay, A. Maillard, P. Georges, D. Rytz, P. Villeval, L. Mc Donagh, F. Salin 2008 *Thematic Meeting of CNRS/CMDO Network* http://cmdo.cnrs.fr/IMG/pdf/Loiseau_CMDO_Palaiseau_nov_08.pdf
[23] N. Kozhaya, M. Ferriol, M. Cochez, M. Aillerie, A. Maillard 2011 *Opt. Mater.* **33** 1621
[24] A. Stone, M. Sakakura, Y. Shimotsuma, G. Stone, P. Gupa, K. Miura, K. Hirao, V. Dierolf, H. Jain 2010 *J. Non-Cryst. Solids* **356** 3059
[25] B.M. Epelbaum 2007 *Shaped Crystals: Growth by Micro-Pulling-Down Technique* eds. T. Fukuda, V.I. Chani (Springer, Berlin) chapter 5 p.93

IOP Conf. Series: Journal of Physics: Conf. Series **879** (2017) 012008 doi:10.1088/1742-6596/879/1/012008

High-temperature internal nitriding of heavy-section steel products

V. Khatkevich, S. Nikulin, S. Rogachev, A. Sergeychev
The National University of Science and Technology "MISIS", Leninsky pr. 4, 119049 Moscow, Russia

csaap@mail.ru

Abstract. A method for the internal nitriding of 0.08C-17Cr-0.5Ti corrosion-resistant steel sheets more than 2.5 mm thick has been suggested. The procedure includes the high-temperature nitriding of thin sheets, their hot pressing, and annealings. After such a treatment, the microhardness magnitude demonstrates the uniform distribution across the sheet section and reaches 280 HV. The formed structure of the material is characterized by a grain size of 0.5 μm and the presence of chromium-nitride particle precipitates. Compression tests showed that the prepared samples are deformed as a monolithic material; no cracks and laminations are formed on the surface.

1. Introduction

The high-temperature internal nitriding process is the saturation of a material with nitrogen. In the course of the process, in contrast to the common surface nitriding process during which nitrides are formed in the surface layer, the deep diffusion of nitrogen (up to through diffusion) takes place, which results in the formation of nitrides over the whole section of a sample. The structure of a material subjected to internal nitriding contains disperse particles distributed over the entire volume of the material, which are thermodynamically stable nitrides of alloying elements and ensure the efficient hardening of the material [1-4].

The internal nitriding is successfully used to uniformly harden thin-wall products, over their volume, made from ferritic corrosion resistant steels [5-8]. In this case, the substantial increase in the static and cyclic strength and high relative elongation at test temperatures of from room temperature to 700 °C take place.

However, the internal nitriding of steel products more than 0.5 mm in thickness is related to the substantial increase in the process time and nonuniform saturation of material with nitrogen [7]. In the present study, a method for the internal nitriding of 0.08C-17Cr-0.5Ti corrosion-resistant steel sheets more than 2.5 mm thick has been suggested. The procedure includes the high-temperature nitriding of thin sheets, their hot pressing, and annealing.

2. Experimental Procedure

Flat 0.08C-17Cr-0.5Ti steel samples 0.5 mm thick in the initially recrystallized state were subjected to nitriding at 1000-1100 °C in pure nitrogen and subsequent vacuum tempering at 700 °C. Such a treatment leads to the formation of nitrided-ferrite structure over the whole section of steel sample,

precipitation of second-phase Cr_2N mainly lamellar particles, and increase in the microhardness from 180 to 320 HV. The average grain size was 7±1 μm.
For hot pressing, a sheet pile consisting of 11 alternating non-nitrided and nitrided steel sheets was formed; the thickness of each sheet is 0.5 mm. Subsequently, the sheet pile was vacuumized and subjected to plastic deformation at 800 °C using a Gleeble System 3800 thermal mechanical simulator. Metallographic analysis of the structure of samples was performed with magnifications of ×100 and ×500 using a Buehler optical microscope and sections etched in an aqueous solution of hydrochloric acid (50 ml concentrated HCL + 50 ml H_2O).
Electron-microscopic studies of the cross-section structure were performed using a JEM-2100 (JEOL) transmission electron microscope and thin films.
The Vickers microhardness was measured using the 100-g load applied for 10 s and a MICROMET 5101 tester equipped with a digital camera and ImageExpert MicroHardness 2 software.
Compression tests were performed using spark-cut rectangular samples 2×2 mm in section and 3 mm in height and an Instron 5966 testing machine. Samples were subjected to compression to a degree of deformation of 50%; the compression direction coincided with the plane of samples layers.

3. Results and Discussion

After pressing, sheets 2.7 mm in thickness, which consists of firmly jointed layers, were obtained; no interfaces are observed on the polished section. Laminar structure of the resulting material was revealed after chemical etching (Fig. 1). The average thickness of individual layers after pressing was 250 μm.

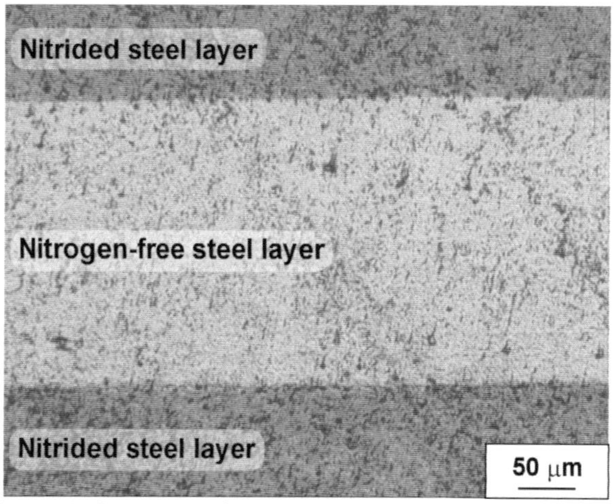

Figure 1. Structure of a cross-section of a pressed sample

Fig. 2 shows the distribution of microhardnes across the thickness of a prepared steel pack subjected to pressing and subsequent heat treatment. The microhardness of nitrogen-free and nitrided steel layers is 180 HV and 340...420 HV, respectively. The short-time heating of the pack at 1000-1100 °C and cooling in air lead to the nitrogen redistribution in nitrogen-free and nitrided layers; in this case, after hardening, the microhardness increases to 550...630 HV. The subsequent vacuum annealing at 700 °C leads to the decrease in microhardness to 280 HV and uniform macrohadness magnitude over the material section. The decrease in the microhardness from 320 to 280 HV (i.e., by 12%) as compared to that for the initial steel state (before pressing) is related to the decrease in the mass fraction of nitrogen in the material because of the nitrogen redistribution between layers during heating.

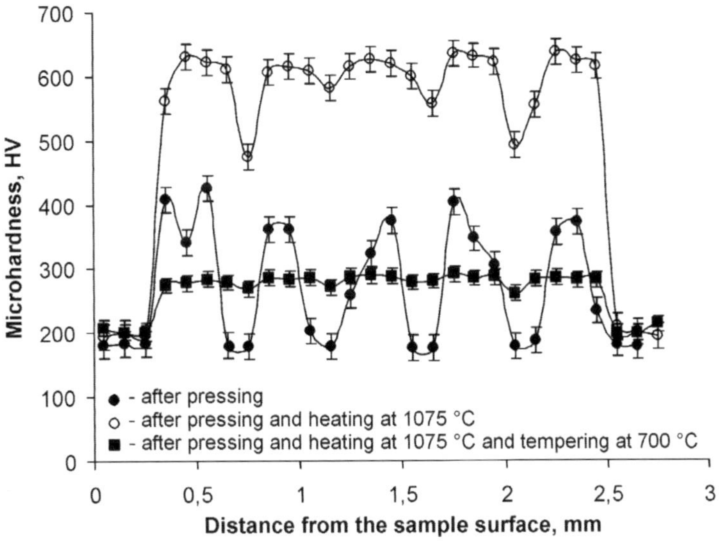

Figure 2. Microhardnes distribution across the steel-pack thickness

According to TEM data, the pressing and subsequent short-time heating at 1075 °C and tempering at 700 °C result in the formation of fragmented structure (in the cross section of a sample) with an average grain size of 0.50±0.05 μm, fine chromium-nitride particles within the grains, and elongated chromium-nitride particles at grain boundaries (Fig. 3).

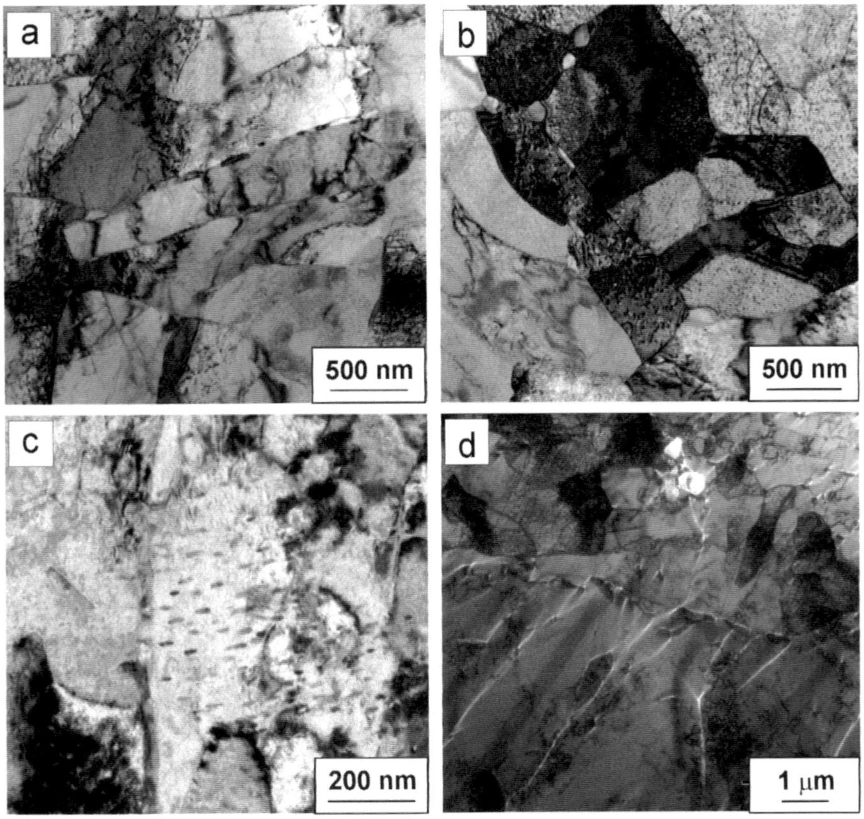

Figure 3. Microstructure (a-c) and (d) interface of steel layers of a sample subjected to pressing, short-time heating at 1075 °C and tempering at 700 °C

The decrease in the average grain size from 7 to 0.5 μm is likely to in part compensate for the softening associated with a decrease in the mass fraction of nitrogen in the material.

Compression tests showed that, after hot pressing, laminations are observed on the surface of samples. Such laminations were also observed in a sample cross section (Fig. 4a). The depth of laminations reaches 30% of the total thickness of the sample.

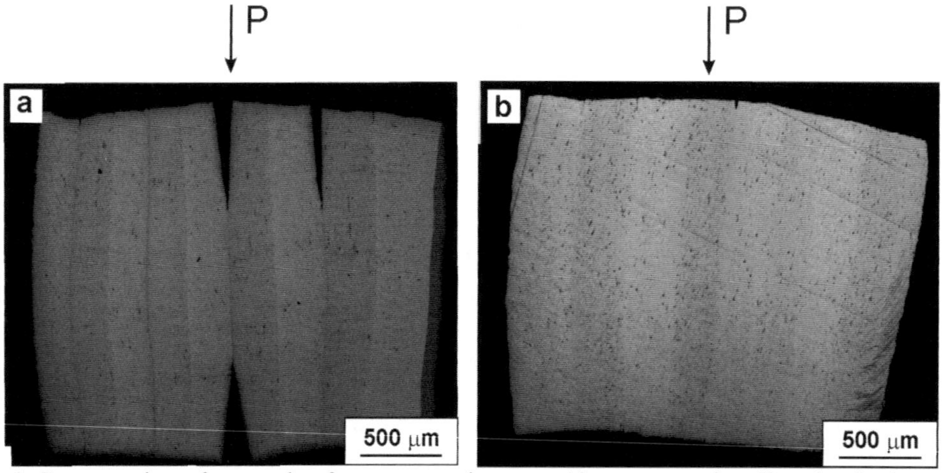

Figure 4. Cross section of a sample after compression tests (the compression direction is shown by an arrow): (a) sample subjected to pressing; (b) sample subjected to pressing and short-time annealing at 1075 °C at tempering at 700 °C

At the same time, samples subjected to hot pressing, short-time annealing at 1075 °C and subsequent tempering at 700 °C were deformed as a monolithic material without the formation of cracks and laminations on the surface. Cross sections of such samples demonstrated only individual laminations no more than 50 μm deep (Fig. 4b).

Thus, in the present study, the possibility of internal nitriding of steel products more than 2.5 mm thick and their substantial hardening was demonstrated.

4. Conclusions

A method for the internal nitriding of 0.08C-17Cr-0.5Ti corrosion-resistant steel sheets more than 2.5 mm thick has been suggested. The procedure includes the high-temperature nitriding of thin sheets, their hot pressing, and annealing. After such a treatment, the microhardness magnitude exhibits a uniform distribution across the sheet section and reaches 280 HV. The formed structure of the material is characterized by a grain size of 0.5 μm and the presence of fine chromium-nitride particle precipitates and elongated chromium nitride particles at grain boundaries. Compression tests showed that the prepared samples deform as a monolithic material, no cracks and laminations being formed on the surface.

Acknowledgment

The work has been performed with financial support of the Ministry of Education and Science of the Russian Federation (Agreement No. 14.581.21.0009 of 03.10.2014, Unique Identifier of the Project No. RFMEFI58114X0009).

References

[1] Petrova L G 2001 *Met. Sci. Heat. Treat.* **43** 11
[2] Lakhtin Y M 1991 *Met. Sci. Heat. Treat.* **33** 124
[3] Petrova L G 2004 *Met. Sci. Heat. Treat.* **46** 18

[4] Nagae M, Takemoto Y, Yoshio T, Takada J, Hiraoka Y 2005 *Mater. Sci. Eng. A* **406** 50
[5] Nikulin S A, Rogachev S O, Khatkevich V M, Rozhnov A B 2013 *Met. Sci. Heat. Treat.* **55** 351
[6] Nikulin S A, Rogachev S O, Khatkevich V M, Rozhnov A B. 2014 *Phys. Met. Metallogr.* **115** 186
[7] Nikulin S A, Rogachev S O, Khatkevich V M, Rozhnov A B, Nechaykina T A 2013 *J. Alloys Compd* **564** 114
[8] Khatkevich V M, Nikulin S A, Rozhnov A B, Rogachev S O 2015 *Met. Sci. Heat. Treat.* **57** 205

Acoustic emission analysis of crack resistance and fracture behavior of 20GL steel having the gradient microstructure and strength

S. Nikulin, A. Nikitin, V. Belov, A. Rozhnov, V. Turilina, V. Anikeenko, V. Khatkevich
The National University of Science and Technology "MISIS", Leninsky pr. 4, 119049 Moscow, Russia

nikulin@misis.ru

Abstract. The crack resistances as well as fracture behavior of 20GL steel quenched with a fast-moving water stream and having gradient microstructure and strength are analyzed. Crack resistance tests with quenched and normalized flat rectangular specimens having different cut lengths loaded by three-point bending with acoustic emission measurements have been performed. The critical *J*-integral has been used as the crack resistance parameter of the material. Quenching with a fast moving water stream leads to gradient (along a specimen wall thickness) strengthening of steel due to highly refined gradient microstructure formation of the troostomartensite type. Quenching with a fast-moving water stream increases crack resistance J_c, of 20GL steel by a factor of ~ 1.5. The fracture accrues gradually with the load in the normalized specimens while the initiated crack is hindered in the variable ductility layer and further arrested in the more ductile core in the quenched specimens.

1. Introduction

Development of gradient microstructures is an efficient way of strengthening and increasing fracture resistance of critical components from carbon steels, e.g. parts of the rail transport (side frames of freight car trucks, etc.). Development of gradient microstructures in steels is possible using heat treatment which includes holding in the austenite temperature range followed by cooling with a fast-moving water stream (so-called spray quenching or jet impingement quenching). This technology is attractive in industry because it provides intensive heat transfer rate between the component and the quenchant and high heat removal capability. Such a heat treatment allows creating the gradient distribution of strength across a component cross section (high strength surface and ductile core). At that, compressive stress is developed on the surface which prevents crack initiation [1-6].

Taking into account the strengthening effect of spray quenching on steel mechanical properties [5, 6], its influence on the static crack resistance may not be obvious, the more so if the gradient nature of the microstructure and its cross-section toughness which are developed by this type of heat treatment are considered. In this regard, the main objective of this work was the study of crack resistance and fracture behavior of 20GL steel quenched with a fast-moving water stream, having the microstructure and strength gradient along the cross section using acoustic emission and fractography.

2. Materials for the study

The study was performed on the specimens cut from the side frames of the freight car trucks made from 20GL steel (Table 1). First, a 700 mm long part was cut out of the side frame as shown on the Figure 1a. The part was subjected to heating in the resistance furnace up to 950-970 °C and then to quenching with a fast-moving water stream during 2 minutes in a special quenching device which allows to cool the part from all sides using multiple pressurized water jets. Then, specimens with a central cut were cut out of the part, oriented in the longitudinal direction of the side frame (see Figure 1a). The same specimens cut out of the side frame in the initial condition were also studied for comparison (normalized at T=930-960 °C, 2 h, cooled in air).

The wall thickness of the side frame which the specimens were cut out from (16-19 mm) did not allow making typical specimens for crack resistance tests, therefore special specimens were prepared and used. Their dimensions are shown on Figure 1b. The height of the specimens constituted b = 13.0±0.1 mm, specimens thickness, t = 10.0±0.1 mm, length, L_1 = 75.0±0.1 mm, cut depth, l = 1, 2, 3, or 4 mm (specimens with different cut length were tested), cut width, e = 2 mm, distance between the supports, L = 50.0±0.1 mm.

Table 1 – Chemical composition 20GL grade steel

C	Si	Mn	V	Al	S	P	Other
0.197	0.336	1.163	0.10	0.035	0.009	0.013	0.14 Cr; 0.092 Ni; 0.129 Cu

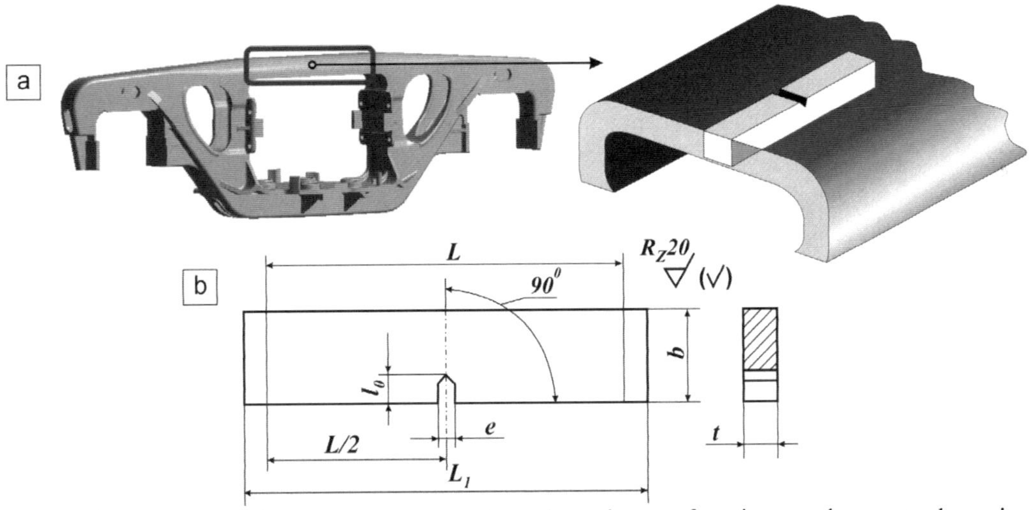

Figure 1. Specimens used for crack resistance tests (a - scheme of cutting out the part and specimen from a side frame; b - specimen geometry and sizes)

3. Methods of study

3.1. Microstructure analysis and hardness measurement

The microstructure was studied on the cross-sections after etching with 4% alcohol solution of nitric acid using Axio Observer D1m Carl Zeiss optical microscope with ×500 magnification. The microstructure of the specimens was studied in different points of the cross-section after quenching with a fast moving water stream (at different depths of the cooled surface).

Hardness was measured on the same specimens across the cross-section using Rockwell's method and Buehler MacroMet 5101T hardness gauge with 1 mm step. C scale and 150 kg loading were used for the quenched specimens which have higher hardness, and A scale and 60 kg loading were used for the normalized specimens. The measured values were then converted to a single scale (HB) using hardness tables for the comparative analysis.

3.2 Mechanical tests and acoustic emission measurement

The tests were performed at room temperature following the three-point bending on the base of 50 mm using the universal Instron 150 LX test machine with active grab travel rate of 0.2 mm/min and maximum force of 150 kN. The tests were performed on flat rectangular specimens having different cut lengths (l=1, 2, 3, or 4 mm). The acoustic emission was continuously registered during the three-point bending tests simultaneously with the "load P – deflection f" curve registration. The acoustic emission (AE) method allows precisely determining the moment of crack initiation and controlling the material failure kinetics. In particular, this method was successfully used in works [7, 8] to study the static and cyclical strength of side frames of freight car trucks. Also, this method allows obtaining quantitative assessments of the crack sizes [9, 10]. In this work, the AE method was used both for determining the moment of crack initiation during specimen loading and for detailed study of mechanisms and kinetics of 20GL steel failure.

AE signals were detected with the CTS-19 piezoelectric transducer installed on the polished surface of a specimen through a layer of glycerol (Figure 2). AE signals after preliminary amplification (30±3 dB in 5 MHz frequency band) were recorded via the PXI-1042Q (National Instruments) modular computer platform controller with up to 10 MHz ADP analogous signal frequency sample in 70 dB dynamic range. The measured parameter was peak amplitude of AE signals. The "load P – deflection f" curves were time-aligned with the acoustic emission diagrams after tests.

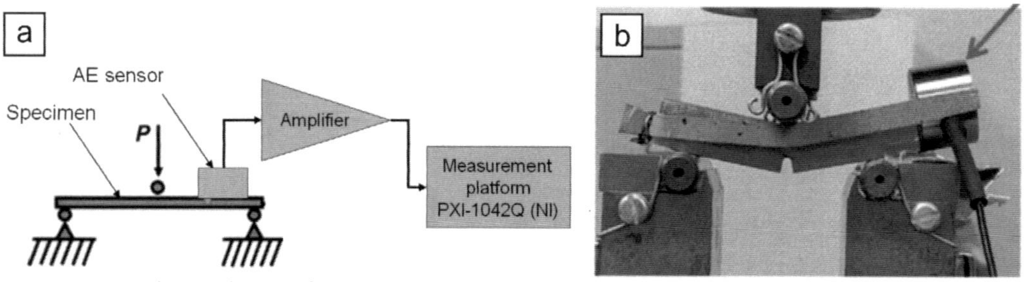

Figure 2. Experimental setup for AE measurement under mechanical three-point bending test (a-scheme: b - view of specimen with transducer attached)

Crack initiation in the cut tip was determined using acoustic emission (AE) signals which exceeded the noise level significantly, with simultaneous recording of the critical load P_c corresponding to this crack initiation. Analysis of the crack propagation mechanism and kinetics was performed using the analysis of AE signals quantity, occurring after the crack initiation and their peak amplitudes (U_{AE}, V). Crack initiation on the surface of the specimens was also observed visually using a binocular during loading and after it was stopped following registration of high-power AE pulses.

3.3 Critical J-integral value calculation

The linear fracture mechanics parameters are usually used to assess the crack resistance of brittle materials, e.g. stress intensity critical factor which does not depend on the specimen thickness (K_{Ic}, MPa·m$^{1/2}$). Non-linear fracture mechanics parameters are used for ductile materials, for instance, the critical J-integral (J_{Ic}, kJ/m^2) [11]. 20GL steel after normalization as well as after quenching with a fast moving water stream features medium strength and ductility so, a non-linear fracture mechanics parameter critical J-integral J_c was used as a parameter of the crack resistance.

IOP Conf. Series: Journal of Physics: Conf. Series **879** (2017) 012009 doi:10.1088/1742-6596/879/1/012009

Begley-Landes method was used to determine the J-integral [12, 13]. According to the method, the value of the J-integral for each selected deflection f is determined taking into account the specimen thickness t using the following formula:

$$J = (1/t)(dA/dl), \tag{1}$$

where dA is the deformation energy by the crack initiation moment, dl, crack (cut) length variation, t, flat rectangular specimen thickness.

Deformation energy A was determined as the area under P–f curve along the region from 0 to f_c. The value of f_c was determined by the acoustic emission impulse. Calibration curves $A(f)$ were plotted for different cut lengths based on the determined values of the deformation energy. J–f curves were plotted using the calibration curves $A(f)$ to determine the crack resistance parameter J_c. J–f curves were further approximated by a linear function according to the following equation:

$$Y = kX + a, \tag{2}$$

where Y is the energy divided by the crack growth value (J/m), X, cut length (0.001-0.004 m), k, specific (divided by the specimen thickness t = 0.01 m) mechanical energy per crack length unit. J_c values were determined on the J–f curves using previously determined f_c values.

3.4 Fractography

The fracture surface after the mechanical tests was studied using the JSM-6610LV (JEOL) scanning electronic microscope with ×(100–2000) magnification. The pattern of fracture surface change along the full thickness of a specimen was determined from the cut to the opposite side of the specimen.

4 Results of the study

4.1. 20GL steel microstructure and hardness

The microstructure of the 20GL steel after normalization and quenching is shown on Figure 3 while the hardness distribution along the specimen cross-section is shown on Figure 4. Normalized specimens had a typical for such steels ferrite-pearlite microstructure along the cross-section with a grain size of 25±5 μm (Figure 3a). Hardness of the normalized specimens along the cross-section constituted ~53 HRA (~7-10 HRC), which corresponds to the value of ~200 HB on the Brinell's scale.

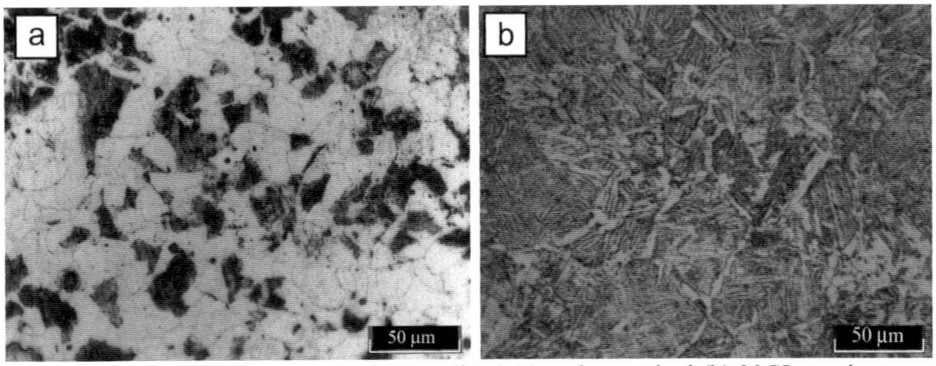

Figure 3. Microstructure of normalized (a) and quenched (b) 20GL steel.

The microstructure of the quenched specimens represented a mix of the martensite-like fine microstructure and proeutectoid ferrite of different morphology located along the grain boundaries and inside of the prior austenite grains (Figure 3b), The volume fraction of the proeutectoid ferrite was growing to some extent while moving away from the quenched surface. The hardness of the specimens after quenching changes from 38 HRC (~500 HB) at the cut tip to 32 HRC (~270 HB) at the opposite side of the specimen (see Figure 2). Decrease in hardness with the distance from the surface can likely be explained not only by the increasing volume fraction of the proeutectoid ferrite but by the change of the fine martensite-like microstructure type as well (the troostomartensite or troostite microstructure prevails near the quenched surface while the central part of the specimen is dominated by the sorbite microstructure).

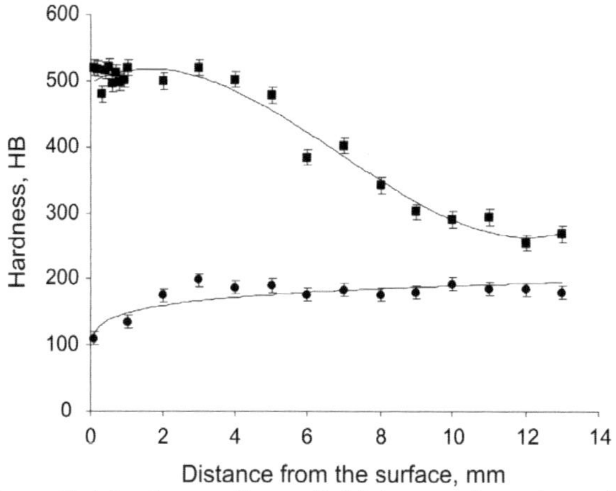

Figure 4. Hardness distribution on the wall thickness of specimen from 20GL steel after quenching (■) and normalization (●)

Thus, the 20GL steel strengthening after quenching with a fast moving water stream is due to formation of the martensite-like fine microstructure. This microstructure changes to some extent with the distance from the quenched surface which promotes the strength gradient formation along the cross-section. The absence of martensite in the microstructure (martensite could form only in a limited portion as a part of the troostomartensite in the surface layer), along with the presence of the proeutectoid ferrite, ensures sufficient ductility of such a microstructure.

4.2. 20GL steel crack resistance

Typical P–f loading curves aligned with AE diagrams registered in crack resistance tests of the 20GL steel after normalization and quenching with a fast moving water stream are shown on Figure 5 (example of specimen with 1 mm cut depth). Significant difference was observed in the mechanical behavior of the normalized and quenched specimens in the following tests irrelevant of the cut length: loading curve deviation from the linear law (corresponding to the material yield strength) and loading drop (corresponding to the reached ultimate tensile strength) were observed in the quenched specimens at significantly higher load as compared to the normalized specimens. Crack initiation registered by large AE impulses was observed in the region corresponding to uniform plastic deformation in both states of 20 GL steel (quenched and normalized).

Deflection values f_c were determined corresponding to the crack initiation moment during tests for each tested specimen. After that, the deformation energy A was determined for each specimen as the area under P–f curve along the region from 0 to f_c. Based on the determined values of the deformation energy, A-f calibration curves were plotted for different cut lengths. They further were used for J–f calibration curves plotting (Figure 6). J–f curves were used to evaluate the crack resistance parameter J_c, by the deflection f_c, corresponding to the specimen deflection at the crack initiation moment.

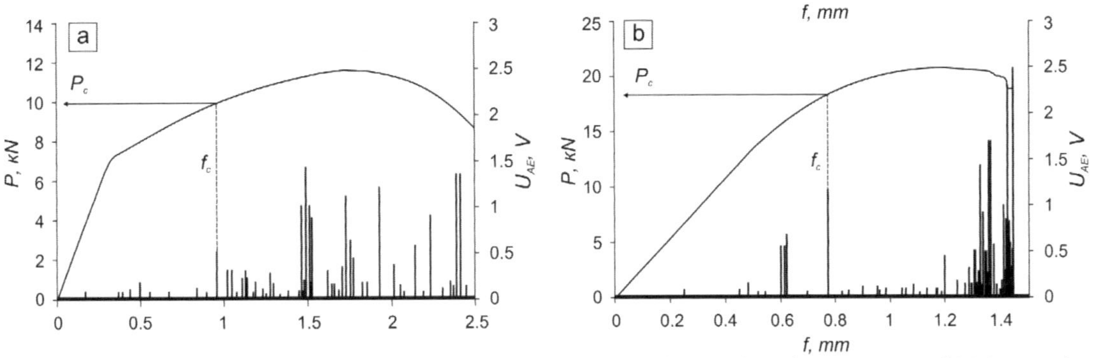

Figure 5. *P–f* loading curves aligned with AE diagrams under crack resistance tests of 20GL steel after normalization (a) and quenching (b).

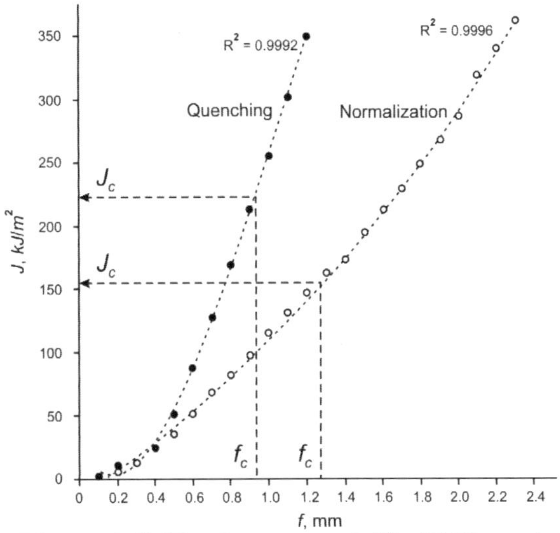

Figure 6. Calibration curves *J–f* for 20GL steel

It is clear from Figure 6 that the 20GL steel static crack resistance after quenching with a fast-moving water stream by J_c parameter constituted 225±15 kJ/m^2 which is 1.5 times higher than in normalized state (150±15 kJ/m^2). Such a behavior of the 20GL steel can likely be explained by the formation of the highly dispersed fine microstructure with a sufficiently high ductility margin. Besides, the energy to crack initiation could be influenced positively by the compressive stresses, which form in the surface layer after quenching.

4.3. Mechanism of crack propagation

Detailed comparative analysis of acoustic emission signals registered throughout the three-point bending test of 1-mm cut length specimens was performed to study the mechanisms of crack propagation in the 20GL steel after quenching and normalization. Acoustic emission diagrams aligned with the loading corves were shown on Figure 5. It is clear from Figure 5a that when loading the normalized steel specimens, the first AE signals (having low amplitude U_{AE}, less than 0.5 V) are registered at ~8 kN load and are likely to be related to the initiation of the material plastic deformation and, possibly, to a small microcrack initiation in the cut tip. One or several AE impulses of a sufficiently large amplitude (U_{AE} =0.7-1.0 V) corresponding to a macrocrack initiation in the cut tip are observed at load of ~10 kN. The initiated surface macrocrack continuously grew afterwards as the load increased. This process is accompanied by dozens of AE impulses with ~0.5 V amplitude.

Starting from the load of $0.9P_{max}$, the crack developed by macro-steps which was accompanied by several high-amplitude ($U_{AЭ}$ = 1.5–2 V) AE impulses and load drop of ΔP ~ 0.3 kN. The crack further propagated by steps at the load drop after Pmax has been reached.

First microcracks on the quenched steel specimen surface are registered at higher loads than in normalized specimens (P~15 kN). The crack initiation process is accompanied by AE impulses with U_{AE} ~0.5 V amplitude (Figure 5*b*). The forming surface cracks 15-45 μm long were observed in the specimen fracture surface after registering AE signals with amplitude over U_{AE}=0.5 V, rapid specimen unloading, thermal crack decoration (holding in a furnace at 150 °C, 40 min) and final specimen failure. Macrocrack initiation in the cut tip is registered by AE impulse of a sufficiently large amplitude (U_{AE} = 1.5 V) at the load of P~18 kN. The cause of the crack arrest in quenched specimens is likely to be the development of a plastic deformation in its top which leads to the crack tip blunting and delay that was observed on the polished surface of the specimen. In case of further loading, the propagation of the main crack by large steps restarts at the maximum load and during load drop as well.

The fractography analysis showed that the 20GL steel in both normalized and quenched state, features fully ductile fracture with typical failure elements - dimples (Figure 7). In both states, non-metal inclusions are observed at the bottom of some dimples. However, the microstructure of the ductile fractures of the specimen after normalization and quenching was somehow different. Thus, a non-uniform dimple fracture is observed in the steel after normalization in which, along with a small deep dimples up to 10 μm in size, a sufficiently large number of large "flat" shallow dimples of ~30-50 μm size containing non-metallic inclusions (Figure 7a) can be observed. The degree of the plastic deformation development around such "flat" dimples is relatively high. The dimple fracture is more uniform in the samples after quenching, the large "flat" dimples formed on non-metal inclusions are rarer, and their size (diameter) is smaller (up to 20 μm). Most part of the fracture surface is covered with a "small-dimple" component with dimple dimension, as a rule, of less than 5-7 μm.

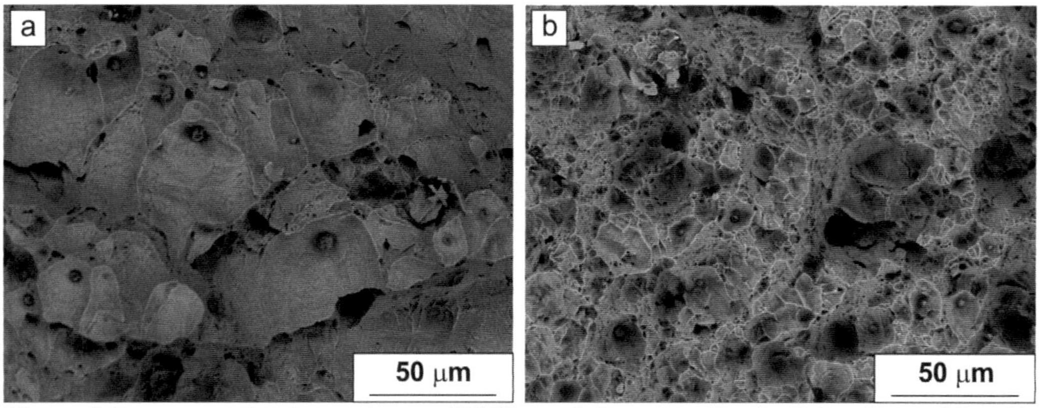

Figure 7. Fracture surface in specimens of 20GL steel after normalization (a) and quenching (b).

The disperse martensite-like microstructure formed during quenching with a fast-moving water stream ensures generally small-dimple pattern of the fracture surface. Large amount of ferrite in the normalized state represented by relatively large grains of about 25 μm in size, causes formation of large ductile fracture dimples and relevant decrease in the total length of the ductile ridges (which are walls of the dimples), all of this leading to decreased energy intensity of the fracture.

The results of the fractography analysis explain the significant difference in the crack resistance assessed by the energy parameter of the material fracture - the critical J-integral J_c. Thus, the higher energy intensity of the fracture process in the quenched specimens as compared to the normalized specimens is confirmed by the more highly dispersed dimple structure of the fracture surface in this state.

5. Conclusion

The study of the microstructure and hardness of the 20GL steel showed that quenching with a fast-moving water stream lead to the formation of the martensite-like fine microstructure and ensures formation of the strength gradient along the cross-section (surface hardness increased by a factor of 2.5, and by a factor of 1.5 in the core as compared to the normalized state).

The crack resistance tests with acoustic emission measurement along with the fractography analysis showed that creation of the gradient strength and microstructure by quenching with a fast moving water stream increased the static crack resistance of 20GL steel ≈by a factor of 1.5, as compared to the normalized state. During three-point bending loading of normalized specimens having uniform strength along the cross-section, the crack grows continuously by steps throughout the test duration while in case of quenched specimens having a strength gradient along the cross-section, the crack decelerates in the ductile core at a certain stage of its development.

Acknowledgment

The work has been performed with financial support of the Ministry of Education and Science of the Russian Federation within implementation of the Federal Target Program "Research and Development in Priority Directions of Advancement of the Scientific and Technological Complex of Russia in 2014 – 2020" (Agreement No. 14.578.21.0020 of 03.10.2014, Unique Identifier of the Project No. RFMEFI57815X0139).

References

[1] Totten G E *et al* 1993 *Handbook of quenchants and quenching technology* (Spray quenching) (ASM Int.) Chapter 7 pp. 239-291
[2] Kobasko N *et al* 2010 *Intensive Quenching Systems: Engineering and Design.* ASTM Stock Number: MNL64.
[3] Ravikumar S V *et al 2013 Steel research int.* **84 /11** 1157
[4] Pola A *et al* 2013 *Journal of Materials Processing Technology* **213** 2247
[5] Nikulin S A, Rozhnov *et al* 2016 *Mater. Lett.* **185** 499
[6] Nikulin S A *et al* 2016 *Met. Sci. Heat Treat.* **57** 678
[7] Stepanova L N *et al 2011 Russian Journal of Nondestructive Testing*, **47/3** 158
[8] Muraviev V V, Muravieva O V 2016 *Deformation and Fracture of materials.* **9** 24 (in Russian).
[9] Khanzhin V G, Shtremel M A 2009 *Met. Sci. Heat Treat.* **51** 250
[10] Khanzhin V G *et al* 1990 *Soviet Journal of Nondestructive Testing.* **26** 256
[11] Lanes J D, Saxena A, Merkle J G 1988 *Nonliner Fractuire Mechanics* (Elastic-Plastic Fracture. Vol.II) (ASTM Int.)
[12] Begley J A, Landes J D 1972 *The J-integral as a fracture criterion* (Crack resistance. P. II) (ASTM STP 514) pp. 1-20
[13] Rice J 1975 *Mathematical methods in fracture mechanics* (Fracture, Vol. 2) Peace pp. 204-335 (in Russian)

IOP Conf. Series: Journal of Physics: Conf. Series **879** (2017) 012010 doi:10.1088/1742-6596/879/1/012010

Investigation of polymer membranes modified by fullerenol for dehydration of organic mixtures

Mariia E Dmitrenko[1], Anastasia V Penkova*[1], Anna I Kuzminova[1], Sergey S Ermakov[1] and Denis Roizard[2]

[1]Institute of Chemistry, St. Petersburg State University, Universitetsky pr. 26, Peterhof, 198504 Saint Petersburg, Russia
[2]Laboratoire Réactions et Génie des Procédés, CNRS, Université de Lorraine, ENSIC, 1 rue Granville, 54000 Nancy, France

a.penkova@spbu.ru

Abstract. This study focuses on the development of novel dense and supported mixed-matrix membranes based on chitosan and poly(2,6-dimethyl-1,4-phenylenoxide) (PPO) with low-hydroxylated fullerenol $C_{60}(OH)_{12}$. These novel membranes containing nano-carbon particles were prepared to reach high membrane performances for further integration in a dehydration process like distillation coupled with pervaporation. SEM microscopy was used to visualize the internal morphology of the membrane. It was found that all membranes were well stable and highly water-selective in spite of the different nature of polymers.

1. Introduction

At the present time the membrane processes are of significant technological and economic importance in the chemical industry. Membrane methods are highly demanded in various industries for purification, concentration and separation of liquid and gaseous mixtures due to the compactness of the equipment, environmental friendliness and low energy consumption. One of the most promising processes for the separation of low-molecular liquid mixtures is the pervaporation. This membrane method is a low-energy, waste-free and environmentally friendly way of separating various mixtures: azeotropic mixtures, mixtures of isomers, and also close-boiling and thermally unstable substances. The operational characteristics of this method make it attractive to the chemical, petrochemical, biochemical, pharmaceutical industry, as an alternative method of separation by modern methods.

Rapid development of membrane technology requires the creation of new materials with improved characteristics. One of the promising methods to get tailoring physico-chemical and transport characteristics of membrane materials is the modification of known polymers by carbon materials, i.e., the creation of membranes with a mixed matrix.

Previous works showed the relevance of the use of fullerene and its water-soluble derivatives to improve the transport properties of pervaporation membranes as a modifier [1,2] and a cross-linking agent [3–5].

Two different types of polymeric materials were chosen as polymers: chitosan, a well-known hydrophilic polymer and polyphenylene oxide, a hydrophobic polymer.

Many studies [6–11] are aimed at the study of chitosan as a membrane material because of its unique physico-chemical characteristics: high affinity for water, good film-formability, mechanical and chemical stability. However, chitosan-based membranes often have low selectivity due to excessive swelling in the water. To reduce the degree of swelling of chitosan in water, various methods of cross-linking the polymer (physical, chemical) are used, using, as a rule, toxic or ineffective cross-linking reagents (sulfuric acid, alkali, glutaraldehyde, acrylic acid, etc.). One of the most promising directions in this area is the modification and functionalization of the polymer by nanoparticles. In this paper, fullerenol $C_{60}(OH)_{12}$ particles were as cross-linking agent. This choice was based on the previous studies on modification of PVA by fullerenol [3–5].

Polyphenylene oxide (PPO) is a glassy polymer with good film-forming and mechanical properties, which has thermal stability and chemical stability in solutions of alkalis, mineral acids and salts, in superheated steam environment, racks under radioactive irradiation and the action of microorganisms. Conversely to chitosan, this polymer does not swell extensively in water or protic solvents. In diffusion processes of separation, PPO exhibits properties as one of the most permeable glassy polymers especially for water. However, this polymer has a relatively low selectivity in the separation processes. Previously, attempts were made to improve the separation factor of PPO using laborious and unsafe methods of chemical modification of this polymer (sulfonation, bromination, alkylation, etc.) [12,13], but these methods have not found industrial applications to date. In the previous studies it was shown that in spite of hydrophobic nature of PPO under pervaporation separation of ethanol-water mixtures the main component of the permeate was water according to the mechanism described in [14]. In this work, the modification of the PPO by a nanoparticle (fullerenol $C_{60}(OH)_{12}$) has been studied with the aim to improve further the transport properties of membranes due to OH-groups of water fullerene particles.

Therefore, the purpose of this study was to create a mixed-matrix membranes, characterizing their structural and transport properties in pervaporation.

2. Experimental part

2.1. Materials

Chitosan with medium molecular weight and poly (2,6-dimethyl-1,4-phenylenoxide) (PPO) were purchased from Sigma-Aldrich (USA) and used as the membrane materials. $C_{60}(OH)_{12}$ (Fullerene Technologies, Russia) was used for polymer modification. A porous membranes based on aromatic polysulfone amide (UPM, pore size 200 Å) and on copolymer of vinylidene fluoride (fluoroplastic F42L) and tetrafluoroethylene (MFFC, pore size 0.05 μm) were purchased from Vladipor (Russia) and a porous membrane based on polyacrylonitrile (PAN, pore size 200 Å) was purchased from Vladipor (Germany) and used as membrane supports. Ethanol and tetrahydrofuran (THF) was purchased from ZAO Vecton (Russia).

2.2. Preparation of chitosan and chitosan-fullerenol membranes

2.2.1. Preparation of composites. The chitosan-fullerenol composites were prepared by mixing solutions of chitosan in 1 wt.% acetic acid aqueous solution and fullerenol in water in amounts that provided the required content of fullerenol in the composite (up to 1 wt.%). The resulting solution was allowed to stand for 3-4 days for interactions to take place between the polymer and fullerenol molecules. Next, the composite solution was sonicated for 40 min.

2.2.2. Preparation of membranes. Dense membranes based on chitosan and fullerenol-chitosan composites with the thickness ~ 40 μm were obtained by casting a 1 wt.% polymer solution on a Petri dish. The solvent was removed by evaporation at 40 °C; the membrane was separated from the substrate and dried in a vacuum oven at 40 °C up to the constant weight. Then the membranes based on parent polymer and its composite were subjected to heat treatment at 140 °C during 100 min for cross-linking.

The formation of the thin selective layer in the supported membrane was achieved by casting 1-wt% chitosan solution with and without fullerenol (1% to the weight of the polymer) onto the surface of the UPM and PAN supports and drying at room temperature to form a selective layer with a thickness of 2 ± 0.3 μm, as determined by scanning electron microscopy. Cross-linking of the selective layer was achieved by heating the membrane to 140 °C for 100 minutes.

The maximum concentration of fullerenol was limited to 1 wt.% in the case of chitosan and 2 wt.% in the case of PPO because fullerenol concentrations exceeding these concentrations resulted in a poor dispersion of fullerenol in solution.

2.3. Preparation of PPO and PPO-fullerenol membranes

IOP Conf. Series: Journal of Physics: Conf. Series **879** (2017) 012010 doi:10.1088/1742-6596/879/1/012010

2.3.1. Preparation of composites. PPO-fullerenol composites containing 2 wt.% fullerenol were obtained by thorough mixing of PPO and fullerenol powders in agate mortar. After the solid-phase interaction the composites were dissolved in chloroform. The PPO/fullerenol solution was intensely stirred and sonified before membranes preparation.

2.3.2. Preparation of membranes. Dense membranes based on PPO and composites PPO-fullerenol with the thickness ~ 50 μm were obtained by casting a 4 wt.% polymer solution on a cellophane surface. The solvent was removed by evaporation at 40°C; the membranes were separated from the substrate and dried in a vacuum oven at 40 °C up to the constant weight

Supported membranes were prepared by casting 4 wt.% PPO (or PPO-fullerenol) solutions in chloroform on the surface of an MFFC support consisting of a fluoroplast F42L layer and polypropylene base. To create a selective layer with 6-8 μm thickness. Then, the composite membrane was dried at room temperature during 1 day and after in a vacuum oven at 40 °C up to constant weight.

2.4. Scanning electron microscopy
SEM micrographs of the fracture surfaces perpendicular to the membrane plane were obtained using a Zeiss Merlin SEM. The homogeneous membranes were submerged in liquid nitrogen for five minutes and fractured perpendicular to the surface. The prepared specimens were observed using SEM at 1 kV.

2.5. Pervaporation experiments
The transport properties were studied using a laboratory cell [5].
A downstream pressure of < 10^{-2} kPa was maintained using a vacuum pump and was controlled by a pressure gauge. The permeate was collected in a liquid nitrogen trap, weighed, and then analyzed via gas chromatography. The gas chromatograph used (SHIMADZU GC-2010) was equipped with a HP-PLOT/U column and a thermal conductivity detector to perform a quantitative analysis of the feed and the permeate composition.

The membrane permeation flux J (kg/m^2h) was determined to be the amount of liquid transported through a unit of the membrane area per hour and was calculated as:

$$J = \frac{W}{A \times t} \quad (1)$$

where W (kg) is the weight of the liquids that permeated through the membrane, A (m^2) is the effective membrane area, and t (h) is the measurement time.

Each measurement was performed at least three times, and the average value was recorded for later analysis.

2.6. Swelling Study
Two types of dry non-porous membranes from PPO and chitosan were immersed in liquid at 20°C and atmospheric pressure for 10 days. At definite intervals, the swollen membranes were taken out from the weighing bottles, carefully wiped with filter papers to remove residue liquid on the membrane surface, and then quickly weighed. The experiment was carried out until the swollen membranes obtained a constant weight that indicated a state of sorption equilibrium. Then, the membranes were placed in a vacuum box at 40 °C for 7 days to control the weight of the dry membranes. The liquids under the study depend of the studied polymer and were water, THF, and ethanol. The degree of the swelling was calculated by the following equation:

$$Sw = \frac{M_s - M_d}{M_d} \quad (2)$$

where Ms is the weight of a swollen membrane in equilibrium state and Md is the weight of a dry membrane.

3. Results and discussion

3.1. Scanning electron microscopy
A detailed study of the inner membrane structure was carried out by scanning electron microscopy (SEM) using the Zeiss Merlin SEM microscope.

Figure 1 shows micrographs of the membrane cross-sections showing the features of the internal structure of chitosan-based membranes and the chitosan-fullerenol composite containing 1 wt.% $C_{60}(OH)_{12}$.

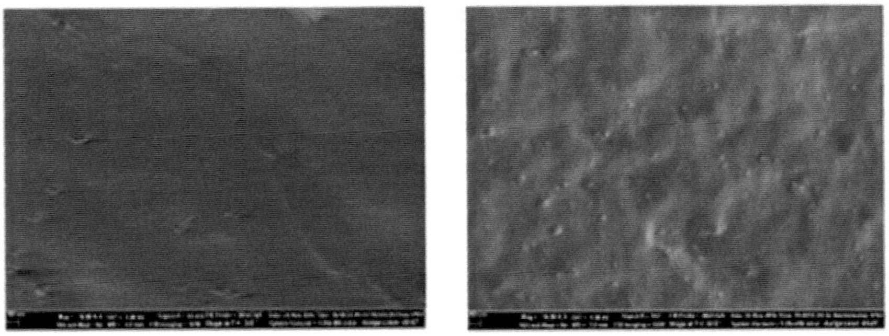

a) chitosan b) chitosan-fullerenol (1%)

Figure 1. SEM micrographs of the cross-section of heat-treated membranes (140 ^{0}C 100 minutes) based on: a) chitosan, b) chitosan-fullerenol (1%)

For membranes based on chitosan and its composite with fullerenol (Figure 1), the following changes are observed: cross-section of a membrane from pure polymer is relatively homogeneous, while for a membrane with 1 wt.% fullerenol surface roughness is observed and inclusions of nanoparticles are visible.

Figure 2 shows micrographs of dense membranes based on PPO and PPO-fullerenol (2 wt.%) composite. As can be seen from the figure 2, the membrane from pure PPO has a roughness structure and heterogeneity, which increases with the introduction of fullerenol into the polymer matrix.

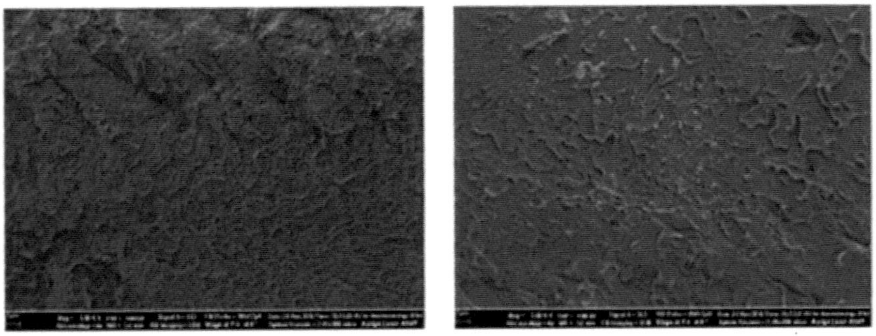

a) PPO b) PPO-fullerenol (2%)

Figure 2. SEM micrographs of cross-section of membranes based on:
a) PPO, b) PPO-fullerenol (2%)

The presented SEM micrographs demonstrated that the properties of the polymer matrix depend not only on the method of film formation, but also on the filler, in this case fullerenol. These changes in morphology under modification of the membranes with fullerenol are directly reflected on other physico-chemical and transport parameters of the membranes, the results of which are presented in next sections.

3.2. Sorption data

The study of the equilibrium swelling of polymeric membranes was carried out for two reasons, the first is the determination of the degree of cross-linking of polymer chains (for chitosan). For this study, the membranes were immersed in water. The second reason is that the first stage of the pervaporation mechanism is sorption, so there is a need to study the sorption characteristics of the membranes with respect to the components of the separated mixtures (ethanol, tetrahydrofuran (THF)) to explain the mechanism of transport of low molecular substances through the membrane. The results are shown in Table 1.

Table 1. Sorption characteristics of dense membranes.

Membrane	Sorption, %		
	THF	Ethanol	Water
chitosan	0,4		354,1
chitosan-fullerenol (1%)	0		330,3
PPO		7,8	0
PPO-fullerenol (2%)		4,1	0

The results presented in Table 1 indicate a decrease in the sorption capacity of fullerenol-modified membranes with respect to solvents as compared to unmodified membranes. The data of the table demonstrate that the degree of swelling in water and tetrahydrofuran of membranes based on chitosan decreases with the introduction of fullerenol into the polymer matrix, which indicates the cross-linking of polymer chains. Water is practically not sorbed by membranes from PPO and PPO-$C_{60}(OH)_{12}$. While the introduction of fullerenol into the PPO matrix reduces the degree of swelling in ethanol to 4.1%. Thus, the obtained data demonstrate different sorption behavior of membranes, which directly affects their transport characteristic. In the case of chitosan, the very low degree of swelling in THF and, conversely, the very high degree of water sorption indicate that both sorption and diffusion phenomena will favor water transport. On the other hand for PPO, the selective transport is expected to arise mainly from the facilitated diffusion of water in the glassy network.

3.3. Pervaporation data

The study of the transport properties of membrane was carried out in the process of pervaporation, which is an actual membrane process for the separation of low-molecular substances, especially for azeotropic, close-boiling and difficult-to-separate mixtures that are difficult to separate by simple separation methods. For the evaluation of membrane materials for industrial scale applications, it is necessary to take into account such properties as the possibility of using a membrane to separate the feed mixture and creating a supported membrane with a thin selective layer to increase productivity. The investigation of dense membranes is necessary to analyze the transport characteristics of the selected polymer and to exclude the influence of the substrate and defects in the selective layer of the supported membranes. The creation of composite supported membranes allows a significant increase in permeability.

3.3.1 Transport properties of membranes based on chitosan and chitosan-fullerenol composite. Pervaporation of tetrahydrofuran (THF) - water mixture. The transport properties of membranes based on chitosan and its composite with fullerenol have been studied by separating the azeotropic composition of a mixture THF-water at 20^0C by pervaporation. Tetrahydrofuran is an important universal aprotic solvent. In the production process, water is a by-product of the reaction, in connection with which its dehydration is necessary. The difficulty in separating the THF-water mixture is that the THF forms an azeotropic mixture with water (5.7 wt.% water), which has a boiling point of 63.4°C, as well as strong hydrogen bonds. Thus, the use of distillation to separate this system is not advisable.

Table 2. Pervaporation of a mixture THF-water azeotropic composition at 20°C using dense and supported membranes based on chitosan and chitosan-fullerenol composite.

Membranes (140^0C 100 min)		Flux (kg/m^2 h)	Permeate, wt.%	
			water	ethanol
dense	chitosan	0.090	97.35	2.65
	chitosan-fullerenol (1%)	0.063	99.34	0.66
supported	chitosan /UPM	0.133	94.05	5.95
	chitosan /PAN	0.125	96.38	3.62
	chitosan-fullerenol (1%)/PAN	0.099	98.37	1.63

Table 2 shows the transport properties for thermally cross-linked (140^0C 100 minutes) dense and supported membranes based on chitosan and chitosan-fullerenol(1%) composite. Dense membranes were studied to explain the mechanism of mass transfer. However, for use in industry, the thickness of the membrane must be reduced that can be achieved by creating supported membranes by depositing a thin selective layer on a porous support. In this paper, industrial ultrafiltration membranes based on polysulfonamide (UPM) and polyacrylonitrile (PAN) were chosen as porous supports. The thickness of supported membranes was about 2 μm. The flux for supported membranes increased in 1.47 times for the membrane on the UPM support and 1.38 fold for the PAN compared to the chitosan-based dense membrane. However, for the supported membrane on the UPM the selectivity with respect to water was decreased (to 94 wt.% of water) in the permeate compared to the dense membrane (97.35 wt.%) while the supported membrane on the PAN has good level of selectivity (96.38 wt.% water in permeate). Thus, the membrane on the PAN support was selected for further modification with a nanoparticle and further study. The introduction of 1 wt.% fullerenol into the chitosan matrix of the supported membrane on PAN significantly increased the selectivity (up to 98.37 wt.% water in the permeate) with a relatively similar flux (Table 2). Thus, a new effective nanocomposite supported membrane for purifying THF from water impurities was obtained.

3.3.2 Transport properties of membranes based on PPO and PPO-fullerenol composite. Pervaporation of a mixture of ethanol-water. The effect of fullerenol $C_{60}(OH)_{12}$ (2 wt.%) on the transport properties of PPO membranes was studied for the separation of water-ethanol mixture. This mixture was chosen as previous studies demonstrated that for ethanol-water mixtures PPO exhibits water selective properties. In this case the modification of PPO by fullerenol could lead to the improvement of transport characteristics due to the change of surface and inner morphology of the PPO membranes. The system under investigation is characterized by azeotropic composition point: 95.57 wt.% ethanol and 4.43 wt.% water at 20°C, 760 mmHg. The obtained results are presented in the table 3.

Table 3. Pervaporation of a mixture ethanol-water with azeotropic composition at 20^{0}C using dense and supported membranes based on PPO and PPO-fullerenol composite

Membranes		Flux (kg/m^2 h)	Permeate, wt.%	
			water	ethanol
dense	PPO	0.014	92.98	7.02
	PPO-fullerenol (2%)	0.028	95.12	4.88
supported	PPO /MFFC	0.181	91.41	8.59
	PPO-fullerenol (2%)/MFFC	0.313	94.11	5.89

It was found that the introduction of fullerenol into the PPO matrix increases the flux (up to 0.028 kg / m^{2}h) and the membrane selectivity (up to 95.12 wt.% water in the permeate) when the azeotropic point of the ethanol-water mixture was separated. The high selective properties established for the PPO-fullerenol(2%) membrane make it promising for participation in real ethanol purification processes. However, the investigated membrane has a significant disadvantage - low productivity, which is largely determined by the thickness of the membrane. The dense membranes based on PPO had a thickness of ~ 50 μm. To increase flux and preserve the mechanical characteristics, supported membranes consisting of a thin selective layer based on PPO and PPO-fullerenol(2%) composite (~ 2 μm) were deposited on a porous substrate that provided mechanical strength and practically no resistance to penetrant transport. The industrial MFFC membrane based on a copolymer of vinylidene fluoride and tetrafluoroethylene (fluoroplastic F42L) on a polypropylene base was used as the support. The choice of this type of support provided, above all, high flux and slightly changed of the separation factor. Thus, the creation of a supported membrane based on the PPO-fullerenol(2%) composite leads to a significant increase in flux (up to 0.313 kg/m^{2}h) with comparatively the same selectivity (94.1 wt.% of water in the permeate) compared with the dense PPO-fullerenol(2%) membrane, which indicates the creation of a highly effective membrane for the purification of ethanol from water.

4. Conclusions

Novel dense mixed-matrix membranes based on chitosan and PPO were developed using low-hydroxylated fullerenol $C_{60}(OH)_{12}$ as selective nanoparticles. In the case of chisotan, fullerenol acts also as a crosslinker. The introduction of fullerenol into the two different polymer matrices favors to rise the transport of water and creation of dehydration membranes with improved pervaporation transport properties. Clearly the use of fullerenol as nano-modifier for chitosan and PPO membrane led to the increase both flux and selectivity. The application of fullerenol as cross-linking for chitosan caused the increase of selectivity with the similar level of flux. Supported membranes having a thin, dense layer were also successfully prepared. These membranes exhibited much higher fluxes – factor 15 to 20 – while keeping almost constant the water selectivity.

5. References

[1] Sudareva N N, Penkova A V, Kostereva T A, Polotskii A E and Polotskaya G A 2012 Properties of casting solutions and ultrafiltration membranes based on fullerene-polyamide nanocomposites *Express Polym. Lett.* **6** 178–88
[2] Penkova A, Toikka A, Kostereva T, Sudareva N and Polotskaya G 2008 Structure and Transport Properties of Fullerene–Polyamide Membranes *Fullerenes, Nanotub. Carbon Nanostructures* **16** 666–9
[3] Penkova A V., Acquah S F A, Sokolova M P, Dmitrenko M E and Toikka A M 2015 Polyvinyl

alcohol membranes modified by low-hydroxylated fullerenol C60(OH)12 *J. Memb. Sci.* **491** 22–7
[4] Penkova A V., Acquah S F A, Dmitrenko M E, Sokolova M P, Mikhailova M E, Polyakov E S, Ermakov S S, Markelov D A and Roizard D 2016 Improvement of pervaporation PVA membranes by the controlled incorporation of fullerenol nanoparticles *Mater. Des.* **96** 416–23
[5] Penkova A V., Acquah S F A, Dmitrenko M E, Chen B, Semenov K N and Kroto H W 2014 Transport properties of cross-linked fullerenol–PVA membranes *Carbon N. Y.* **76** 446–50
[6] Uragami T, Matsuda T, Okuno H and Miyata T 1994 Structure of chemically modified chitosan membranes and their characteristics of permeation and separation of aqueous ethanol solutions *J. Memb. Sci.* **88** 243–51
[7] Uragami T, Kato S and Miyata T 1997 Structure of N-alkyl chitosan membranes on water-permselectivity for aqueous ethanol solutions *J. Memb. Sci.* **124** 203–11
[8] Ghazali M, Nawawi M and Huang R Y M 1997 Pervaporation dehydration of isopropanol with chitosan membranes *J. Memb. Sci.* **124** 53–62
[9] Volkov V I, Skirda V D, Vasina E N, Korotchkova S A, Ohya H and Soontarapa K 1998 Self-diffusion of water-ethanol mixture in chitosan membranes obtained by pulsed-field gradient nuclear magnetic resonance technique *J. Memb. Sci.* **138** 221–5
[10] Ren J and Jiang C 1998 Transport Phenomena of Chitosan Membrane in Pervaporation of Water-Ethanol Mixture *Sep. Sci. Technol.* **33** 517–35
[11] Qunhui G, Ohya H and Negishi Y 1995 Investigation of the permselectivity of chitosan membrane used in pervaporation separation II. Influences of temperature and membrane thickness *J. Memb. Sci.* **98**
[12] Shih C-Y, Chen S-H, Liou R-M, Lai J-Y and Chang J-S 2007 Pervaporation separation of water/ethanol mixture by poly(phenylene oxide) and sulfonated poly(phenylene oxide) membranes *J. Appl. Polym. Sci.* **105** 1566–74
[13] Khulbe K C, Matsuura T, Lamarche G and Kim H J 1997 The morphology characterisation and performance of dense PPO membranes for gas separation *J. Memb. Sci.* **135** 211–23
[14] Polotskaya G A, Penkova A V, Toikka A M, Pientka Z, Brozova L and Bleha M 2007 Transport of small molecules through polyphenylene oxide membranes modified by fullerene *Sep. Sci. Technol.* **42** 333–47

IOP Conf. Series: Journal of Physics: Conf. Series **879** (2017) 012011 doi:10.1088/1742-6596/879/1/012011

Processing of magnesium bearing materials by dry chlorination - An overview of kinetic aspects

N Kanari, F Diot, E Allain and J Yvon

Université de Lorraine, UMR 7359 CNRS, CREGU, GeoRessources Laboratory, 2 rue du Doyen Roubault, TSA 70605, Vandoeuvre-lès-Nancy Cedex, F-54518, France

E-mail: ndue.kanari@univ-lorraine.fr

Abstract. This paper is focused on the reactivity of Cl_2 towards MgO in presence of a reducing (CO) and oxidizing (O_2) agent at temperatures up to 1025 °C. Kinetics parameters are determined by thermogravimetry under isothermal conditions. The overviewed results showed that the carbochlorination (Cl_2+CO) occurred at temperatures lower than oxychlorination (Cl_2+O_2). The reaction of MgO with Cl_2+CO from 425 °C to 600 °C proceeds with an activation energy (E_a) of about 49 kJ/mol, while the reaction rate does not longer depend on the temperature between 600 and 675 °C. The reaction of MgO with Cl_2+O_2 at 850-1025 °C was characterized by a value of E_a of 214 kJ/mol indicating that the overall rate of the process is controlled by the rate of chemical reaction. Oxychlorination can also be used for removing the impurities contained in the industrial magnesium raw materials. Further, the behavior of the MgO in chlorine atmosphere leads to give an insight about the reaction mechanisms useful for the selective chlorination of complex ores and concentrates containing magnesium oxide.

1. Introduction

Although magnesium is abundant in the earth crust, its natural resources for a commercial purpose include only a few number of minerals (magnesite, dolomite, brucite, carnallite, olivine,...) as well as salt pits, magnesia-rich brine and seawater. Reduction of magnesia and electrolysis of molten magnesium chloride are two principal routes for the production of magnesium [1, 2]. The thermal processes (mostly silicothermic and carbothermic reductions) are performed at high temperature and under a reduced pressure to promote the vaporization of the obtained magnesium. Concerning the anhydrous chloride route, the magnesium chloride is generated during chlorination by Cl_2+C of MgO bearing materials and/or from brine dehydration. Therefore, the electrolysis of molten $MgCl_2$ producing liquid $MgCl_2$ is performed at 700-800 °C and as it could be expected, chlorine (Cl_2) is generated at the anode from the oxidation of chloride ion (Cl^-).

Use of the dry chlorination process for the transformation of magnesium oxide and more complex magnesium substances into magnesium chloride seems to be a good alternative route due to the high reactivity of chlorine and to the fact that chlorine can be recycled during metallic magnesium recovery. A literature review showed that several studies [3-9] are devoted to the chlorination of materials containing magnesium compounds by using various chlorinating agents. One may mention the work by Ishii et al. [3] that studied the chlorination of MgO, olivine (Mg_2SiO_4), protoenstatite ($MgSiO_3$) and talc [$Mg_3Si_4O_{10}(OH)_2$] as well as FeO, Fe_2O_3 and SiO_2 between 25 °C and 1000 °C using differential thermal and thermogravimetric analysis (DTA and TGA). According to these authors, the reactivity of MgO towards chlorine, in absence of carbon, is negligible at 640 °C. Two peaks were perceived during the chlorination of MgO in presence of carbon. The first is exothermic and observed at 500 °C corresponding to the formation of $MgCl_2$. The second peak is endothermic and detected at about 700 °C. It is attributed to the fusion of $MgCl_2$. The reaction rate is fast at the beginning of the reaction that afterwards proceeds slowly and is stopped completely at 55 % of MgO conversion. The authors suggested that $MgCl_2$ formed on the surface of the MgO particle acts as a barrier for further chlorination. The kinetics parameters of chlorination of MgO and Fe_2O_3 contained in a chromite concentrate using HCl+C between 1000 °C and 1200 °C are given by Vil'nyanskii and Martirossyan [4]. The apparent activation energy of the MgO chlorination is about 118 kJ/mol and the apparent reaction orders with respect to HCl are 0.98, 0.99 and 0.99 at 1000 °C, 1100 °C and 1200 °C, respectively. The chlorination of MgO by HCl in a potassium chloride melt between 850 °C and 940 °C has been studied by Stupina et al. [5], they proposed that the rate-controlling step of the MgO chlorination is the diffusion of HCl to MgO surface. The same phenomenon was also underlined by the same authors [6] during the chlorination of MgO by HCl in fused KCl and NaCl between 780 °C and 1000 °C. Further literature details regarding to the chlorination of magnesium oxide were summarized earlier [7, 8]. Recently, Cecchi et al. [9] examined the production of $MgCl_2$ from raw chrysotile tailings. They concluded that temperature is the determining factor for both the recovery of Mg (85%) and for limiting impurities.

This paper gives a summary of our kinetics results obtained during dry chlorination of the magnesium oxide under a reducing (CO) and oxidizing (O_2) atmosphere.

2. Materials and methods

The sample used for the experimental tests was a magnesium oxide powder with a purity of 98 pct. X-ray diffraction (XRD) analysis revealed the presence of MgO as the only crystallized phase in the sample. The thermogravimetric (TG) chlorination tests were realized with 40 to 100 mg of sample, using the experimental setups described in [7]. Its main device is a CAHN 1000 microbalance with a sensitivity of 10 µg. The results are mostly expressed as the evolution of the reaction extent as a function of the chosen experimental conditions and the methodology applied for the kinetic study [10] consists of the following items:

- thermogravimetric analysis (TGA) using non isothermal conditions,
- comparison of the reaction rate with that of the volatilization of the reaction products,
- determination of adequate gaseous mixture flow rate,
- effect of the proportions of the reactive gases on the reaction rate,
- determination of the apparent reaction orders with respect to the reactive gases,
- evaluation of the apparent activation energy of the process (effect of temperature),
- data processing of the experimental data to define the geometry of the reaction interface (shape factor),
- analysis of the solid products (SEM, microprobe, XRD, specific surface area, etc.),
- indication of the possible rate controlling step.

Only some typical data will be described here illustrating the steps for the chlorination of MgO. Detailed results can be found in our previously published papers [7, 8, 10].

3. Results

3.1. Kinetics of the carbochlorination of MgO

The first isothermal tests for this kinetics study were performed in order to check the effect of Cl_2+CO gas velocity on the carbochlorination rate. Using this gas velocity will avoid starvation and will minimize the mass transfer phenomena. A series of isothermal tests was carried out at 600 °C using total gas flow rates of 8 to 60 L/h and keeping the Cl_2/CO molar ratio equal to 1. Figure 1 (a) illustrates the evolution of MgO reaction extent "X" as a function of reaction time. Figure 1 (b) traces the evolution of the reaction rate (expressed as % X/min) as a function of the gas velocity ($0.05 \leq X \leq 0.40$). This figure shows that the reaction rate increases as the gas flow rate augments up to 40 L/h (linear velocity of gases: Vg = 33 cm/min). For higher gas velocities, the chlorination rate does not depend on the gas flow rate. Consequently, a gas flow rate of 48 L/h (Vg ≈ 40 cm/min) was used for further tests of the MgO carbochlorination with Cl_2+CO.

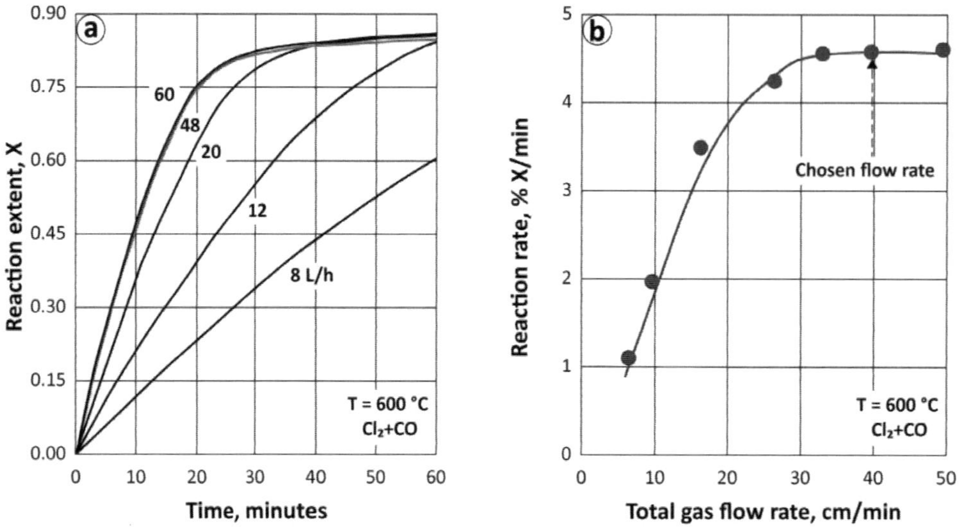

Figure 1. (a) Isotherms of the MgO carbochlorination at 600 °C at various gas flow rates; (b) evolution of the reaction rate as a function of the linear gas velocity.

A careful examination of figure 1 (a) shows that the maximum conversion of MgO into $MgCl_2$ tends to an asymptote of about X = 0.85 indicating that the reaction of Cl_2+CO with MgO is almost stopped. To give an insight about this phenomenon, several boat isothermal tests were performed for a reaction time of two hours and the results are grouped in figure 2. Treated residues were water-leached to dissolve formed $MgCl_2$ in order to measure the carbochlorination degree of MgO. Besides, the reaction rate is also calculated from the mass losses up to about 700 °C. The reaction extents obtained by these two methods are almost equal (figure 2).

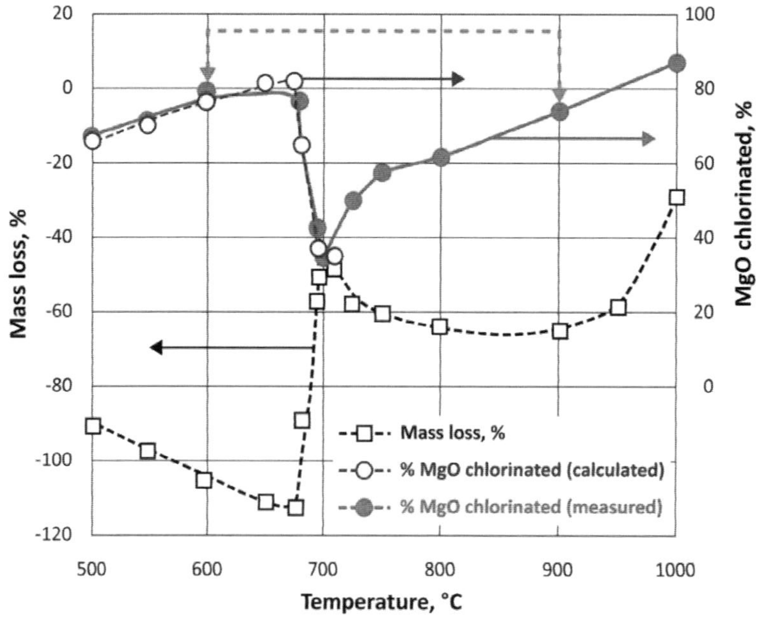

Figure 2. Carbochlorination of MgO in the horizontal set-up during 2 hours.

This figure shows that about 67 % of MgO are chlorinated at 500 °C and reaction extent increases with temperature and reaches 82% at 675 °C. Therefore, the reaction rate drops and only 35 % of the sample are chlorinated at 700 °C. This is probably due to the fusion of $MgCl_2$ leading to the formation of a liquid layer that decreases the mass transfer between MgO and the carbochlorinating gas mixture. Temperatures as higher as 900 °C are required to achieve the same rate obtained at 675 °C. Beyond 900 °C, magnesium chloride is sufficiently volatile and is recovered by cooling the gas phase. At 1000 °C, about 85 % of MgO are chlorinated with a reaction time of two hours and $MgCl_2$ and MgO are identified by XRD in the obtained residues. This confirms that chlorination of MgO is incomplete even at 1000 °C. Moreover, the presence of $MgCl_2$ in the treated residues suggests that the carbochlorination rate of MgO is higher than the volatilization rate of $MgCl_2$.

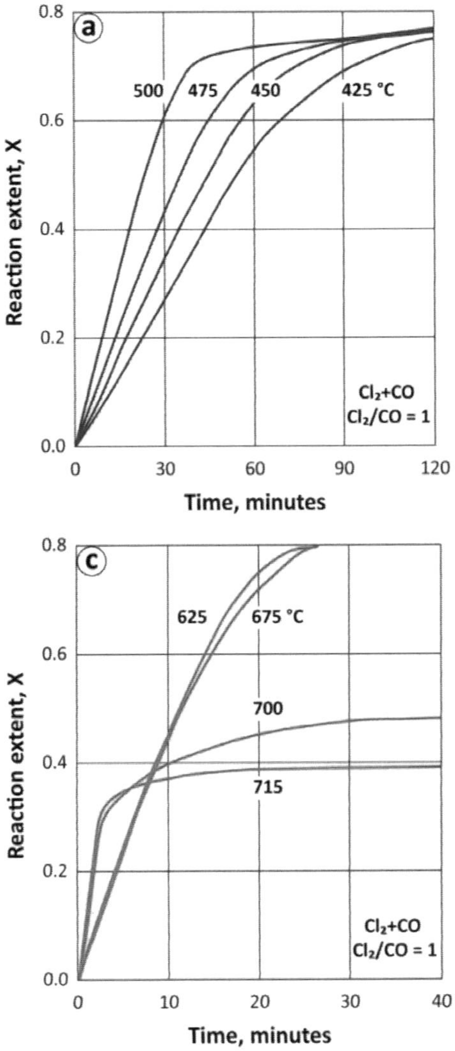

Figure 3. Evolution of the reaction extent as a function of the reaction time during MgO carbochlorination:
(a) between 425 °C and 500 °C,
(b) between 525 °C and 625 °C,
(c) between 625 °C and 715 °C.

The effect of temperature on the carbochlorination of MgO was investigated from 425 °C to 715 °C by thermogravimetry and the obtained isotherms are grouped in figures 3. This figure shows that a reaction extent of about 75 % requires about 120 and 20 minutes at 425 °C and 600 °C, respectively. As in previous case, figure 3 also reveals that the conversion of MgO into $MgCl_2$ reaches a maximum of about 85 %. One may speculate that the $MgCl_2$ formed on the surface of the MgO particle acts as a barrier against further progress of the reaction. Note that the molar volume of $MgCl_2$ (41.04 cm^3/mol) is much higher than that of MgO (11.26 cm^3/mol).

As obviously demonstrated in figure 3 (c), for a reaction time of 20 minutes, the reaction extent decreases as the temperature increases from 675 °C to 715 °C which is provoked by the fusion of $MgCl_2$ (M.p. = 714 °C). Note also that the high exothermic nature of the carbochlorination reactions will increase the temperature in the reaction zone allowing the fusion of the generated $MgCl_2$.

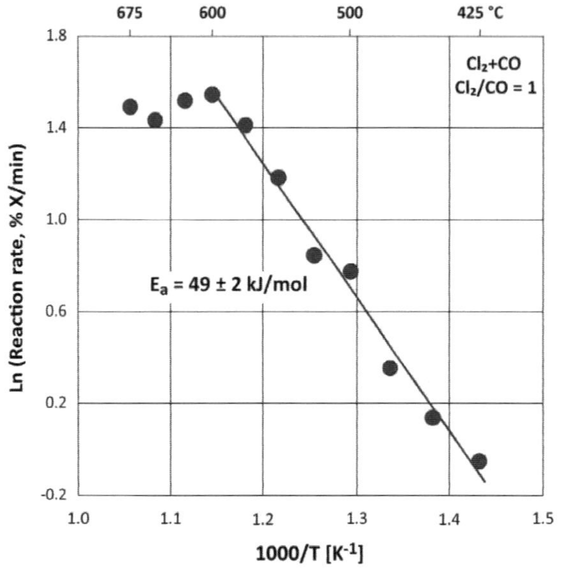

Figure 4. Arrhenius plot for the carbochlorination of MgO by Cl_2+CO.

The Arrhenius plot of MgO carbochlorination ($0.05 \leq X \leq 0.40$) is given in figure 4. The apparent activation energy is 49 ± 2 kJ/mol between 425 °C and 600 °C. This value suggests that the kinetics of MgO carbochlorination is affected by diffusion. Beyond 600 °C, the reaction rate does not depend on the temperature. As mentioned above, the exothermic nature of carbochlorination reaction could lead to the partial fusion of $MgCl_2$, consequently leading to a layer formation surrounding the MgO particles and thereby decreasing the mass exchange with the chlorinating gas mixture. Due to the hygroscopic nature of $MgCl_2$, it was difficult to examine the carbochlorination residues by microscopy, while the XRD confirmed the formation of $MgCl_2 \cdot 6H_2O$.

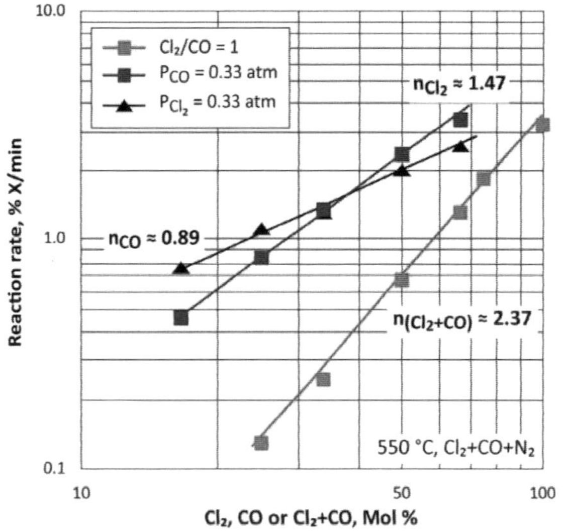

Figure 5. Effect of gases composition on the carbochlorination rate of MgO at 550 °C.

The effects of Cl_2+CO, Cl_2 and CO partial pressures on the carbochlorination rate of MgO at 550 °C are compared in figure 5. The apparent reaction orders with respect to Cl_2+CO, Cl_2 and CO are about 2.37, 1.47 and 0.89, respectively. Obviously, the effects of reactive gases follow the sequence: $n(Cl_2+CO) > nCl_2 > nCO$ and these reaction orders satisfy the relation: $n(Cl_2+CO) \approx nCl_2 + nCO$.

IOP Conf. Series: Journal of Physics: Conf. Series **879** (2017) 012011 doi:10.1088/1742-6596/879/1/012011

3.2. Kinetics of the oxychlorination of MgO

The chlorination of MgO by chlorine produces magnesium chloride which is characterized by its melting and boiling points of 714 °C and 1412 °C, respectively. As the TGA technique is used to measure the reaction rate, it was mandatory to insure that the rate of $MgCl_2$ volatilization is higher than that of its formation. One possible way to decrease the reaction rate of MgO with chlorine is to increase the partial pressure of oxygen in the system. The impact of the oxygen potential on the reaction rates during the chlorination of the iron oxides was also investigated earlier [11-13].

To check this effect on the MgO interaction with chlorine, a series of experiments was performed at 950 °C, using a Cl_2+O_2 gas mixture with a molar ratio of $Cl_2/(Cl_2+O_2)$ from 0.33 to 1.00. Only mass losses of sample were observed during the oxychlorination using a Cl_2+O_2 mixture containing less than 90 % Cl_2 suggesting that the observed mass loss is a direct measure of the extent of the overall oxychlorination reaction. For this reason, the kinetics parameters of MgO oxychlorination were studied using the Cl_2+O_2 gas mixtures having chlorine content lower or equal to 80 %.

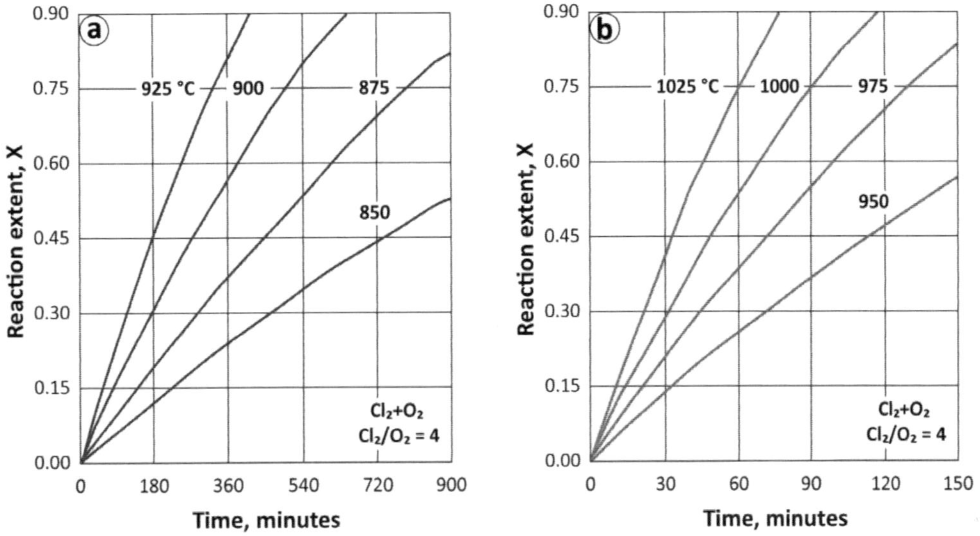

Figure 6. Isotherms of oxychlorination of magnesium oxide for: (a) T ≤ 925 °C and (b) T ≥ 950 °C.

The temperature effect on MgO oxychlorination was studied in isothermal conditions between 850 °C and 1025 °C. An oxychlorinating gas mixture of Cl_2+O_2 with a molar ratio $Cl_2/O_2 = 4$ was used. Figure 6 gives the evolution of the reaction extent 'X' as a function of time for different isotherms. More than 24 hours were required to reach a value of X of about 0.80 during the experiment at 850 °C, while the same reaction extent at 1025 °C was obtained for a reaction time of about 1 hour.

The data of figure 6 were used to determine the effect of the temperature on the reaction rate for a reaction extent between 0.05 and 0.40. Figure 7 gives the results in the Arrhenius plot. A value of the apparent activation energy of MgO oxychlorination close to 214 kJ/mol was deduced from this plot. Such a high value suggests that overall rate of MgO oxychlorination is mostly controlled by the rate of the chemical reaction.

The effect of the partial pressures of the reactive gases on the oxychlorination of MgO is studied at 950 °C using a $Cl_2+O_2+N_2$ gas mixture with the optimal gas flow rate. For the Cl_2+O_2 impact, the (Cl_2+O_2) partial pressure is varied from 0.25 to 1.00 atm by keeping the Cl_2/O_2 molar ratio constant and equal to 4. To follow the influence of the pCl_2, the oxygen partial pressure is kept as a constant and equal to 0.33 atm, while chlorine partial pressure is varied from 0.20 to 0.67 atm. Similarly, the influence of the O_2 in the reaction rate is measured at $pCl_2 = 0.33$ atm by changing the oxygen content between 20 % and 67 % in the $Cl_2+O_2+N_2$ gas mixtures. Obtained results are summarized in figure 8 as the evolution of the reaction rate versus reactive gas content using a natural logarithm scale.

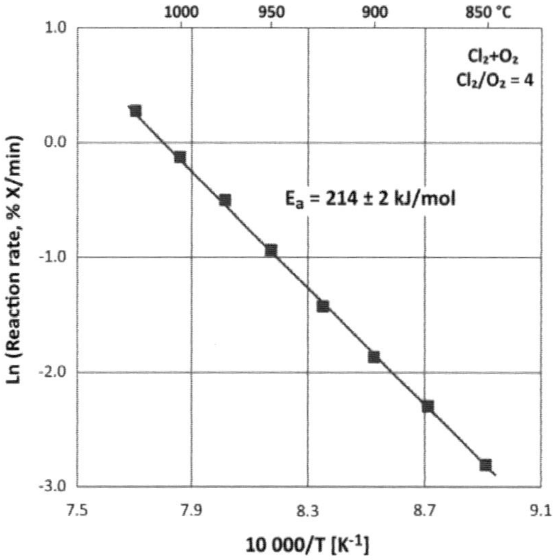

Figure 7. Arrhenius plot for the oxychlorination of MgO by Cl_2+O_2.

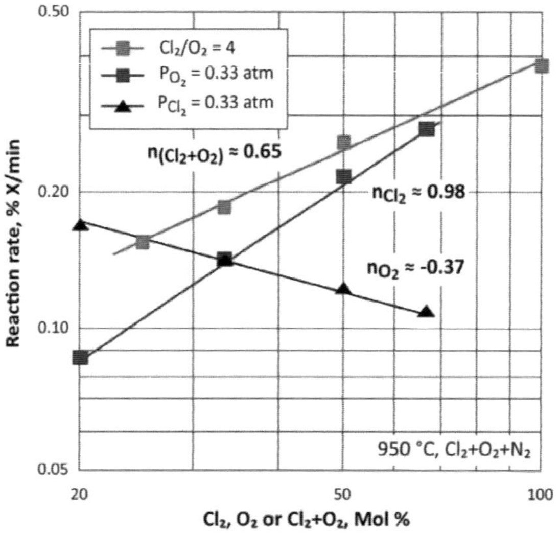

Figure 8. Effect of gases composition on the oxychlorination rate of MgO at 950 °C.

Examination of figure 8 shows that the reaction rate of MgO oxychlorination is almost proportional to the pCl_2 since the apparent reaction order with respect to chlorine is close to one. The apparent reaction order with respect to O_2 is about -0.37 indicating that the oxygen has a negative effect on the chlorination of MgO. The significance of the effects of reactive gases, expressed as apparent reaction orders, follows the sequence: $nCl_2 > n(Cl_2+O_2) > nO_2$ and the apparent reaction orders are related by the following relationship: $n(Cl_2+O_2) \approx nCl_2 + nO_2$.

The low reactivity of Cl_2+O_2 towards MgO can be applied for the removal of impurities such as iron oxides from the magnesium bearing materials. These results are also used [14, 15] to understand the oxychlorination mechanisms of chromite [(Fe^{2+}, Mg)(Cr, Al, Fe^{3+})$_2O_4$] in order to separate selectively its main constituents.

As reported recently [16, 17], the selective removal of the iron from chromite, resulting to Cr/Fe increase, is achieved by chlorination and the selectivity is enhanced by adding sodium chloride at temperatures lower than or equal to 720 °C. Other investigations used chlorination for the purification of talcs [18, 19] at about 900 °C allowing an important extraction of iron without magnesium loss.

4. Conclusions

Kinetics of the MgO reaction with Cl_2+CO was studied under isothermal conditions. The carbochlorination process from 425 °C to 600 °C proceeded with an activation energy of about 49 kJ/mol, while the initial reaction rate does not longer depended on the temperature between 600 °C and 675 °C. The carbochlorination rate is dropped down at about 700 °C due to the fusion of the generated magnesium chloride and the reaction is not complete for two hours at 1000 °C. The apparent reaction orders with respect to Cl_2+CO, Cl_2 and CO at 550 °C are about 2.37, 1.47 and 0.89, respectively.

The oxychlorination of MgO with Cl_2+O_2 starts at temperatures higher than 800 °C and the reaction rate is lower than that of the $MgCl_2$ volatilization. The reaction of magnesium oxide with Cl_2+O_2 (Cl_2/O_2 = 4) between 850 °C to 1025 °C was characterized by an apparent activation energy of about 214 kJ/mol indicating that the overall rate of the process is probably controlled by rate of chemical reaction. The reaction rate of MgO oxychlorination at 950 °C had an almost linear relationship with the partial pressure of chlorine. The reaction orders, at this temperature, with respect to Cl_2+O_2 and O_2 were 0.65 and -0.37, respectively.

This kinetic study was explored to explain the observed phenomena during the chlorination of raw materials such as chromite concentrate. Thermodynamic studies were also performed to complement the kinetic studies to achieve a selective separation of valuable metals (Cr, Mg, Fe, Al) contained in these complex materials.

References

[1] Rongti L, Wei P and Sano M 2003 *Metall. Mater. Trans. B* **34B** 433
[2] Tian Y, Xu B Q, Yang B, Yang C B, Qu T, Liu D C and Dai Y N 2015 *J. Magnes. Alloys* **3** 149
[3] Ishii T, Furuichi R and Kobayashi Y 1974 *Thermochim. Acta* **9** 39
[4] Vil'nyanskii Ya E and Martirossyan V A 1973 *Arm. Khim. Zh.* **26** 881
[5] Stupina A M, Bezvoritnii V A and Bezukladnikov A B 1974 *Zh. Prikl. Khim.* **47** 2544
[6] Stupina A M, Bezvoritnii V A and Bezukladnikov A B 1974 *Tsv. Met.* **12** 35
[7] Kanari N and Gaballah I 1999 *Metall. Mater. Trans. B* **30B** 383
[8] Kanari N, Gaballah I and Allain E 1999 *Metall. Mater. Trans. B* **30B** 1009
[9] Cecchi E, Mercier G and Bergeron M 2009 *Int. J. Miner. Process* ***93*** 278
[10] Kanari N, Mishra D, Mochón J, Verdeja L F, Diot F and Allain E 2010 *Rev. Metal. Madrid.* **46** 22
[11] Kanari N, Allain E and Gaballah I 1999 *Thermochim. Acta* **335** 79
[12] Kanari N, Mishra D, Filippov L, Diot D, Mochón J and Allain E 2010 *Thermochim. Acta* **497** 52
[13] Kanari N, Mishra D, Filippov L, Diot D, Mochón J and Allain E 2010 *Thermochim. Acta* **506** 34
[14] Kanari N, Gaballah I and Allain E 2001 *Thermochim. Acta* **371** 75
[15] Kanari N, Gaballah I and Allain E 2001 *Thermochim. Acta* **371** 143
[16] Shen S-B, Bergeron M and Richer-Laflèche M 2009 *Int. J. Miner. Process.* **91** 74
[17] Shen S-B, Hao X-F and Yang G-W 2009 *J. Alloy. Compd.* **476** 653
[18] Orosco R P, Ruiz M del C, Barbosa L I, González J A 2011 *Int. J. Miner. Process.* **101** 116
[19] Orosco P, Ruiz M del C, González J 2013 *Thermochim. Acta* **554** 15

IOP Conf. Series: Journal of Physics: Conf. Series **879** (2017) 012012 doi:10.1088/1742-6596/879/1/012012

Improvement of calcium mineral separation contrast using anionic reagents: electrokinetics properties and flotation

Z. Lafhaj[1], L.O. Filippov[1,2*], I. V. Filippova[1,2]

[1]Université de Lorraine, Laboratoire GeoRessources UMR 7359 CNRS-Cregu ENSG, 2 rue du doyen Marcel Roubault, BP 40, F-54501, Vandœuvre-lès-Nancy, France
[2]National Universitie of science and technology « MISiS », Leninsky pr.,4, 119049 Moscow, Russia.

Corresponding author: lev.filippov@univ-lorraine.fr

Abstract. The flotation separation of salt type calcium minerals is problematic, due to the similarities in their same active Ca^{2+} related site for interaction with anionic collectors and similar physicochemical characteristics such as solubility, zero-point charge, surface speciation and Ca-site density. The work was performed to achieve effective and selective separation of the calcium-minerals using pure minerals samples: orange calcite with Mg impurities, optic calcite with impurities level and an apatite. The pure samples surface was examined using techniques sensitive near-surface like infrared spectroscopy (FTIR) and chemical composition was obtained by ICPMS. The isoelectric point (IEP) and point of zero charge (PZC) in electrolyte were recorded using electrophoresis method at different ionic strengths of the solution. Mechanisms of charge development at the mineral-water interface are discussed. The time of contact as important parameter for the charge equilibrium was deduced from kinetics study and fixed to 30 minutes. The difference in the values obtained between IEP and PZSE can be explained by the presence of a specific adsorption of cations and anions on the surface. The effect of pure anionic collectors such as oleic and linoleic acid were studied. At low pH, both collectors lead to a good recovery for the calcites. The flotation recovery of optic calcite at pH 9 with sodium oleate is higher than with sodium linoleate. At alkaline pH, apatite showed a better recovery with sodium linoleate.

1. Introduction

The separation of salt type minerals such as apatite and calcite from silicates has been successfully carried out by flotation methods. However, the flotation is difficult because of their similar surface properties and high surface reactivity with their conventional collectors such as fatty acid and alkyl sulphate [1] [2]. The surface chemistry of calcium minerals plays an essential role in the interaction between collectors with the mineral surface.

Fatty acids as collectors provides a successful recovery of phosphate via flotation. Indeed, oleic acid is the most common and traditionally used collector for Ca minerals. Additionally, the use of oleic acid may require high temperatures, resulting in high processing cost [3], [4]. Several researchers reported that sarcosine anion selectively adsorbs on the apatite but not on calcite neither dolomite [5]. Due to their high cost and complex synthesis, these collectors, despite they showed improved performance for phosphate flotation, have a limited use in industry. The most extensively used types of anionic collectors are long-chain fatty acids and their salts, mainly oleic acid or its soap, sodium oleate.

Many studies, over the recent years, used the mixture of the reagents of different molecular carboxylic with non-ionic reagents i.e. aliphatic alcohols to enhance the selectivity in flotation separation of salt-type minerals [2], [6]. The critical micelle concentrations of the mixtures are higher than that of the surfactants alone) [7]. An important step forward in delineating the physical chemistry of flotation systems was bringing the concepts of the electrical double layer and electro kinetic potentials used in the interpretation of flotation phenomena [8].

The Point of Zero Charge (PZC) and isoelectric point (IEP) are two electrokinetics parameters describing the relation between solution pH and surface charge. Accordingly, the close correspondence between the IEP and the PZC of the solid suggests that they are governed by the same equilibrium which can be calculated from thermodynamic solubility data [9]. In addition, the PZC can be determined from the flotation response of the solid when treated with physically adsorbed anionic and cationic collectors. The zero-point of charge of calcite and apatite were investigated here by measuring the electrophoretic mobility solution equilibrium and flotation response of solution pH.

The study of electro kinetic behaviour of calcium minerals was expanded here to several calcite and apatite samples with different origin. This investigation aims at discovering if the trend of zeta *potential* changes would follow the same pattern despite the different origin, and to find the relationship between the zeta potential and the textural properties and the impurities presence in the sample of calcite and apatite. Additionally, we studied the improvement of calcium minerals separation contrast. The floatability of calcite and apatite using oleic and linoleic acid and their mixture with different ratio have been investigated by flotation tests.

2. Materials and methods

2.1. Samples

The calcium-bearing minerals used in this study were optic calcite, orange calcite, and apatite from Madagascar. The mineral samples were ground in agate mortar to achieve <5 μm size fraction for zeta potential measurement. The size fraction -106+38 μm for flotation tests was obtained with a tungsten annular grinder. Collectors used in this paper were 98% pure sodium oleate (Ol) and sodium linoleate (Li) purchased from Fluka-Sigma-Aldrich.

2.2. Characterization

The composition of the samples was determined by FTIR spectroscopy, X-ray diffraction and ICP-MS for trace elements and results are presented in Table 1.

Table 1. Chemicals analysis of Ca-minerals with ICP-MS.

					%				
	CaO	**P_2O_5**	**MgO**	**SiO_2**	**K_2O**	**Na_2O**	**Al_2O_3**	**LOI**	**Total**
Apatite Madagascar	53.45	37.73	-	3.89	-	0.05	-	0.61	95.76
Calcite optic China	55.44	-	0.27	0.06	-	-	-	43.54	99.31
Calcite orange Mexico	54.05	-	1.23	0.07	0.01	0.14	-	43.32	98.82

The influence of impurity ions on the surface charge and zeta potential values was also investigated. Apart from calcium carbonate as the principal component, other common minerals such as calcium sulfate, calcium phosphate, magnesium hydroxide and silica could influence the potential. Nucleation and crystallization of calcium carbonate in hard water seems evident and can be influenced by magnetic field [10], [11]. Despite, the corresponding mechanisms are still unknown, they can be grouped into four basic categories [12]: intra-atomic effects (e.g. changes in electron configuration), contamination effects (due to magnetically-enhanced dissolution), inter-molecular/ionic effects and interfacial effects (e.g. distortion of the double layer). The major impurities were silica in the apatite sample and magnesium in the calcite orange sample with minor magnesium contents in all the other samples. The apatite sample also contain some quantities of cerium, thorium, lanthanum, neomidium and serium, in that order. Serium is also found in the calcite optic, and especially in the calcite orange. All these impurities may have some influence in the surface properties of these samples.

2.3. Zeta potential measurement

Analysis was conducted with a Zeta meter IV instrument from CAD. 30 mg of the minerals (particles size <5 μm) were first conditioned with 300 ml of KCL (10^{-2} and 10^{-3} M) for 30 min and 20 ml of the supernatant liquid were transferred into the measurement cell to measure the electrophoretic mobility of the particles as a function of pH. The pH of the solution was adjusted with small amounts of HCl and NaOH addition (1%). The zeta potential was calculated using the Smoluchowski equation.

2.4. Flotation tests

Flotation tests using pure minerals were performed in a MINIMET 180 mL cell using 3g of +38-106 μm size fraction samples for each test. The liquid-solid ratio was 1.7%. The pulp was aerated, and the froth product was collected for a total time of 5 min. After completion of the flotation test, the floated and sunk products were filtered, dried at 40 °C, weighed and analysed. The fatty acid collectors were also used as a mixture at sodium oleate: sodium linoléate mass ratios of 1:0, 2:1, 1:1, 1:2 and 0:1 but with the same total concentration of collectors.

3. Results and discussion

3.1. *FTIR analysis*

The two calcites have fundamental vibration absorption characteristics CO_3^{2} groups around 1440 cm^{-1} , 876 cm^{-1} and 713 cm^{-1}. Other most intense bands at 2983 cm^{-1}, 2875 cm^{-1}, 2512 cm^{-1} and 1796 cm^{-1} correspond to harmonic vibrations and/or combinations of CO_3^{2-} groups (Fig. 1).

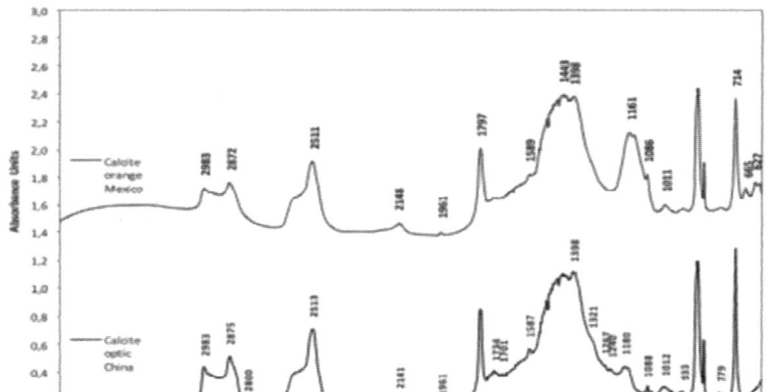

Figure 1. FTIR spectra of calcites samples.

The band at 713 cm^{-1} may correspond both to the presence of calcite and the aragonite. The vibration to 1084 cm^{-1} is, in turn, specific to aragonite. The massive cantered around 1440 cm^{-1} correspond to asymmetric stretching vibrations of CO_3^{2-}. The Mexico orange calcite presents additional bands at 1164, 1143 and 669 cm^{-1} which correspond to the presence of sulphate ions (SO_4^{2-}). vibration bands at 935 cm^{-1} and 849 cm^{-1} present in the orange calcite correspond to specific calcite bands. The orange calcite also shows a band at 775 cm^{-1} which corresponds to quartz.

3.2. *Zeta potential*

The studies show that the IEP of calcite varies with sample history and solution composition. Electrophoretic mobility measurements indicate that the potential-determining ions are not H^+ and OH^-, but rather Ca^{2+} and CO_3^{2-} (or HCO_3^- or $H_2CO_3^0$). This apparent conflict is resolved by a slight modification of the electrical double layer (EDL) model [13]. Thus, the sign and the value of zeta potential should impact the particle adhesion to a solid surface. First, for comparison purpose, in Figure 2 shows the change in zeta potential of calcite optic and orange, and apatite from Madagascar as a function of time in KCl 10^{-2} M for an initial pH of 9.0.

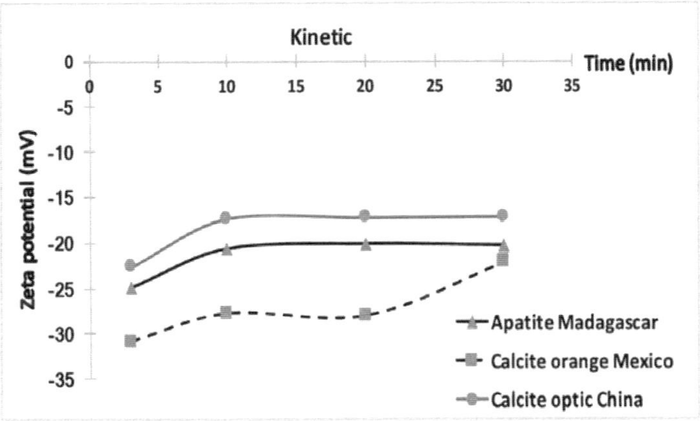

Figure 2. Variation of zeta potential of calcites and apatite with mixing in KCL 10^{-2} M at pH 9.

The zeta potential of the minerals at pH 9.0 is negative with the zeta potential of calcite orange being larger than that of the apatite from Madagascar. The zeta potential appears to decrease in magnitude in the first 10 minutes of conditioning. Figure 3 shows that the pH of the solution containing calcite and apatite increases in the first 30 min of conditioning before plateauing at longer times. Somasundaran and Agar [14] reported a similar increase in pH with conditioning time of calcite until an equilibrium pH value was reached when the initial pH values were acidic or neutral, while pH was found to decrease to a similar equilibrium pH value when the initial pH values were basic. This equilibrium pH value at 8.2 was associated to the point of zero charge of calcite. For calcite at low pH values, carbonate in solution or at the surface consumes proton ($CO_3^{2-} + H^+ \Leftrightarrow HCO_3^-$) resulting in a pH increase while at high pH values calcium in solution or at the surface consumes hydroxyl ions ($Ca^{2+} + OH^- \Leftrightarrow CaOH^-$) resulting in a pH decrease. In all the suspensions, the pH increased on the time scale. The only conclusion that at present might be drawn is that the measured zeta potentials are still not equilibrium ones and they are not determined by the pH of the bulk phase. Thus, they are determined rather by concentrations of Ca^{2+} and CO_3^{2-} in the solution and especially by the latter ones. Moreover, pH 8.2 is the experimental isoelectric point of calcite [14]. At pH values, lower than the IEP, the solution will become more alkaline by addition of calcite. At high pH values, the reverse is true, and the pH of solution will decrease.

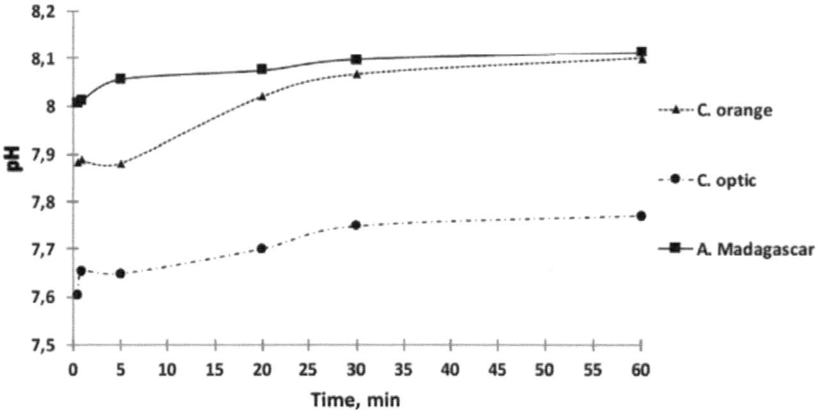

Figure 3 Changes of pH versus time for calcites and apatite in 10^{-2} M KCl.

The ionic impurities affect the zeta potential of freshly precipitated $CaCO_3$ in comparison with that of pure carbonate. The presence of Fe^{2+} keeps the potential positive during 2 h [15], while SO_4^{2-} causes it negative. The effect of Mg^{2+} on the zeta potential is less clear, thought it can form isomorphic replacement. It is possible to suggest the similar behaviour of Mg^{2+} as for Fe^{2+} because of close ionic radius to Ca^{2+}. Thus, the effect of Mg^{2+} and SO_4^{2-} ions released in solution from the impurities on the zeta potential of calcium carbonate probably results from their specific adsorption on the hydrolysed surface, comparing respective zeta potential and pH changes.

Indeed, the sites on the surface of the calcite may be likened to two types of primary hydration sites/protonation: $\equiv CaOH^0$ and $\equiv CO_3H^0$ products by the reactions of protonation and hydration of the surface sites of calcite in aqueous medium, control the calcite surface reactivity [16]. Other point can be determined by potentiometric titrations, it is the point without salt effect (PZSE). According to the theory, all charges curves obtained for different ionic strengths intersect at a point corresponding to PZSE which is considered as PZC. Indeed, the point of zero charge, the number of positive groups is equal to negative groups; therefore, the surface is generally neutral. However, a change in ionic strength or an electrolyte concentration change must leave the invariant surface charge for this pH. It is different for the other pH since in this case the surface charge will end up strongly screened by the ions against the electrolyte and therefore the surface charge is measured will change. Indeed, when the solution pH is more basic than the PZSE, negative ions and complexes will predominate at the interface and the surface will be negatively charged. In a similar way, when a solution pH is more acidic than the PZSE, positive species will predominate and the surface will be positively charged. The PZSE obtained from the potential measurements may differ from the IEP if any specific adsorption of ion occurs (Figure 4).

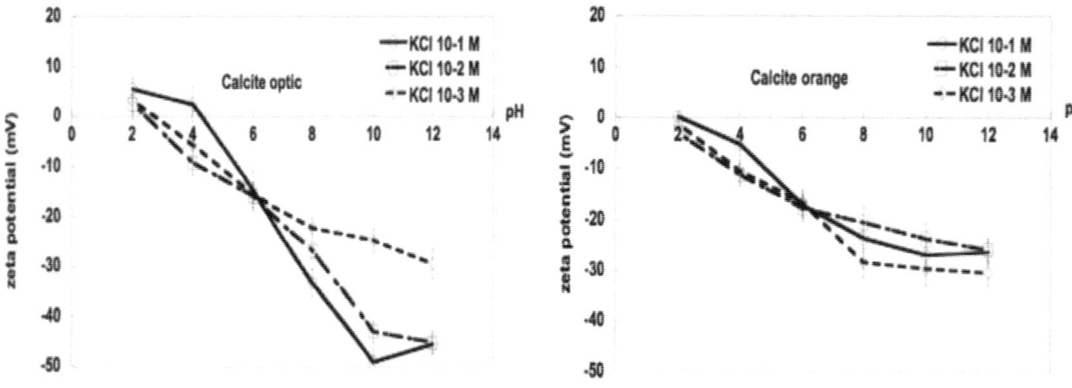

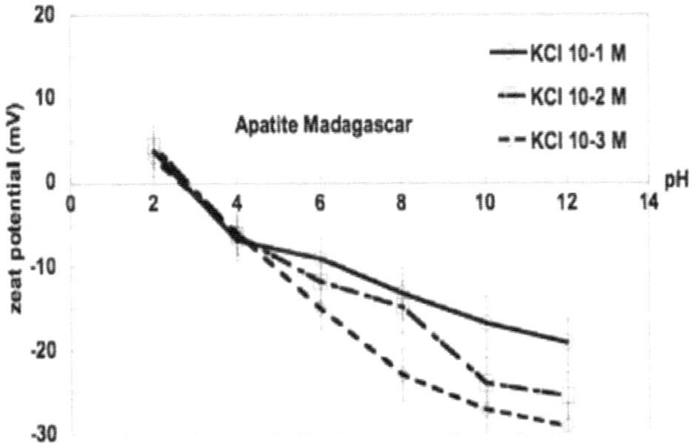

Figure 4. Zeta potential dependence of calcites and apatite on the pH. Supporting electrolyte is 10^{-1}, 10^{-2} and 10^{-3} M KCl.

Thus, this difference can be observed if equilibrium has not yet been attained. Therefore, Ca^{2+} and HCO_3^- are the principal potential determining ions, then a PZSE higher than the IEP (as observed here for calcite and apatite) could be explained, because the calcium ion, being bivalent, would be specifically adsorbed.

3.3. *Flotation tests*

Flotation recovery was studied in the pH range of 5-11 for all minerals at a collector concentration of 10^{-5} M and results are presented in Figure . With increasing pH, calcite recovery decreases while apatite recovery increases when using oleate as the collector. The high mineral recovery obtained is the result of collector adsorption on the mineral surface because recovery of these minerals in the absence of collector is low (<10%) and attributed to entrainment.

In the equilibrium stability diagram of oleic acid, neutral oleic acid is the more stable species at pH values below 8 in both the aqueous and liquid form while at higher pH values negatively charged oleate species as monomer or dimer are dominant. As the concentration of calcium dissolved from calcite increases with decreasing pH values, this should result in more calcium-oleate complex formation ($Ca(OL)_{2,s}$) at the solution-mineral interface followed by its precipitation on the mineral surface, which should therefore increase calcite flotation as pH decreases. Figure 5 shows that the trend in apatite recovery with pH is the same in the presence of oleate or linoleate. It is postulated that the presence of two double bonds and more precisely the stronger chain-chain interaction in the linoleate dimer than in the oleate dimer, increases the interaction with calcium ions in solution (lower solubility of the calcium-(linoleate)$_2$ complex).

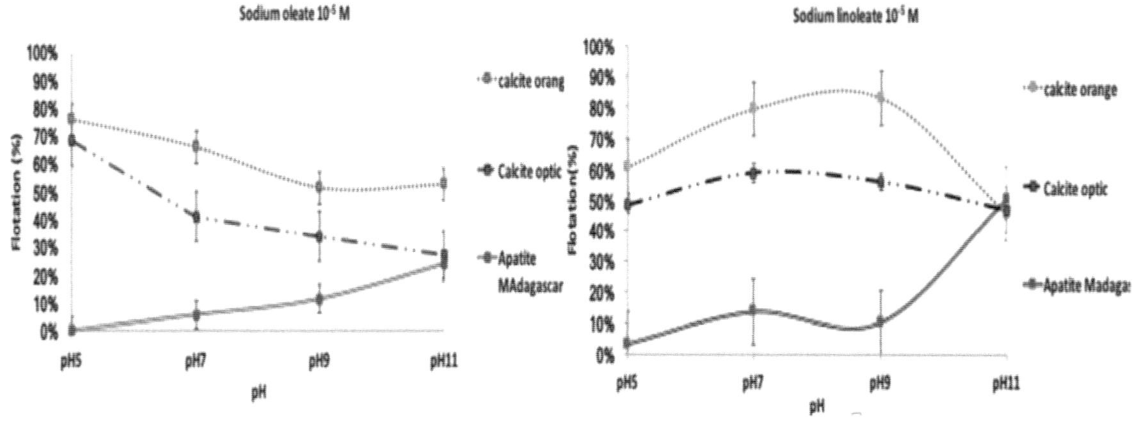

Figure 5. Flotation recovery as a function of pH, in the presence of 10^{-5}M sodium oleate (top) or linoleate (bottom).

This mechanism is also supported by the fact that the dimer, $(linoleate)_2H^-$, is more stable at an intermediate pH value where calcite recovery is at its maximum value (Figure), there are obviously considerable differences in flotation efficiency between them. At low pH, both the collectors lead to a good recovery for the calcites, while with optic calcite and orange calcite the use of sodium oleate shows a better flotation recovery than with linoleate. At higher pH, the trend is reversed showing better recovery with sodium linoleate. The decrease of flotation with calcite below pH 9.0 is due to its dissolution. At pH values below 8 oleate exists in solution mainly in the form of neutral oleic acid molecules and precipitated oleic acid.

However, at pH 9, flotation of calcite is lower which may be due to saturation of the surface, which can cause a decrease in adsorption strength of the collectors. But according to [16], at pH above 8.2, the $CaCO_3^-$ group mainly composes the surface of calcite; and the collector cannot fix and calcite floats less. Thus, at this pH, the use of sodium linoleate alone is more efficient than the oleate alone.

The only differences between surface properties of calcite and apatite are at the level of dissolution of the mineral content, and the proportion of species and group located on their surface. The Ca^{2+} site density on the surface of these minerals is almost identical. The most common chemical species on the surface of these two minerals are Ca^{2+} and HCO_3^-. Then, Ca^{2+} is more important in proportion to pH 8 on the surface of the calcite and HCO_3^-, on the surface of the apatite. According to the species distribution diagram of oleic acid [19], we can note that the concentrations of oleic ion monomer and dimer remain almost constant above the precipitation pH and decrease sharply below it

4. Conclusion

Following of electrokinetics studies the calcites and apatites separation were discussed in this paper by using anionic collectors. The close correspondence between the IEP and the PZSE determined for calcite and apatite shows different values. The zeta potential experiments implied that chemical adsorption was achieved in the surface of calcium minerals. The surface charge of calcite and apatite became more negative after adsorption of sodium oleate and sodium linoleate. The maximum flotation recovery of calcite and apatite was obtained at low and high pH, respectively. The results are consistent with the oleic and linoleic acid hydrolyses to form ions ($RCOO^-$) at high pH and molecules (RCOOH) at acid and neutral pH which defines the differential behaviour of two calcium minerals.

5. References

[1] S. K. Mishra 1982 Electrokinetic properties and flotation behaviour of apatite and calcite in the presence of sodium oleate and sodium metasilicate *Int. J. Miner. Process.* **9** 59–73.

[2] I. V. Filippova, L. O. Filippov, A. Duverger, and V. V. Severov 2014 Synergetic effect of a mixture of anionic and non-ionic reagents: Ca mineral contrast separation by flotation at neutral pH *Miner. Eng.* **66–68** 135–44.
[3] H. Sis and S. Chander 2003 Reagents used in the flotation of phosphate ores: a critical review," *Miner. Eng.* **16** 577–85.
[4] G. Ying-yong, G. Shun-peng, X. Zeng, and Y. Yu 2008 Double reverse flotation process of collophanite and regulating froth action *Trans. Nonferrous Met. Soc. China* **18** 449–53
[5] J. Gordeijev and P. Hirva 1999 Theoretical studies on the interaction of oleoyl sarcosine with the surface of apatite *Surf. Sci.* **440** 321–26.
[6] L. Filippov, A. Duverger, I. Filippova, H. Kasaini, and J. Thiry 2012 Selective flotation of silicates and Ca-bearing minerals: The role of non-ionic reagent on cationic flotation *Miner. Eng.* **36–38** 314–23.
[7] K. H. Rao, J. M. Cases, P. De Donato, and K. S. E. Forssberg 1991 Mechanism of oleate interaction on salt-type minerals *J. Colloid Interface Sci.* **145** 314–29.
[8] A.Gaudin, D. Fuerstenau 1955 Quartz flotation with anionic collectors *Trans AIME* **202** 66–72.
[9] G. Parks and P. de Bruyn 1962 The zero point of charge of oxides *Phys Chem* **66** 967–73.
[10] E. Chibowski and L. Hołysz 1995 Changes in zeta potential of TiO2 and CaCO3 suspensions treated with a radiofrequency electric field as measured with a Zeta Plus instrument. *Colloids Surf. Physicochem. Eng. Asp.* **105** 211–20.
[11] K. Higashitani, H. Iseri, K. Okuhara, A. Kage, and S. Hatade 1995 Magnetic Effects on Zeta Potential and Diffusivity of Nonmagnetic Colloidal Particles *J Colloid Interface Sci* **172** 383–88.
[12] J. S. Baker and S. J. Judd 1996 Magnetic amelioration of scale formation *Water Res.* **30** 247–60.
[13] S. L. S. Stipp 1999 Toward a conceptual model of the calcite surface: hydration, hydrolysis, and surface potential *Geochim. Cosmochim. Acta* **63** 3121–31.
[14] P. Somasundaran and G. E. Agar 1967 The zero point of charge of calcite *J. Colloid Interface Sci.* **24** 433–40.
[15] E. Chibowski, L. Hotysz, and A. Szcześ 2003 Time dependent changes in zeta potential of freshly precipitated calcium carbonate *Colloids Surf. Physicochem. Eng. Asp.* **222** 41–54.
[16] O. S. Pokrovsky, J. Mielczarski, O. Barres, and J. Schott 2000 Surface Speciation Models of Calcite and Dolomite/Aqueous Solution Interfaces and Their Spectroscopic Evaluation *Langmuir* **16** 2677-88.
[17] D. Fuerstenau 1956 Streaming Potential Studies on Quartz in Solutions of Aminium Acetates in Relation to the Formation of Hemi- micelles at the Quartz-Solution Interface **60** 981–85.
[18] D. W. Fuerstenau and Pradip 2005 Zeta potentials in the flotation of oxide and silicate minerals," *Adv. Colloid Interface Sci.* **114–115** 9–26.
[19] P. Somasundaran, K. P. Ananthapadmanabhan, and I. B. Ivanov 1984 Dimerization of oleate in aqueous solutions *J. Colloid Interface Sci.* **99** 128–35.

IOP Conf. Series: Journal of Physics: Conf. Series **879** (2017) 012013 doi:10.1088/1742-6596/879/1/012013

Rubber-like materials derived from biosourced phenolic resins

G. Amaral-Labat[1,2], L.I. Grishechko[1,3], G.F.B. Lenz e Silva[2], B.N. Kuznetsov[3,4], V. Fierro[1], A. Pizzi[5,6] and A. Celzard[1]

[1] Institut Jean Lamour - UMR Université de Lorraine – CNRS n°7198. ENSTIB, 27 rue Philippe Séguin, CS 60036, 88026 Épinal Cedex, France
[2] University of São Paulo, Department of Metallurgical and Materials Engineering PMT-USP, Avenida Mello Moraes, 2463-Cidade Universitária, CEP 05508-030, - São Paulo, SP – Brazil
[3] Institute of Chemistry and Chemical Technology, SB RAS, 50/24, Akademgorodok Street, Krasnoyarsk 660036, Russia
[4] Siberian Federal University, 79, Svobodny Avenue, Krasnoyarsk, 660041, Russia
[5] LERMAB, Université de Lorraine, 27 rue Philippe Séguin, CS 60036, 88026 Épinal cedex, France
[6] Dept. of Physics, King Abdulaziz University, Jeddah, Saudi Arabia

Email: alain.celzard@univ-lorraine.fr

Abstract. The present work describes new gels derived from cheap, abundant and non-toxic wood bark extracts of phenolic nature, behaving like elastomers. Especially, we show that these materials might be used as rubber springs. Such amazing properties were obtained by a quite simple synthesis based on the autocondensation of flavonoid tannins in water at low pH in the presence of a plasticizer. After gelation and drying, the materials presented elastic properties that could be tuned from hard and brittle to quite soft and deformable, depending on the amount of plasticizer in the starting formulation. Not only the materials containing the relevant amount of plasticizer had stress-strain characteristics in quasi-static and cyclic compression similar to most commercial rubber springs, but they presented outstanding fire retardance, surviving 5 min in a flame at 1000°C in air. Neither flame propagation nor drips were noticed during the fire test, and the materials were auto-extinguishable. These excellent features make these materials potential substitutes to usual organic elastomers.

1. Introduction

Natural rubber is a well-known material produced at the level of several million tons every year and used for many centuries, as the first uses were reported in the ancient Aztecs' and Mayans' culture [1]. Despite its quite valuable properties, its limited supply at some periods of the 20th century – or at least the existing gap between production and fast-growing world demand – encouraged the development of synthetic rubbers, despite many of them were technically not as good as natural rubber [2]. Whether natural or not, rubbers are elastomers obtained after a crosslinking step, called vulcanization in this context, and after which the polymer chains have enough mobility to reorganize for redistributing an applied stress. The crosslinks allow that the material recovers its initial shape and dimensions once the

IOP Conf. Series: Journal of Physics: Conf. Series **879** (2017) 012013 doi:10.1088/1742-6596/879/1/012013

stress is removed. Most elastomers are based on polyenes, mainly as copolymers based on isoprene or butadiene, although polyurethane and polysiloxane also deserve being called "rubbers" [3].

In the present work, rubber-like materials were prepared and characterized from plant-derived polyphenols. Whereas such raw materials are famous for giving hard resins from which adhesives can be obtained mainly for wood products [4], getting soft and elastic solids was unexpected. This could be achieved, however, by using the relevant plasticizer in relevant amounts so that neither the material remained brittle nor the crosslinking was prevented. In the following, the main steps describing the synthesis of these new rubber-like materials are presented, as well as a selection of physical and chemical properties.

2. Materials preparation and characterization

2.1. *Materials preparation*

The preparation method of rubber-like phenolic materials is described below. It reports how simple the synthesis is, as it only requires 3 very common and cheap compounds and nothing else but a minimalist laboratory equipment: a mixer, an oven, and glass tubes.

2.1.1. Chemicals. One of the originalities of the present work is using condensed (flavonoid) tannin as the base of the elastic resin from which the materials were prepared. Condensed tannin was supplied by the company SilvaChimica (Italy) and was used without purification. It is a natural extract from black wattle trees, sold under the name Fintan OP, and obtained by leaching barks with a warm aqueous solution of sodium bisulphite at 1 wt.%. After spray-drying of the concentrated solution, a light brown powder is obtained which is soluble in water, see figure 1. A typical oligomer constituting about 80% of the tannin extract is also shown in figure 1 [5]. It can be easily crosslinked with aldehydes, such as formaldehyde [5], thereby leading to hard resins. However, this kind of tannin is also prone to autocondensation, provided that the pH is low enough [6].

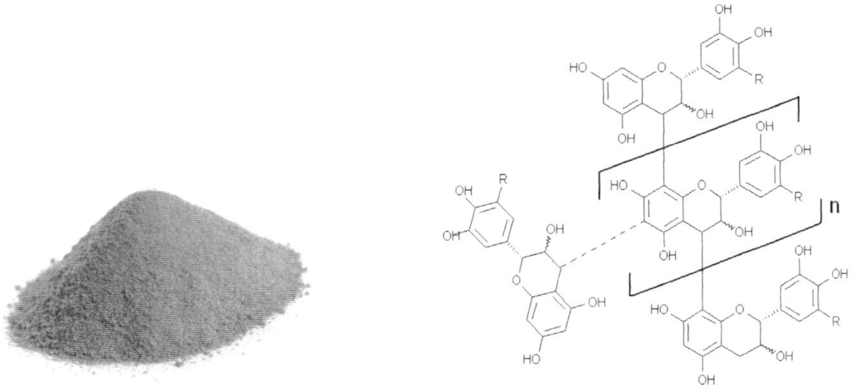

Figure 1. Black wattle bark extract (left) and corresponding oligomer (right), in which *n* typically ranges from 1 to 9.

Low pH was reached by use of para-toluenesulphonic acid (p-TSA), having the advantage of being both a strong and solid acid. This allowed significant changes of pH with negligible dilution. Finally, ethylene-glycol (EG) was used as plasticizer.

2.1.2. Synthesis. Tannin (T) was first dissolved in distilled water (W), and the solution was stirred magnetically until complete homogenization. The stirring was maintained when adding successively para-toluenesulphonic acid (p-TSA) followed by ethylene-glycol (EG). The corresponding formulations are given in table 1, which shows that various ethylene-glycol to water (EG/W) weight ratios were tested while keeping a constant amount of T and p-TSA. The final, brown solutions were introduced in glass tubes which were hermetically closed afterwards and let for curing in an oven at 85°C for 5 days. The time spent in the oven was enough to produce the gelation of the solutions. Then the glass tubes were broken, the materials were recovered in the form of long cylinders of diameter 1 cm, and left for one full month in air. During this second period, the gels dried and shrank until a constant diameter was reached. Based on the observed dimensional changes, it is assumed that the properties stabilised after this time so that the samples were ready for characterisation. The names of the samples were given according to their EG/W weight ratio, as shown in table 1.

Table 1. Formulations of rubber-like materials and their bulk density.

Samples	EG/W (g/g)	EG/W (wt. %)	T/pTSA (g/g)	(T+pTSA)/total) (wt. %)	Bulk density (g cm^{-3}) ± 5%
TEG(0)	0/15	0	5/0.8	28	1.49
TEG(20)	3/12	20	5/0.8	28	1.43
TEG(33)	5/10	33	5/0.8	28	1.45
TEG(47)	7/8	47	5/0.8	28	1.42
TEG(60)	9/6	60	5/0.8	28	1.46

2.2. Materials characterization

The materials were characterized through a limited number of methods, in order to describe their main characteristics in terms of mechanical properties and fire retardance. As far as rubber-like materials are concerned, many other intrinsic properties might have been investigated as well, and will be in the near future, but the present paper only aimed at presenting such new materials and their principal features.

2.2.1. Bulk density. The materials were cut into cylindrical blocks of carefully measured diameters and heights, typically 8-10 mm and 7-11 mm, respectively. 2 to 5 samples of the same kind were investigated, and the average density was then calculated.

2.2.2. Mechanical tests. Compression tests were carried out as follows in the quasi-static regime, using an Instron 5944 universal testing machine equipped with a 2 kN load cell. First, samples of identical dimensions (typical diameter 8-10 mm and typical height 5-6.5 mm, so as to maintain a constant diameter/height ratio) were compressed with a rate of 2 mm/min until they collapsed completely or broke into pieces. From the corresponding stress-strain curves, elastic modulus and compression strength were determined. The former was estimated from the slope of the initial, linear part of the curve, whereas the latter was assumed to correspond to the upper limit of the elastic regime, i.e., to the point at which a strong change of slope is noticed.

A second set of samples was then used for cyclic compression tests. Unlike the former experiments for which the strain was not limited, the compression was carried out until 10% strain at the same compression speed of 2 mm/min. 50 successive compression-decompression tests were performed, applying a recovery time of 1 min between each cycle.

2.2.3. Fire tests. One sample was put directly into the flame of a propane burner fed by an excess of air, more exactly in the zone wherein the temperature is the highest, about 1000°C, i.e., just at the tip of the blue cone. The sample weight was measured before and after 5 min spent in the flame.

3. Results and discussion

3.1. Materials aspect and general features

The materials were all black and shiny, irrespective to their initial formulation. Figure 2 illustrates the case of TEG(33) sample, for which the elastic character is easy to appreciate as soon as the gel is gently pressed between the fingers. The method by which the materials were prepared suggests that they can be obtained without limit of size and in any shape, the latter only depending on the characteristics of the mold.

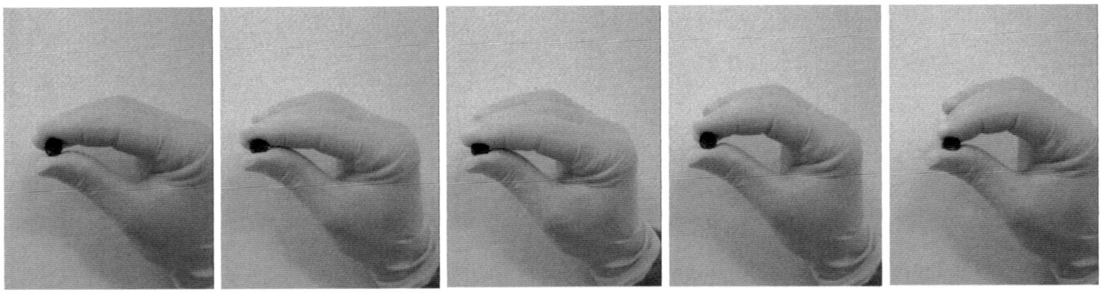

Figure 2. Sequence (from left to right) of compression-release stresses applied to TEG(33) sample between fingers, showing the ease by which it can be deformed while recovering immediately.

The density values reported in table 1, 1.45 g cm^{-3} ± 3% on average, were very close to each other. This suggests that, at fixed content of solid phase, water and ethylene glycol were both equally incorporated and took part to the samples bulk, and that EG replaced water without producing any noticeable shrinkage or swelling.

3.2. Mechanical properties

As already suggested by figure 2, all materials except TEG(0) were very soft and elastic. This is shown in figure 3, where the compressive stress was plotted as a function of strain. A clear trend can be observed from TEG(0) to TEG(60), corresponding to an increasing amount of plasticizer in the formulation. As TEG(0) did not contain ethylene glycol at all, it behaved as most other tannin-based resins so far, i.e., presented a hard and brittle behaviour [7]. This is evidenced by the steep, first part of the curve, corresponding to a much higher modulus than for the other materials, followed by a brittle failure. In contrast, the other materials became softer and softer when the amount of plasticizer increased. Thus, the general shape of the curves presented by TEG(20) to TEG(47) are typical of what is known for commercial rubber springs, and with similar values of both stress and strain [8]. It is then expected that these materials can be used for the same applications. TEG(60) is, however, probably too weak, as suggested by its rather flat curve shown in figure 3.

The aforementioned trends also apply to the compressive strength, defined as the stress at the change of slope of the curves, i.e., at around 10-15% strain. The higher was the EG content, the lower was the strength. The corresponding values of modulus and strength are gathered in table 2. TEG(60) was logically the softest material, and corresponded to the limit of this range of materials. Formulations presenting higher amounts of ethylene glycol indeed did not gel, from where the range of compositions formerly presented in table 1.

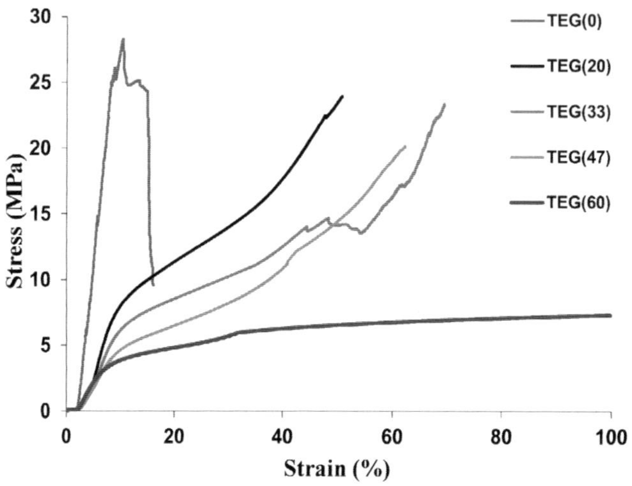

Figure 3. Stress-strain characteristics of rubber-like materials submitted to compression at a loading rate of 2 mm/min and without limitation of strain.

Table 2. Elastic moduli and compressive strengths derived from figure 3.

Samples	Elastic modulus (MPa)	Compressive strength (MPa)
TEG(0)	381	25.0
TEG(20)	141	10.7
TEG(33)	101	6.4
TEG(47)	68	5.2
TEG(60)	72	4.1

TEG(33) sample before and after compression test is shown in figure 4. Even after having been compressed at a strain as high as 40%, it completely recovered its initial height, although around 10 min were required for this. Such severe deformation, far beyond what is requested for rubber springs (usually limited from 15% to 25% strain [9]), produced an irreversible change of diameter. Interestingly, the latter was lower after than before the test, suggesting a negative Poisson ratio.

Figure 4. TEG(33) sample before (left) and 10 min after (right) compression tests up to 40% strain at 2 mm/min.

However, lower strains such as those usually encountered in rubber spring applications did not produce any dimensional change, and the same sample recovered completely again after many stresses applied up to 10% strain. Figure 5 thus shows a series of 50 successive compression-decompression curves for the same TEG(33) material. As can be seen, the cycles became nearly impossible to separate beyond cycle #30.

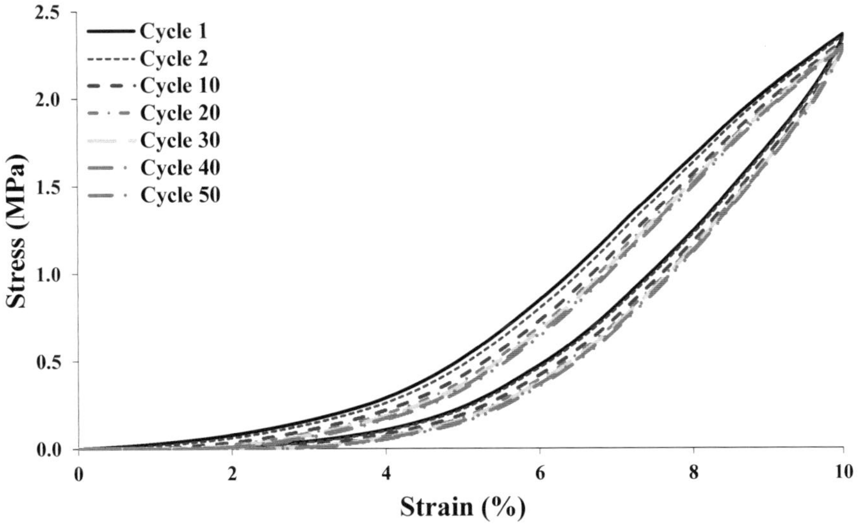

Figure 5. Successive compression-release cycles for TEG(33) up to 10% strain at 2 mm/min, with 1-min recovery between each cycle.

From the hysteresis loops presented in figure 5, the energy loss coefficient was calculated as the area below the loading curve minus the area below the unloading curve [10]. This parameter corresponds to the ease with which a material can recover and hence retain its elastic properties. The values given in figure 6 indicate that the energy that can be dissipated is quite high, as it should be expected for a rubber spring.

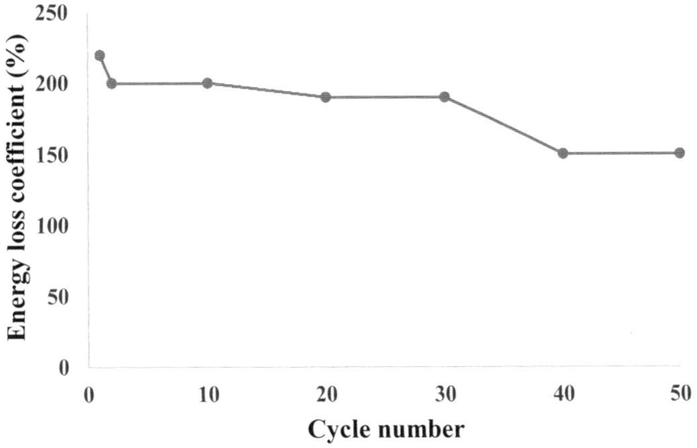

Figure 6. Energy loss coefficient derived from figure 5.

IOP Conf. Series: Journal of Physics: Conf. Series **879** (2017) 012013 doi:10.1088/1742-6596/879/1/012013

3.3. Fire retardance

The fire retardance of these materials was found to be outstanding. Such feature is already known for phenolic resins in general, which are indeed used as foundry sand resins in the metal casting industry, or as classical matrix in ablative composites or friction materials [11]. When heated, these materials indeed tend to decompose into glassy carbon whereas a very low amount of poorly flammable gas is released. During such degradation process, a significant amount of heat is absorbed, and almost no flame is produced. Whereas this phenomenon had to be expected because of the highly aromatic character of the crosslinked phenolic resin [12], observing it with rubber-like materials still is amazing.

Figure 7 thus presents TEG(33) sample submitted to a burner at 1000°C. No flame appeared at the surface of the sample all along the test, but only a little smoke was emitted at the beginning (see the orange colour after 1 min in figure 7). When taken out of the burner, the red-glowing sample auto-extinguished immediately. After 5 min, the weight loss was around 65% and, as expected, the material completely lost its elastic character and became very hard. However, it survived, as shown in the last photo of figure 7. In so severe conditions, any other usual organic elastomer would have burnt completely with flames and, in some cases, with release of faming drips.

Figure 7. Top raw: view of TEG(33) sample submitted to the flame of a propane burner at 1000°C at times t = 0, 1 min, 2 min, 3.5 min, and 5 min (from left to right). Bottom: view of the same sample at 5 min just after removing the flame.

4. Conclusion

The first rubber-like materials based on a natural source of phenolic resin, flavonoid tannin, were reported. They were prepared from an ultra-simple mixing of commercial wattle bark extract in water with ethylene glycol and para-toluenesulphonic acid, subsequently let for curing for 5 days at 85°C and dried in room conditions for one month.

These new materials proved to behave as true elastomers in compression, whereas their chemical structure had nothing to do with classical vulcanizates. In the present case, the oligomers crosslinked by autocondensation, as no crosslinking agent such as formaldehyde was requested. The presence of ethylene glycol (EG), acting as internal plasticizer, allowed observing various mechanical behaviors, from hard and brittle in the complete absence of EG to ultra-soft with the highest possible amount of EG. Therefore, a range of materials could be obtained just by playing with the concentration of EG in the formulation, all the other parameters being kept constant. The material with the medium content of EG was found to be able to dissipate a significant amount of energy when used as rubber spring, and its stress-strain characteristics were indeed quite similar to those of commercial elastomers used for the same application. Moreover, the reversibility of the compression-release curves was complete after 30 cycles up to 10% strain.

Besides being new biosourced rubber-like materials, the latter also presented an outstanding fire retardance related to their phenolic nature. The sample with the medium content of EG lost it elastic properties but survived after having been submitted to a flame at 1000°C during 5 min. No ignition could be observed, and the material was auto-extinguishable.

Such advantageous properties definitely demonstrate the interest for these new materials for springs, and further studying their properties is worth doing in the near future.

Acknowledgements
The authors gratefully acknowledge the financial support of the CPER 2007–2013 "Structuration du Pôle de Compétitivité Fibres Grand'Est" (Competitiveness Fibre Cluster), through local (Conseil Général des Vosges), regional (Région Lorraine), national (DRRT and FNADT) and European (FEDER) funds. We also thank the Région Lorraine, especially one of us (L.I.G) for her grant through the Lorraine-Russia ARCUS cooperation program.

References
[1] Backhaus R A 1998 *Biopolymers from Renewable Resources* ed D L Kaplan (Berlin Heidelberg: Springer) chapter 13 pp 323-354
[2] Jones K P and Allen P W 1992 *Natural rubber: biology, cultivation and technology* ed M R Sethuraj and N M Mathew (Asterdam: Elsevier) chapter 1 pp 17-19
[3] Morton M 2010 *Rubber Technology* ed M. Morton (Dordrecht: Kluwer Academic Edition)
[4] Pizzi A 2003 *Handbook of Adhesive Technology* ed A. Pizzi and K L Mittal (Basel: Marcel Dekker) chapter 27 pp 573-587
[5] Pizzi A 1983 *Wood adhesive chemistry and technology* ed A. Pizzi (New York and Basel: Dekker)
[6] Meikleham N and Pizzi A 1994 *J. Appl. Polym. Sci.* **53** 1547
[7] Szczurek A, Fierro V, Pizzi A, Stauber M and Celzard A 2014 *Ind. Crop. Prod.* **54** 40
[8] Wood LA 1977 *J Res Natl Bur Stand* **82** 57
[9] Polurethane Products Ltd. Datasheet 2013 *Engineering Property - Urethanes in Compression* (http://www.polyprod.com/)
[10] Zhao W, Li Y, Wang S, He X, Shang Y, Peng Q, Wang C, Du S, Gui X, Yang Y, Yuan Q, Shi E, Wu S, Xu W and Cao A 2014 *Carbon* **76** 19
[11] Pilato L 2010 *Phenolic Resins: A Century of Progress* ed. L Pilato (Berlin Heidelberg: Springer)
[12] Celzard A, Fierro V, Amaral-Labat G, Pizzi A and Torero J 2011 *Polym. Degrad. Stabil.* **96** 477

Modelling the physical properties of glasslike carbon foams

M. Letellier[1], J. Macutkevic[2], D. Bychanok[3], P. Kuzhir[3], C. Delgado-Sanchez[1], H. Naguib[4], S. Ghaffari Mosanenzadeh[4], V. Fierro[1] and A. Celzard[1]

[1] Institut Jean Lamour - UMR Université de Lorraine – CNRS n°7198. ENSTIB, 27 rue Philippe Séguin, CS 60036, 88026 Épinal Cedex, France
[2] Department of Radiophysics, Vilnius University, Sauletekio 9, 10022 Vilnius, Lithuania
[3] Research Institute for Nuclear Problems BSU, 11 Bobruiskaya Str., 220030 Minsk, Belarus
[4] Department of Materials Science and Engineering, University of Toronto, 5 King's College Road, Toronto, ON M5S 3G8, Canada

Email: alain.celzard@univ-lorraine.fr

Abstract. In this work, model alveolar materials – carbon cellular and/or carbon reticulated foams – were produced in order to study and to model their physical properties. It was shown that very different morphologies could be obtained whereas the constituting vitreous carbon from which they were made remained exactly the same. Doing so, the physical properties of these foams were expected to depend neither on the composition nor on the carbonaceous texture but only on the porous structure, which could be tuned for the first time for having a constant pore size in a range of porosities, or a range of pore sizes at fixed porosity. The physical properties were then investigated through mechanical, acoustic, thermal and electromagnetic measurements. The results demonstrate the roles played by bulk density and cell size on all physical properties. Whereas some of the latter strongly depend on porosity and/or pore size, others are independent of pore size. It is expected that these results apply to many other kinds of rigid foams used in a broad range of different applications. The present results therefore open the route to their optimisation.

1. Introduction

The research presented here lies in between physics of heterogeneous media, materials chemistry and engineering science. The goal was to prepare model cellular materials, with the main objective of solving a number of open questions in physics and allowing either to validate existing models, to propose more refined versions, or to observe new phenomena. If this type of studies has already been partially attempted in the past (see [1]), it has never been possible so far to produce model alveolar materials whose structural parameters can be independently and significantly varied at constant composition. The originality of the present studies is then based on the use of vitreous carbon for developing rigid foams with perfectly controlled morphologies and composition, enabling both rigorous physical studies and their optimization for specific performances in terms of mechanical, thermal, acoustic, and electromagnetic properties. Ultimately, it should be possible to optimize these rigid foams depending on the planned applications.

Herein, the leading idea was therefore to produce model cellular materials from phenolic-furanic resins based on different formulations and preparation routes, allowing a strict control of the porous structure. Since such materials presented different chemical compositions possibly influencing the resultant physical properties to investigate, all were next pyrolysed at 900°C for producing carbon monoliths having the same final composition and the same carbonaceous texture. As a result, a broad range of porous materials could be obtained for the first time, based on which the influence of porous structure on mechanical, acoustic, thermal and electromagnetic properties could be investigated with an unprecedented accuracy.

2. Materials preparation and characterization

2.1. Materials preparation

Most carbon foams prepared here were obtained after pyrolysis of rigid cellular foams derived from phenolic-furanic (thermoset) resins. If this type of foam is not completely new [2–6], many new formulations were produced herein, and for the first time, the development of a variety of structures formerly impossible to achieve by other processes was carried out. Over 60 cellular foams made of vitreous carbon with different structures were produced through the pyrolysis at 900°C of foamed phenolic polymers.

It is known for long [6] that the density and the cell size of such foams can be varied through the amount of foaming agent. In the present study, significant changes in the structure were also obtained by the addition of various plasticizers, surfactants and copolymers. Hence, 14 types of formulations giving different structures (called STD, PMDI, T1, ...) were developed, and the amount and the nature of the foaming agent were varied to expand the range of available structures. The properties were then studied on the basis of these 14 families of formulations. Indeed, it was found to be more rigorous, and easier (in terms of repeatability) to model the behaviour of each set of materials and then compare the sets with each other. The details about the formulations and their impact on the final porous structure have been abundantly discussed elsewhere [7], and therefore will not be recalled here.

Reticulated carbon foams were also produced by templating of commercial reticulated polyurethane foams and using the same pyrolysis process as for cellular foams. Such materials not only allowed obtaining other types of structures but provided an appropriate basis of comparison with cellular foams. Indeed reticulated foams represent the only type of vitreous carbon foams currently on the market.

2.2. Materials characterization

2.2.1. Porous structures. The carbon foams were first characterized in terms of apparent density measurements, and cell size. The average cell size was determined for all foams from the quantitative analysis of images obtained with a scanning electron microscope (SEM) and an X-ray microtomograph.

2.2.2. Nature of the carbon. Given the chemical nature of the precursors, phenolic-furanic resins, the resultant carbon after pyrolysis was always glasslike, as expected. A number of carbon foams prepared from the most different formulations were then studied by elemental analysis (chemical composition), helium pycnometry (skeletal density) and Raman spectroscopy. The relative density was calculated from the values of bulk and skeletal density.

2.2.3. Physical properties. As for mechanical properties, a comparative study based on quasi-static compression tests carried out with or without plates glued to the samples' faces was performed on a large number of samples. The tests were done with an Instron 5944 universal testing machine equipped with a 2 kN head. The permeability and the sound absorption of 10 mm-thick samples were measured with a lab-made air permeameter and a two-microphone impedance tube (between 800 and 6300Hz), respectively. The thermal conductivities of the foams were determined at room temperature using the hot disk method, using a Hot Disk TPS 2500 S apparatus. Finally, the electromagnetic

IOP Conf. Series: Journal of Physics: Conf. Series **879** (2017) 012014 doi:10.1088/1742-6596/879/1/012014

properties were measured over a broad frequency range from static regime to near infrared using a complete set of RLC meters and network analysers, waveguides and THz spectrometers.

3. Results and discussion

3.1. Carbon foams' general features

Figure 1 shows only a few examples of carbon foams obtained in this work. Two are cellular foams, i.e., comprised of more or less spherical cells connected with each other through more or less open circular windows, whereas the two other ones are reticulated foams, i.e., are only based on struts; in other words, reticulated foams don't have pore walls. The important feature to remember is that, thanks to so many formulations, foams with different pore sizes could be obtained at constant bulk density, i.e., constant porosity, whereas the opposite also applied based on foams having a fixed pore size but different porosities.

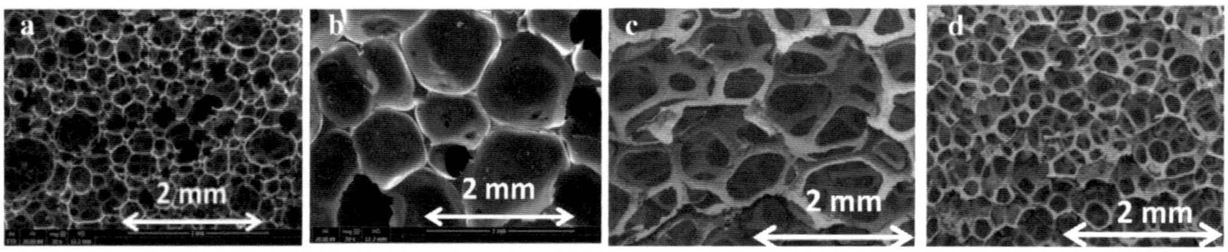

Figure 1. Examples of SEM pictures of carbon foams prepared from different formulations and methods: (a-b) cellular foams; (c-d) reticulated foams.

Such essential feature is presented in figure 2. For each type of formulation, a regression between density and cell size could be obtained. The ranges of bulk densities ($0.015 \le d \le 0.333$ g cm^{-3}) and average cells sizes ($100 \le D \le 5000$ μm) obtained here were the broadest ever investigated within the same piece of work.

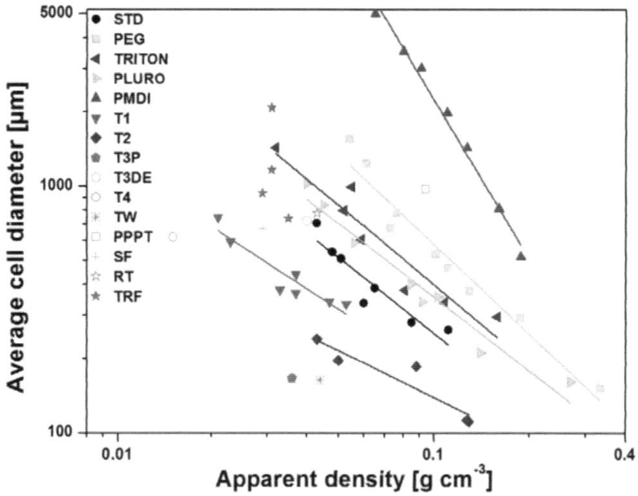

Figure 2. Average cell sizes determined from SEM pictures (and confirmed by micro-computed tomography) as a function of bulk density, for the different foams tested.

It was also shown from elemental analysis that the carbon content of these foams was high and quite comparable, 91.5 ± 2.5%. The skeletal densities measured with helium pycnometry were all approximately equal to 1.98 g cm^{-3}. The Raman spectra of the 3 foams having the most different formulations were very similar again and had, after fits and deconvolutions of the bands, exactly the same D1/G intensity ratio. For the characterizations carried out in this work, we considered that these carbons had nearly identical chemical composition, and that the differences of properties measured thereafter were thus only due to the differences of porous structures.

3.2. Physical properties

3.2.1. Mechanical properties. The compression tests showed that gluing plates to the samples' surfaces did not change the value of the measured maximal compressive stress, but allowed determining much more accurately the compressive modulus. Indeed, without plates, the applied stress was localized on the only struts in contact with the plateau of compression and led to a progressive rupture of the structure, contrarily to the global rupture occurring with plates. The compressive modulus increased with the apparent density for all foams but with different trends (see figure 3). Without plates, the modulus was much lower and could be described with the same law, namely that of Gibson and Ashby (G&A) for open-cell foams [8], for all the foams studied.

However with plates, and whereas a similar relation was also relevant for reticulated foams, the relation of G&A for closed-cell foams was used for cellular foams. The latter relationship took the fraction ϕ of solid only contained in the struts into account, which was then determined for different groups of foams in figure 3. It is the first time that the modulus of cellular open-cell foams was shown to exhibit similar behaviour to that of closed-cell foams.

As for the maximal compressive stress, it increased with density for all the foams according to the law of G&A for open-cell foams. However, it was found that cellular foams having relatively thick cell walls exhibited higher maximal stress than reticulated foams of the same density. Moreover, unlike the prediction of the theory of Weibull [9], no effect related to the cell size could be observed for these foams.

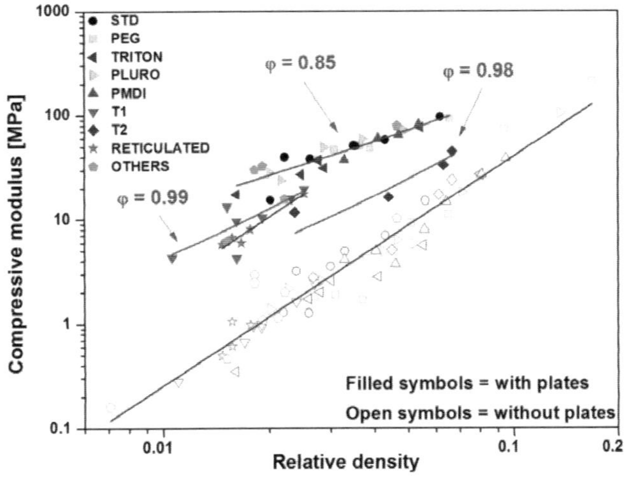

Figure 3. Compressive modulus as a function of relative density of the foams tested with and without plates. Different equations coming from G&A [8] were fitted. φ represents the fraction of solid (see text).

3.2.2. Acoustic properties. The determination of the permeability allowed showing that the latter property was not directly linked to the porosity but increased with the average cell size for all the foams studied. For the reticulated foams, the permeability was the highest and, using the data of figure 1, it could be concluded that the permeability directly increased with the cell size. The sound absorption of these foams was low (see figure 4) and, based again on the data of figure 1, decreased with the increase of cell size. As for the cellular ones, the permeability was relatively low and

IOP Conf. Series: Journal of Physics: Conf. Series **879** (2017) 012014 doi:10.1088/1742-6596/879/1/012014

depended on the equivalent windows diameter (also called hydraulic diameter, not given here but which can be calculated according to [10]). For these cellular foams, the sound absorption was higher and largely increased with the hydraulic diameter. An optimum of permeability was found in the range $1 \times 10^{-9} - 3 \times 10^{-9}$ m^2, leading to the highest values of sound absorption coefficient.

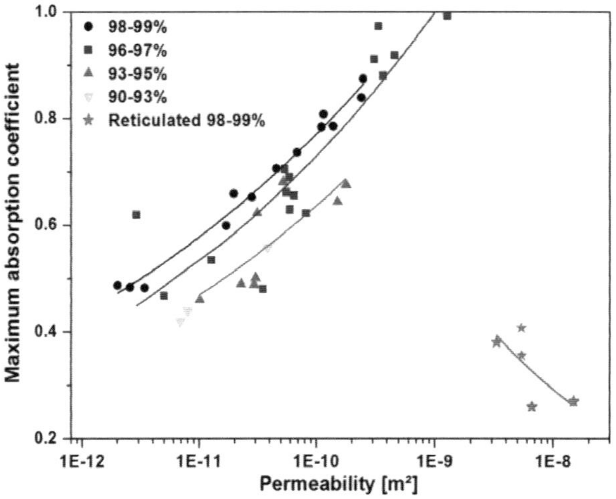

Figure 4. Maximum sound absorption coefficients as a function of permeability for different classes of porosity (expressed as vol. % in the legend, see text).

3.2.3. Thermal properties. The thermal conductivities of 68 carbon foams were measured at room temperature and found to range from 0.041 to 0.18 W m^{-1} K^{-1}, for respective bulk densities within the range 0.029 – 0.264 g cm^{-3} (see figure 5). The model of Glicksman [11], based on a simple geometry of the porous medium, fitted these results pretty well. The fraction of solid contained only in the struts, ϕ, was found to be 0.86 for the cellular foams by neglecting the effects of radiation (i.e., very close to ϕ = 0.85 determined in mechanical studies, so in excellent agreement). However, the measured conductivities of several foams, especially the lightest ones, were higher than the values determined by this model because the effects of the radiations were no more negligible, even at room temperature, for foams of low densities and / or large cell sizes.

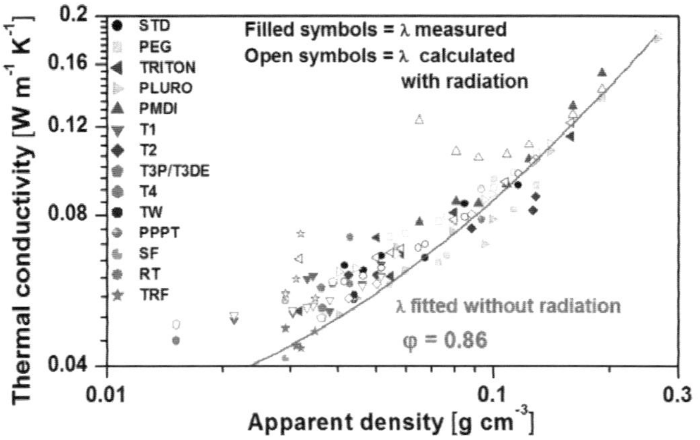

Figure 5. Fit and calculation of the model of Glicksman [11] for the thermal conductivities, λ, measured at room temperature by the Hot Disk method, for the different foams tested.

The model of Glicksman was then used by taking into account the radiation contribution (without the contribution of the cell walls to the extinction coefficient) for calculating the thermal conductivities from the values of density and cell size. The values obtained from this second model are in good agreement with the experimental values of the high-density foams but not with those having low densities and large cell sizes. This result proves that the contribution of the cell walls needs to be measured and should be taken into account for a more accurate description of the experimental data.

3.2.4. Electromagnetic properties. Only a few results are reported here. It was found that the bulk density is the dominant parameter until the end of the microwave range. Similarly to what occurs in a percolating system, the real part of the permittivity and the electrical conductivity increased with bulk density according to power laws whose exponents decreased with frequency (figure 6). The conductivity in the static regime continuously increased with temperature according to Mott's law for disordered conductors (see [12]), whose coefficients changed from a typical value of 2D material to that of 3D material when the thickness of the cell wall increased. This is the first time that such behaviour was observed in rigid foams.

No change of electromagnetic properties was found by varying other structural parameters such as cell size or interconnectivity. At low frequencies, the carbon foams behaved similarly to metals with high impedances, tangential losses and reflection, and became conductors with losses at a frequency which depended on the density (in the microwave range). Reflection decreased and absorption increased with the decrease of density or the increase of frequency. When the frequency increased further, the carbon foams switched from a more capacitive to a more resistive behaviour, the impedance and tangential losses decreased slowly, and the absorption then became the main mechanism.

The cellular foams finally tended to behave as black bodies at high frequencies (above the terahertz range).

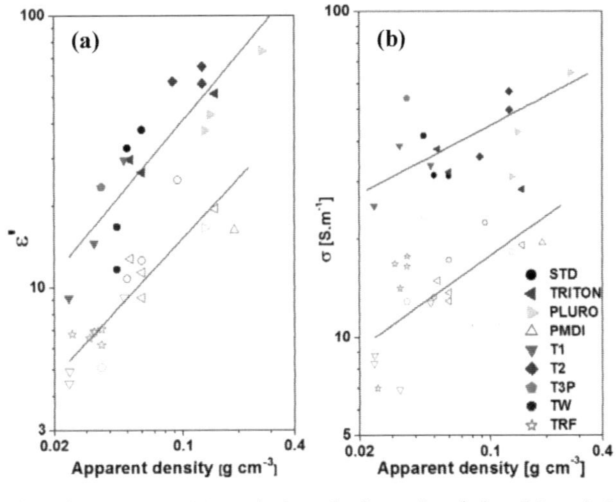

Figure 6. Dielectric constant (a) and electrical conductivity (b) at 8.64 GHz (filled symbols) and at 26 GHz (open symbols), as a function of apparent density, for the different foams tested.

4. Conclusions

In this work, model alveolar materials – carbon cellular and/or reticulated foams – were produced in order to study and to model their physical properties. Cellular carbon foams were obtained by pyrolysis of rigid phenolic-furanic foams presenting many different structures. Reticulated foams were also prepared from polymeric commercial foams through a template method. It was shown that very different morphologies could be obtained whereas the constituting vitreous carbon from which they were made was the same. The physical properties of these foams were investigated through mechanical, acoustic, thermal and electromagnetic measurements. The following could be concluded.

Concerning mechanical properties: (1) the compressive strength increases with the thickness of cell walls; (2) the famous Gibson and Ashby (G&A) law for open cell foams applies to reticulated foams and to cellular foams measured without plates glued to their faces; (3) the G&A law for closed cell foams (in which the fraction of solid only contained in the struts was determined and taken into account) applies to cellular foams measured with plates glued to their faces; (4) brittle foams with perforated cell walls have modulus as high as that of closed cell foams; (5) no influence of cell size on modulus and compressive strength is noticed.

Concerning acoustic properties: (1) at fixed cell size, the permeability of reticulated foams is about 100 times higher than that of cellular foams; (2) the permeability of reticulated foams increases according to a power law of cell size with an exponent 2 times higher than that of cellular foams; (3) the cell size controls the average sound absorption coefficient of all foams and an optimum of permeability is clearly observed at which the sound absorption coefficient is the highest.

Concerning thermal properties: (1) Glickman's model best fits the results of conductivity vs bulk density, using a fraction of solid only contained in the struts which was found to be identical to that determined by mechanical tests; (2) but the model is not accurate enough for foams having low density and/or high cell size if the radiation contribution is not taken into account.

Concerning electromagnetic properties: (1) the real part of the permittivity and the electrical conductivity are controlled by bulk density up to the THz range; (2) the foams are reflective at low frequency and turn to absorptive at high frequency; (3) no influence of cell size is noticed; (4) cellular foams nearly behave as black bodies in the THz – infrared range.

These conclusions were not all supported by the data presented in this paper, which is definitely much too short for its content, but are robust and expected to apply to many other kinds of rigid foams. The results demonstrate that bulk density and cell size are the main parameters controlling all physical properties, and are therefore the keys of their optimisation for any given application.

Acknowledgements

This study was funded by the Institut Universitaire de France (IUF) and by the Lorraine Region (France). The authors also acknowledge the financial support of the CPER 2007-2013 "Structuration du Pôle de Compétitivité Fibres Grand'Est" (Competitiveness Fibre Cluster), through local (Conseil Général des Vosges), regional (Région Lorraine), national (DRRT and FNADT) and European (FEDER) funds. The work was also partly funded by the Lithuanian-French program Gilibert, project Nr. TAP LZ 02/2013, by FP7-PEOPLE-2013-IRSES-610875 NAmiceMC, Belarus-CNRS project BRFFI F13F-004 (Belorussian and French teams) and the France-Canada Research Fund (FCRF).

References

[1] Maheo L, Viot P, Bernard D, Chirazi A, Ceglia G, Schmitt V and Mondain-Monval O 2013 *Compos. Part B Eng.* **44** 172
[2] Tondi G, Fierro V, Pizzi A and Celzard A 2009 *Carbon* **47** 1480
[3] Tondi G, Blacher S, Léonard A, Pizzi A, Fierro V, Leban J M and Celzard A 2009 *Microsc. Microanal.* **15** 384
[4] Zhao W, Pizzi A, Fierro V, Du G and Celzard A 2010 *Mater. Chem. Phys.* **122** 175
[5] Zhao W, Fierro V, Pizzi A, Du G and Celzard A et al 2010 *Mater. Chem. Phys.* **123** 210
[6] Li X, Basso M C, Braghiroli FL, Fierro V, Pizzi A and Celzard A 2012 *Carbon* **50** 2026
[7] Letellier M, Szczurek A, Basso M C, Pizzi A, Fierro V, Ferry O and Celzard A 2017 *Carbon* **112** 208
[8] Gibson L J and Ashby M F 1999 *Cellular Solids: Structure and Properties* (Cambridge: Cambridge University Press)
[9] Brezny R and Green D J 1990 *Acta Metall. Mater.* **38** 2517

[10] Letellier M, Fierro V, Pizzi A and Celzard A 2014 *Carbon* **80** 193
[11] Glicksman L R 1994 *Low Density Cellular Plastics* ed N C Hilyard and A Cunningham (Amsterdam: Springer) chapter 5 pp 104-152
[12] Letellier M, Macutkevic J, Paddubskaya A, Plyushch A, Kuzhir P, Ivanov M, Banys J, Pizzi A, Fierro V and Celzard A 2015 *IEEE Trans. Electromagn. Compat.* **57** 989

High-performance reagent modes for flotation recovery of platiniferous copper and nickel sulfides from hard-to-beneficiate ores

T N Matveeva and V A Chanturiya

Institute of Comprehensive Exploitation of Mineral Resources, Russian Academy of Sciences (IPKON RAS), Kryukovskiy tupic 4, 111020 Moscow, Russia

tmatveyeva@mail.ru

Abstract. The paper presents the results of the recent research performed in IPKON Russian Academy of Sciences that deals with development and substantiation of new selective reagents for effective flotation recovery of non-ferrous and noble metals from refractory ores. The choice and development of new selective reagents PTTC, OPDTC, modified butylxanthate (BXm) and modified diethyl-dithiocarbamate (DEDTCm) to float platiniferous copper and nickel sulfide minerals from hard-to-beneficiate ores is substantiated. The mechanism of reagents adsorption and regulation of minerals floatability is discussed. The study of reagent modes indicates that by combining PTTC with the modified xanthate results in 6 – 7 % increase in the recovery of copper, nickel and PGM in the flotation of the low-sulfide platiniferous Cu-Ni ore from the Fedorovo-Panskoye deposit. The substitution of OPDTC for BX makes it possible to increase recovery of Pt by 13 %, Pd by 9 % and 2 – 4 times the noble metal content in the flotation concentrate.

1. Introduction

Platiniferous copper-nickel ores are complicated mineral raw material with thin impregnations of valuable components, close mutual intergrowth of sulfide minerals and high content of pyrrhotite in sulfide ores. Low-sulfide Cu-Ni ores are characterized by low content of copper and nickel, relatively high PGM content and gangue minerals with high floatability, like talc and other silicates.

PGM deposits are divided into three types, depending on the ratio of the sum of PGM and sulfides content in the ore [1]:

- rich-sulfide copper-nickel ores containing pyrrhotite and chalcopyrite (total sulfides content is more than 15% and the average PGM content is 1 ppm or less);
- platinum disseminated ores (sulfides content is 10 - 15% and the average content of total PGMs is 10 ppm and up);
- platinum low-sulfide ore (sulfides content is 1 - 3% and the average content of PGMs is more than 10 ppm). The proportion of disseminated ores in the overall balance of the PGMs currently stands at 70% and will increase further.

PGM Cu-Ni ores are hard-to-beneficiate raw material, because these ores are usually complex, the largest platinum-ore provinces and regions of Russia are located in the harsh climate zones, such as the Arctic zone. The prevalence of minerals of platinum metals and rich solid solutions of PGM in pentlandite, chalcopyrite and pyrrhotite predetermines the use of gravity-flotation beneficiation circuits with the release of heavy platinum minerals in the gravity concentrate (for example, at the plant "Pechenganikel") and removing the solid soluble platinum group metals into the combined sulfide concentrate.

Most of the disseminated and low-sulfide platinum ores are characterized by fine impregnation of PGM minerals inside Cu and Ni sulfides, and a large variety of PGM mineral forms [2], resulting in high losses of fine particles of noble and non-ferrous metals into waste products. In processing circuits these specific features create certain difficulties when choosing flotation reagents and time-processing efficient reagent regimes [3].

One of the main trends in the development and rational use of Pt-Cu-Ni raw material is based on increasing the recovery of non-ferrous and precious metals and reducing production costs. It can be achieved by maximizing the selectivity of the flotation process, by controlling the parameters of the composition, creation of new effective reagents for the extraction of platinum-containing Cu-Ni sulfide minerals.

Current challenges in the design and application of effective flotation reagents for the recovery of Cu, Ni and PGMs are to produce selective reactants based on directional introduction into their molecular structure of the donor-acceptor substituents, and the ion-complexing agent, and a combination of ionic and non-ionic collectors that physically and chemically absorb on the mineral surfaces.

The methodological basis of intensification of selective flotation and the development of new flotation reagents originate from the fundamental studies of Russian and foreign scientists such as, Plaksin, Chanturiya, Kakowski, Mitrofanov, Okolovich, Ryaboy, Bogdanov, Solozhenkin, Abramov, Mantsevich, Ivankov, Vigdergauz, Gaudin, Taggart, Polling, Forssberg, O'Connor, Nagarai and others.

In our study we followed the main principle that the ability of new reagents to form water-insoluble complexes with the noble metals on the surface of minerals in flotation is the basis for selecting the compounds as collectors for extracting precious metals from hard-to-beneficiate ores. When we selected a class of reagents we also took into account their availability, simplicity of synthesis, environmental safety.

Under the guidance of the academician V.A. Chanturiya, in IPKON RAS, new class of collectors – cloud point polymers with specific functional groups of atoms like thioamine, phosphine and aminosulfide have been designed and tested for recovering platinum into nickel-pyrrhotite concentrate. These reagents owing to their complexing activity towards Cu, Ni and Pt produced high recovery of these metals but their synthesis is sufficiently expensive and complicated [4].

The main aim of the paper is the scientific substantiation of new complexing agents, produced from the conventional collectors of butyl xanthate and diethyl-dithiocarbamate, and their combinations, and on this basis to develop highly efficient reagent regimes to extract platinum-sulfide minerals from refractory ores.

2. Objects and methods

In this paper cyclic propylene tri-thiocarbonate (4-methyl-1, 3-dithiolan-2-thione) (PTTC) and oxypropyl ether diethyl-dithiocarbamate acid (OPDTC) were used as new selective flotation reagents towards Cu-Ni and PGMs minerals. Due to their ability to form a coordination bond with noble metals these compounds produce sparingly soluble compounds of non-ionic molecule structure and water-repelling properties [5, 6].

The use of S-dithiocarbamic acid esters, dithiocarbamates and cyanate-ethylene compounds in the flotation of copper-nickel and copper-molybdenum ores is known to have positive effects [7]. The other advantage of PTTC and OPDTC is that these reagents may be prepared and added into flotation cell as individual compounds or incorporated into modified collector solutions composed of xanthate or dithiocarbamate as the products of their reaction with propylene chlorhydrin. This technique was worked out in IPKON RAS by T.A. Ivanova (Ph.D., senior research associate) [8] and the authors highly appreciated her assistance in this study. The advantage of using modified reagent solutions directly before being added into the flotation cell instead of adding them as separate components is that this technique avoids additional dissolution procedure.

The mineral fractions of pyrrhotite, chalcopyrite and pentlandite were from the Monchegorsk ore and the sample of low-sulfide ore was from the Fedorova Pana deposit in Russia. The chemical composition of these mineral samples is presented in table 1. The pentlandite sample incorporated the impurities of pyrrhotite and chalcopyrite, and the chalcopyrite sample had about 16.6 % of silicates.

Table 1. Chemical composition of mineral samples.

Mineral sample	Content							
	Au	Pt	Pd	Ni	Cu	Fe	S	Oxides
	ppm			%				
Pentlandite (pyrrhotite, chalcopyrite)	0.15	0.9	7	16.3	4.7	42.9	32.6	3.2
Pyrrhotite	—	—	—	0.03	0.04	59.0	38.1	2.8
Chalcopyrite (silicates)	0.15	0.9	1.2	0.6	26.4	28.1	28.3	16.6

Qualitative and quantitative analyses of the aqueous solutions of the reagents studied PTTC, OPDTC, BXm and DEDTCm before and after their contact with ground mineral samples (-0.1 + 0.063 mm), as well as the organic (hexane) extracts of aqueous and solid (mineral) phases after reagent-mineral contact were performed by UV-spectrophotometry using a SHIMADZU UV-1700 spectrometer. The amounts of reagents adsorbed on the surface of sulfides were obtained from the difference between the initial and final concentrations of the reactants in the liquid phase, taking into account the amount of reactant desorbed into the organic solvent from the surface of the mineral.

Mineral polished sections were studied by scanning electron microscopy (SEM, LEO 1420VP microscope) with X-ray microanalysis (EPMA, energy dispersive spectrometry INCA Oxford 350), optical microscopy (OM, Olympus BX51 microscope), laser microscopy (KEYENCE VK-9700) and X-ray diffraction (Rigaku D/MAX-2200). Mathematical statistics methods were used for the analysis of the results.

3. Results and discussion

The efficiency of flotation reagents applied for sulfide minerals extraction using both traditional and new collectors is largely dependent on the formation of certain hydrophobic compounds on the surface of the sulfides.

The results of tests on the adsorption of PTTC, OPDTC and BX on mineral samples by UV extractive spectroscopy have shown that cyclic propylene trithiocarbonate (PTTC) incorporated into the modified butylxanthate solution and OPDTC incorporated into the modified diethyl-dithiocarbamate solution, have shown different values of adsorbed reagent onto the surface of Cu-Ni-Pt sulfide minerals [9]. A maximum PTTC adsorption of $23x10^{-6}$ g/g was measured on pentlandite with chalcopyrite and PGM impurities and of $4.6x10^{-6}$ g/g on chalcopyrite. There was no PTTC absorbed on pyrrhotite which is free of Cu, Ni and PGM impurities (table. 2).

OPDTC showed the same trend of adsorption on the surface of the Pt-Cu-Ni sulfides with highest adsorption on pentlandite ($73x10^{-6}$ g/g), medium adsorption on chalcopyrite ($20.7x10^{-6}$ g/g) and lowest adsorption on pyrrhotite ($4.6x10^{-6}$ g/g) (see table 2). For comparison, the adsorption of butylxanthate (BX) on all studied samples under identical concentrations was much higher than those of PTTC and DEDTCm with adsorption of $1.33x10^{-3}$ g/g on pentlandite (100% of initial BX), $1.01x10^{-3}$ g/g on chalcopyrite (79.6% of initial BX), and 46.8 % adsorption on pyrrhotite .

IOP Conf. Series: Journal of Physics: Conf. Series **879** (2017) 012015 doi:10.1088/1742-6596/879/1/012015

Table 2. PTTC, OPDTC and BX adsorption on mineral samples (UV extractive spectroscopy).

Mineral sample	Adsorption					
	PTTC		BX		OPDTC	
	10^{-6} g/g	% init.	10^{-3} g/g	% init.	10^{-6} g/g	% init.
Pentlandite (pyrrotite, chalcopyrite)	23	31.6	1.33	100	73	16.5
Chalcopyrite (silicates)	4.6	6.3	1.01	79.6	20.7	4.7
Pyrrhotite	0	0	0.66	46.8	4.4	1.0

Such a high adsorption of BX on pyrrhotite may explain why it is difficult to separate Cu-Ni sulfides from pyrrhotite using this collector. The designed reagents of PTTC and OPDTC demonstrated higher selectivity to Cu and Ni sulfides compared to pyrrhotite, which makes them potential collectors for selective flotation of Cu-Ni-Pt minerals.

The study of newly formed films of combined collector DEDTCm (that incorporates OPDTC and DTC), on polished sections of chalcopyrite by ASEM (LEO 1420VP INCA Oxford 350) and ALM (KEYENCE VK-9700) indicated that after contact with the reagent solution characteristic molecular shape of adsorbed collector appeared on the Cu-Ni mineral surface. For example, figure 1 and 2 demonstrate the presence of DEDTCm on chalcopyrite.

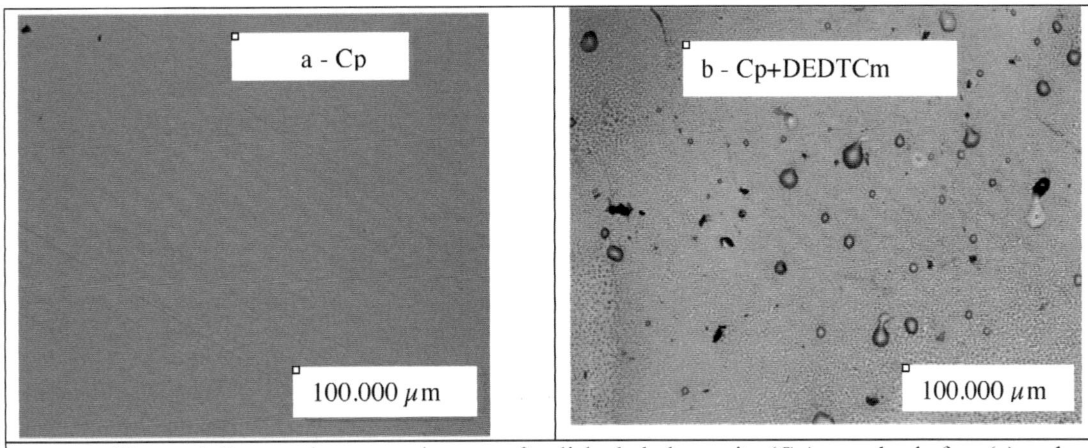

Figure 1. Laser microscopy images of polished chalcopyrite (Cp) samples before (a) and after (b) contact with DEDTCm (OPDTC+DTC) solution (C = 100 mg/l, t = 5 min).

One can see the convex neoplasms that fixed onto very small chemically adsorbed species of the reactant (figure1b and figure 2a). Adsorbed species have differential shape sizes: the largest species is about 60.5 μm in diameter and about 4.0 μm in height, and the smallest species are less than 1.0 μm in diameter and less than 0.1 μm in height (figure 2d, 2c). It was found that the amount of adsorbed species depended on the initial reagent concentration, and that they firmly anchored on the surface and did not dissolve in water after subsequent washing.

IOP Conf. Series: Journal of Physics: Conf. Series **879** (2017) 012015 doi:10.1088/1742-6596/879/1/012015

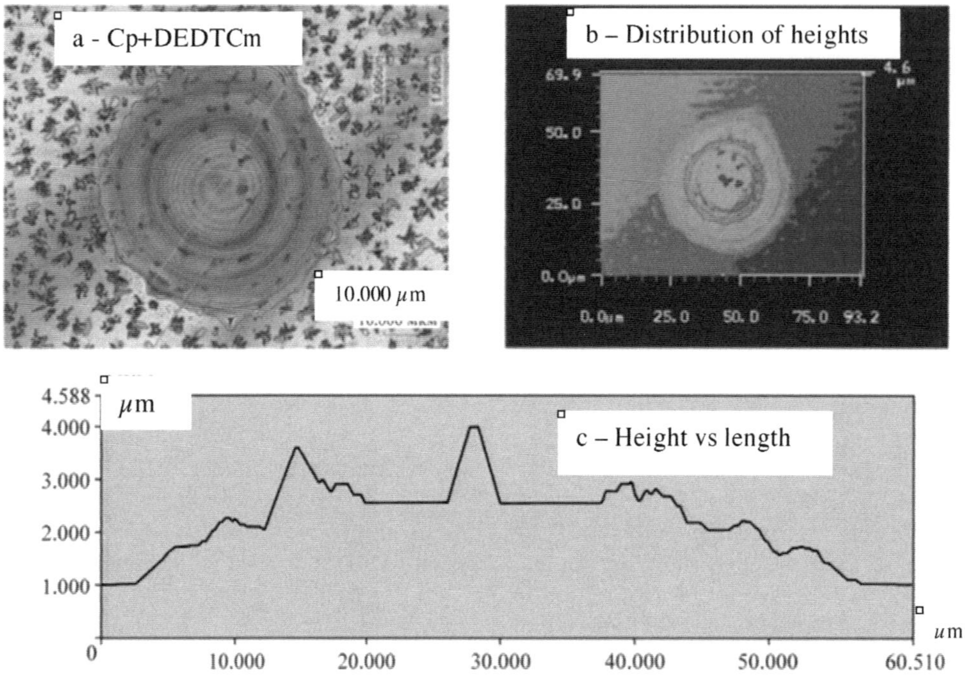

Figure 2. Laser microscopy images of polished chalcopyrite (Cp) samples after DEDTCm (OPDTC+DTC) treatment (C = 100 mg / l) (a), distribution of heights (b) and size of the reagent species on Cp surface (c).

Flotation tests on a sample of low-sulfide platiniferous copper-nickel ore of the Fedorova-Pana deposit showed the effectiveness of the modified xanthate solution, in which cyclic PTTC was included, instead of the conventional collector (butyl xanthate). Ore sample was ground to a particle size of 80% - 0.074 mm and floated in a single-stage operation. Using modified xanthate together with dithiophosphate in the proportion of 1:1 improved the quality of Cu - Ni concentrate and provided a total gain of recovering copper and nickel of 6 - 7% [10].

Laboratory tests of new reagent schemes of combination of OPDTC and BX were also performed on a Pt-Cu-Ni low-sulfide ore sample from the Fedorova-Pana deposit ground to 80% - 0.074 mm.

Using butyl xanthate (BX 100 g/t), isopropyl dithiophosphate (DTC 50 g/t), tannin depressor (300 g/t) and methyl-isobutyl carbinole as a frother (MIBC 60g/t) in a conventional reagent scheme produced a recovery of PGMs not exceeding 72 – 74% (Table 3).

Table 3. Results of low-sulfide ore flotation (Fedorova-Pana field) using OPDTC and BX (PGMs extraction)

Reagents, g/t	Product	Yield, %	Content, ppm		Recovery, %	
			Pd	Pt	Pd	Pt
BX-100, DTP-50	Concentrate	13.33	3.88	1.02	73.97	72.33
Depressor-300	Tails	86.67	0.21	0.06	26.03	27.67
MIDC-60	Ore	100	0.7	0.19	100	100
OPDTC-150	Concentrate	7.59	10.14	4.24	82.23	85.31
Depressor-300	Tails	92.41	0.18	0.06	17.77	14.69

MIBC-60	Ore	100	0.93	0.38	100	100

Table 4. Results of low-sulfide ore flotation (Fedorova-Pana field) using OPDTC and BX (Cu and Ni extraction)

Reagents, g/t	Product	Yield, %	Content, ppm		Recovery, %	
			Cu	Ni	Cu	Ni
BX-100, DTP-50	Concentrate	13.33	0.56	0.43	83.09	71.65
Depressor-300	Tails	86.67	0.017	0.026	16.91	28.35
MIDC-60	Ore	100	0.09	0.08	100	100
OPDTC-150	Concentrate	7.59	1.018	0.8	85.84	76.02
Depressor-300	Tails	92.41	0.014	0.021	14.16	23.98
MIBC-60	Ore	100	0.09	0.08	100	100

Application of OPDTC instead of combination of BX and DTP helped to achieve a higher flotation performance. Compared to the conventional scheme,Pt recovery increased by 13%, Pd recovery increased by 9% and the content of noble metals in the flotation concentrate increased 2 - 4 times (table 3).

Recovery of Cu and Ni was about 83 and 71 %, respectively under the conventional reagent scheme (Table 4). When the new reagent OPDTC was used, the recovery of non-ferrous metals increased by 2.7 – 4.6 % and the content of Cu and Ni in the flotation concentrate increased 2 times.

4. Conclusions

The new reagents-collectors, PTTC, OPDTC, modified BX, and modified DEDTCm were obtained and tested for the recovery of PGM, Cu and Ni from refractory ores.

Experimental studies of the physical and chemical effect of the new reagents on the surface of chalcopyrite, pentlandite and pyrrhotite using analytical methods helped to determine the conditions for the formation of adsorption layer of new reagents-collectors and shows that these reagents selectively adsorbed on the surface of micro- and nanoparticles of Pt, Cu and Ni metals and provide improved recovery of PGMs, Cu and Ni by 2-13% in ore flotation.

It is found that the cyclic PTTC, being a part of the modified xanthate solution, exhibits selectivity with respect to the Pt-containing Cu and Ni sulfide minerals and increases the floatability of these minerals in the flotation of a low-sulfide ore from the Fedorova Pana deposit. Using modified xanthate in conjunction with dithiophosphate at a ratio of 1:1 led to better quality of the Pt - Cu - Ni concentrate and provided a total increase in Cu, Ni and PGMs recovery by 6 - 7%.

The selectivity of OPDTC as additional collector to xanthate in the flotation of Pt-containing minerals is confirmed by its selective adsorption on the surface of copper and nickel sulfides, which provides 82 - 85% recovery of PGM metals from the Pt-Cu-Ni ore.

Designed reagent modes with new selective reagents (PTTC, OPDTC, BXm and DEDTCm) are recommended for the flotation of Cu, Ni and PGMs sulfide minerals from refractory ores.

References

[1] Laverov N P and Distler V V 2003 *Geology of Ore Deposits* vol 45 number 4
[2] Grokhovskaya T L, Bakaev G F, Sholohnev V V et al 2003 *Geology of Ore Deposits* vol 45 number 4
[3] Beloborodov V I, Zakharova I B, Mukhina T N, Marczewska V V, Kulakov A N 2007 *Ore* 6
[4] Chanturiya V A and Nedosekina T V 2008 *Proc. 24th Int. Miner. Proc. Congress (Beijing)*
[5] Agarvala U and Bhaskara Rao P 1967 *Inorg. Nucl. Chem. Lett.*, 3 (6)
[6] Byrko V M 1984 Dithiocarbamates (Moscow: Nauka)

IOP Conf. Series: Journal of Physics: Conf. Series **879** (2017) 012015 doi:10.1088/1742-6596/879/1/012015

[7] Shubov L Y, Ivankov S I and Shcheglova N K 1990 Flotation reagents in mineral raw material enrichment processes Book 1 (Moscow: Nedra)
[8] Ivanova T A, Matveeva T N, and Gromova N K 2010 Gornyi Jurnal 12
[9] Matveeva T N, Ivanova T A and Gromova N K 2007 *Tsvetnye metally* 12
[10] Matveyeva T N, Ivanova T A and Gromova N K 2009 Proc. XIII Balk. Miner. Proc. Congress (Bucharest: Romania) ed by Krausz S I.

Acknowledgments
The authors wish to acknowledge the assistance in this study from colleagues in IPKON RAS, T.A. Ivanova, Ph.D., senior research associate, N. K. Gromova, research associate, and L B Lantsova, senior engineer.

Intensification of the Reverse Cationic Flotation of Hematite Ores with Optimization of Process and Hydrodynamic Parameters of Flotation Cell

Poperechnikova[1,3] O Yu, Filippov[2,3] L O, Shumskaya[1] E N, Filippova[2,3] I V

[1]IVS, Zheleznovodskaya str.11A, Saint-Petersburg, 199155, Russia
[2]Université de Lorraine, GeoRessources, ENSG, 2 rue Doyen Marcel Roubault, BP10162, F-54505, Vandœuvre-lès-Nancy, France.
[3]National University of Science and technology "MISIS", 117049 Moscou, Russia

Email: lev-filippov@univ-lorraine.fr

Abstract. The demand of high grade iron ore concentrates is a major issue due to the depletion of rich iron-bearing ores and high competitiveness in the iron ore market. Iron ore production is forced out to upgrade flowsheets to decrease the silica content in the pelettes. Different types of ore have different mineral composition and texture-structural features which require different mineral processing methods and technologies. The paper presents a comparative study of the cationic and anionic flotation routes to process a fine-grain oxidized iron ore. The modified carboxymethyl cellulose was found as the most efficient depressant in reverse cationic flotation. The results of flotation optimization of hematite ores using matrix of second-order center rotatable uniform design allowed to define the collector concentration, impeller rotation speed and air flowrate as the main flotation parameters impacting on the iron ore concentrate quality and iron recovery in a laboratory flotation machine. These parameters have been selected as independent during the experiments.

1. Introduction

Currently the mining industry faces the challenge of rebellious iron ore processing. Flotation is one of the most used and efficient methods to treat oxidized ferruginous quartzite when it comes to low-grade iron ore (less than 40%) and it can produce high grade concentrates. Specialists of SP ZAO IVS has developed and patented a hematite ore flotation technology [1]. Hematite ores are different by mineral composition of host rocks and texture-structural features, which thus requires a thorough selection of flotation reagents. Consequently, the processing conditions of hematite ores are dictated by their composition and mineral dissemination, and the selective efficiency of reagents applied.

2. Collector selectivity in reverse cationic and anionic flotation routes.

The quartz and carbonate minerals (dolomite, calcite and siderite) cause iron concentrate dilution. The majority of minerals, that can be referred to as oxides and silicates, have strong flotation properties in the presence of anionic (fatty acid, etc.) and cationic type collectors. However, the floatability of minerals using anionic collectors depends on the type and quantity of compounds present on the mineral surface which can vary widely from one mineral to the other. As a result, these minerals are characteristic of a strong dependence of their floatability on mineral composition and preparation for flotation separation [2].

Oxides and silicates can be divided into two different sub-classes for the interaction of their surface with anionic collectors [3].

The minerals in the first sub-class have a large amount of active surface species to interact with anionic collectors and can be floated without activators. Elimination of polyvalent metal compounds from their surface can improve the adsorption of anionic collectors (hematite, magnetite, goethite and others);

The minerals in the second sub-class include quartz, calcite, dolomite and others. The presence of polyvalent metal compounds (cations, hydroxides, oxides, etc.) on their surface can activate their floatability.

To sum up, the flotation performance of each mineral in an ore depends on the type and quantity of polyvalent metal compounds present at the mineral surface which can passivate (first sub-class) or activate (second sub-class) flotation [4].

Flotation comparative experiments have been conducted using different collectors for iron ore flotation such as monoetheramine and modified fatty acids. In both cases, hydrolyzed starch has been applied for iron depression and polyalkylene glycol as a froth control agent. The metallurgical results for iron concentrates obtained as non-floated (cell) products as well as the full material balance for both flotation routes are presented in Table 1. The Fe recovery is high by 9.3% with cationic route while the anionic flotation with modified fatty acids as collector leads to very low silica content of 0.88 % in the iron concentrate.

Table 1 – Impact of collector type on waste rock minerals reverse flotation performance; monoamine ether and modified fatty acids are used in the cationic and anionic reverse flotations, respectively.

Collector	Product	Yield	Grade, %		Recovery, %	
		%	Fe_{total}	SiO_2	Fe_{total}	SiO_2
Monoamine ether	Tailings	50.80	15.62	75.01	19.45	95.63
	Fe concentrate	49.20	66.80	3.54	80.55	4.37
	Ore	100.00	40.80	39.85	100.00	100.00
Modified fatty acids	Tailings	57.77	20.12	67.88	28.75	99.06
	Fe concentrate	42.23	68.20	0.88	71.25	0.94
	Ore	100.00	40.42	39.59	100.00	100.00

Based on data of the mineralogical analysis of iron concentrate (Table 2) obtained using optical microscopy (Nikon, reflected light, parallel Nicol prism), the content in hematite is practically the same in the concentrate, 68.5 and 69.0 %, respectively for the reverse cationic and anionic flotation. The magnetite grade in the iron concentrate of the anionic flotation is increased up to 26.0% while the grade of non-metallic minerals is decreased to 1.2%.

Hematite and magnetite liberation in both flotation tails is almost similar and equals 70% and 60%, respectively. The grain size of particles in the ore and those of the non-metallic minerals in flotation tails does not exceed 40 μm. The main mineral aggregates consist of magnetite and hematite and hematite particles form aggregates with non-metallic minerals.

Table 2 Mineral composition of the iron concentrates generated by reverse cationic and anionic flotation routes using optical microscopy (using reflected light and Nicol prism //)

Cationic route	Anionic route
Fe_{total} = 66.8%, SiO_2 = 3.54%, CaO = 0.4%	Fe_{total} = 68.2%, SiO_2 = 0.88%, CaO = 0.2%
Hematite = 68.5%	Hematite = 69.0%
Magnetite = 22.8%	Magnetite = 26.0%
Goethite = 3.0%	Goethite = 3.8%
Carbonates = 0.8%	Carbonates = 0.2%
Non-metallic minerals = 4.9%	Non-metallic minerals = 1.2%

Figure 1 shows comparative diffractograms of iron ore concentrates. It is clearly seen that flotation cell product do not contain reflexes of carbonate minerals as calcite and dolomite for anionic route. For the carbonate minerals, siderite ($FeCO_3$) is present in the concentrate obtained using the cationic route but almost absent in the concentrate of the anionic flotation. Also, there is practically no quartz in the concentrate obtained with the anionic route.

Therefore, an iron content of up to 68.2% in the non-floated product, which is iron concentrate, using the anionic route is obtained by removing the non-metallic minerals within the froth product. In the other hand a redistribution of hematite and magnetite content in the iron concentrate (cell product) was produced by increasing the latter, and redirecting siderite to froth product (tailings) (Figure 1).

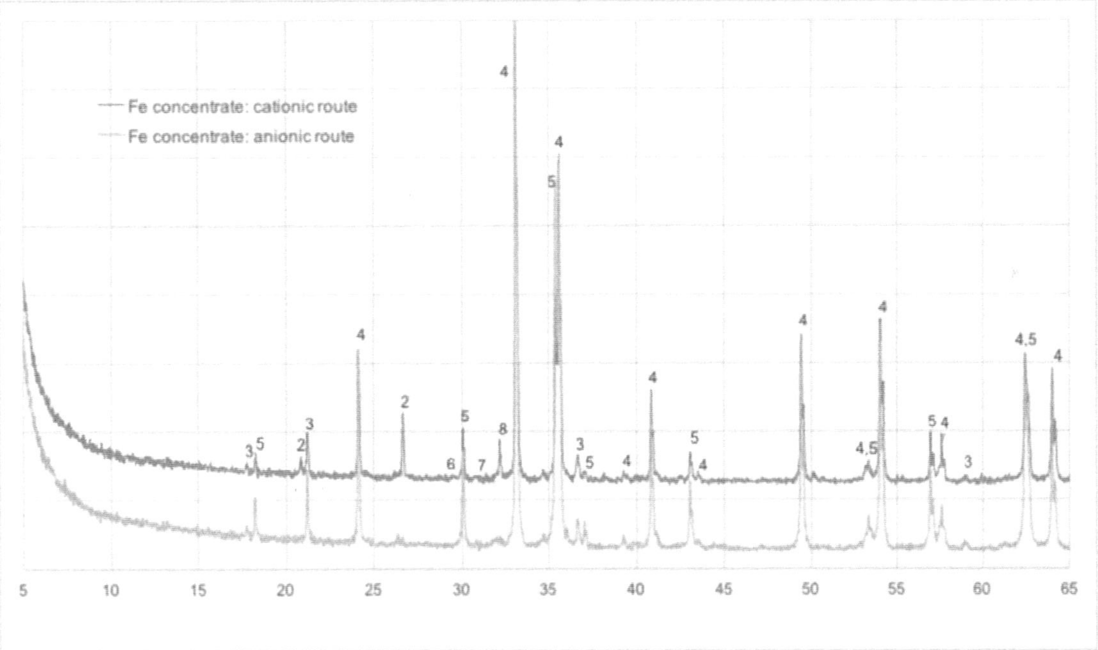

Figure 1 – Comparative diffractograms of iron concentrates indicating main mineral phases. The blue and pink diffractograms are for the cationic and anionic routes, respectively: 1 – mica; 2 –quartz; 3 – geothite; 4 – hematite; 5 – magnetite; 6 – calcite; 7 - dolomite; 8 – siderite.

However, the magnetite proportion is less in the concentrate obtained by reverse cationic flotation. Moreover, the amounts of quartz and non-metallic minerals are significantly higher, 3.5 and 4.9 %, respectively with the reverse cationic flotation, which limits the use of this concentrate for further processing in electrometallurgy. Thus, we suggest that the reduced flotation performances using the reverse cationic flotation are related to the less efficient depression of magnetite and the poor flotation behavior of the Fe-containing silicates minerals such as amphiboles in the presence of starch as depressant [9]. Among the iron containing silicates the actinolite from the family of amphibole was detected by the optical microscopy. Thus, a new depressant was studied in this work to achieve the requirements for iron concentrate by reducing the SiO_2 content below 2% with the reverse cationic flotation route.

3. Study of hematite alternative depressants

Reverse cationic flotation of waste rock is associated to iron oxide depression. The deepressant efficiency and property have been studied for several decades. Liu et al. (2000) have investigated the polysaccharide adsorption mechanism on mineral surface. Polysaccharides (such as starch, dextrin, guar gum and other) are simple sugars used as depressants of iron oxides and, in particular, hematite [5-7].

Recently, the presence of amphiboles in iron ores have been highlighted as one of the main issues in the flotation of oxidized ferruginous quartzite [8, 9]. Fillipov et al. (2013) suggested a starch adsorption mechanism on the Fe-Mg amphiboles surface by mean of hydrogen bonds and followed by formation of complex compounds on minerals surface [10]. Recent studies on the relationships between crystallographic properties of calcic amphiboles (greenschist facies) and flotation properties support the fact that floatability of ferruginous silicates is considerably decreased in the presence of amine used as a collector and starch [11]. As was demonstrated by Severov et al [11] the floatability of amphiboles at a given pH is related to the ratio tetrahedral Si^{4+}/Σoctahedral cations, substitutions be tween Al^{3+} cations and Si^{4+} in tetrahedral coordination and Mg^{2+}ions contents. As a results the floatability drops in the presence of starch for those amphiboles which have more Mg ions in their crystal structures. Moreover, starch adsorption on the surface of ferruginous silicates (including amphiboles) can be explained by layering where the ferruginous cations come out on the surface, and adsorb the starch. This means that the task is to minimize the depressant impact on Fe containing silicates to obtain a high quality iron concentrate with low silica content. It has been shown that an increase of depressant concentration increases iron yield, but in turn, it decreases iron in the concentrate as starch is adsorbed on the surface of ferruginous silicates. It should be noted that starch, used for iron depression, has amylopectin that tends to form agglomerates with ferruginous silicates, and as a result, low grade concentrate (Figure 2) [10-12].

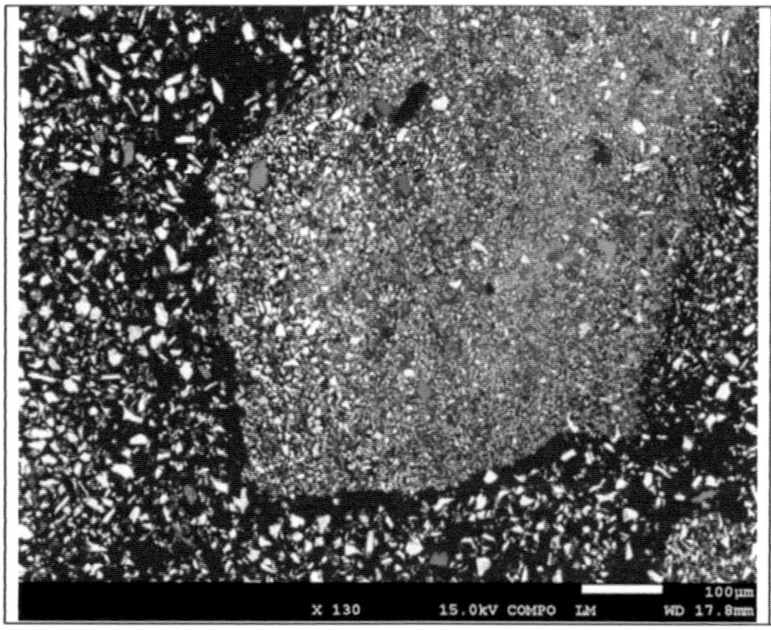

Figure 2 – Iron ore concentrate obtained using hydrolyzed potato starch depressant

Several studies have been focused on the search of more selective depressants than starch [14]. As part of this study, the following depressants have been tested: guar gum, modified carboxyl methyl cellulose (CMC) and caustic potato starch.

The experimental flotation results shown in Figure 3 indicate that at 58% Fe recovery the best Fe grades have been obtained using the modified CMC (DLM XP) followed by starch, CMC and guar gum.

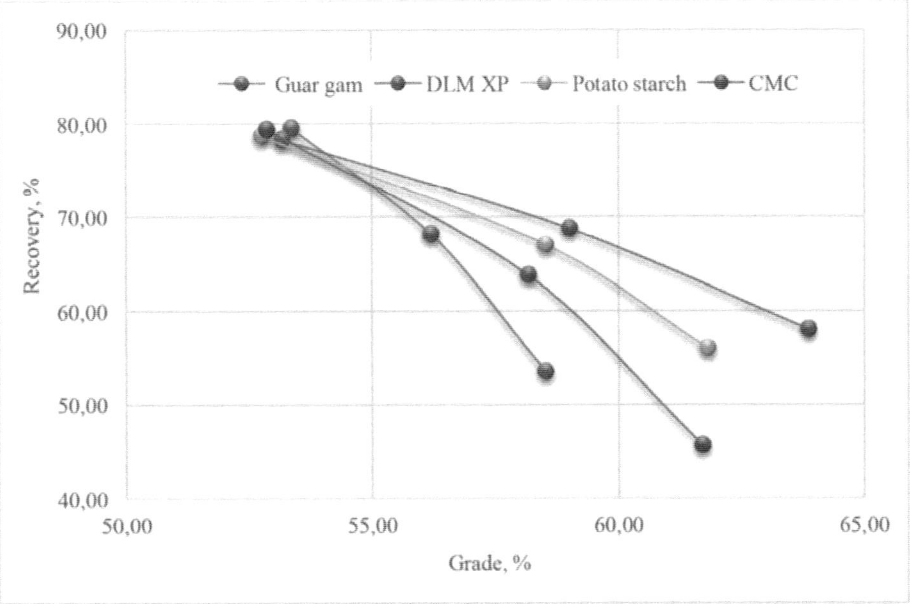

Figure 3 – Iron recovery into concentrate for different types of depressants

4. Intensifying the reverse cationic flotation of hematite ores using an optimization of process and hydrodynamic parameters of flotation cell

The main aim of this second part of the study is to optimise the operational flotation conditions in the cell and the concentrations of collector and depressant for maximum flotation performances.

A second order design of experiment matrix [15] has been used to investigate the effects of multiple factors such as ore properties– laboratory flotation cell operating parameters – standard deviations of iron and silicon dioxide separation. Factorial experimental design approach to optimize the flotation performance is not very often seen as worth mentioning to explain this approach in [16].

Optimization of oxidized ferruginous quartzite flotation has been studied using a laboratory pneumatic mechanical flotation cell designed by RIVS that allows an independent variation of impeller rotation speed and air flowrate.

Three independent factors or variables have been selected in this study: impeller rotation speed, air flowrate and monoetheramine concentration (X1, X2 and X3). Table 3 presents the experiment design conditions according where factors are varied within a lower limit of -1.633 and an upper limit of +1.633, and a principal factor of 0.

Table 3 – Independent variables domain (X_i) and its values

Factor	Value	Unit	-1.633	-1	0	+1	+1.633
X_1	Impeller rotary speed	rpm	800	1033	1400	1767	2000
X_2	Air flowrate	l/min	3	4.36	6.5	8.64	10
X_3	Monoetheramine concentration	%	0.5	1.27	2.5	3.73	4.5

IOP Conf. Series: Journal of Physics: Conf. Series **879** (2017) 012016 doi:10.1088/1742-6596/879/1/012016

The results of flotation tests using center composite design can be seen in Table 4. Under the experiment design conditions, 20 tests have been conducted including eight points of two-level variability interval, six limiting points and six center of design points. All limiting points have factor central value, except factors that have variable domain as ±ε where ε means difference between difference between the principal factor of 0 and the lower/upper limit of factor.

Table 4 – Independent factor matrix to implement optimum second order rotatable design

Test n°	Factor 1	Factor 2	Factor 3	Response 1	Response 2
	X_1	X_2	X_3	Y_1	Y_2
	Impeller rotary speed	**Air flowrate**	**Monoetheramine concentration**	**Fe content in cell product**	**SiO_2 recovery in froth product**
	rpm	**l/min**	**%**	**%**	**%**
1	1033	4.36	3.73	57.8	77.45
2	1767	4.36	1.27	57.1	74.93
3	1033	8.64	1.27	56.2	75.34
4	1767	8.64	3.73	58.9	80.55
5	1400	6.5	2.5	58.8	78.71
6	1400	6.5	2.5	57.8	77.77
7	1033	4.36	1.27	58.6	79.46
8	1767	4.36	3.73	57.2	76.56
9	1033	8.64	3.73	57.5	79.28
10	1767	8.64	1.27	61.3	87.52
11	1400	6.5	2.5	57.9	80.96
12	1400	6.5	2.5	59.2	83.06
13	800	6.5	2.5	49.9	52.71
14	2000	6.5	2.5	58.6	81.05
15	1400	3	2.5	60.3	84.14
16	1400	10	2.5	59.2	82.6
17	1400	6.5	0.5	56.4	73.65
18	1400	6.5	4.5	62.3	87.39
19	1400	6.5	2.5	59.2	82.54
20	1400	6.5	2.5	60.2	83.96

Second power polynomials for each response function are the following:

$$\mathbf{Y_1 = 58.82 + 1.40X_1 + 0.59X_3 + 1.07X_1X_2 - 1.63X_1^2 + 0.44X_2^2}$$
$$\mathbf{Y_2 = 81.01 + 4.07X_1 + 1.43X_3 - 4.78X_1^2 + 2.36X_1X_2 + 1.40X_2^2}$$

When considering the influence of the three factors X_1, X_2 and X_3 on the separation efficiency of oxidized ferruginous quartzite (iron and silicon dioxide) using reverse cationic flotation, the following separation criterion has been revealed: a maximum recovery of silicon dioxide into the froth product is obtained at maximum iron content in concentrate. Figure 5 shows that both impeller rotation speed and monoetheramine concentration have simultaneous impact on the reverse flotation of oxidized ferruginous quartzite.

Study of standard deviations dependency on impeller rotation speed has shown that optimum indexes (as iron grade in concentrate of 59% and silicon dioxide recovery into froth product of 81%) are achieved at a rotation speed of 1600 rpm and that a speed decrease or increase results in sharp decline of qualitative-quantitative values.

Low concentration of monoetheramine produces low silicon dioxide recovery and as a consequence, low grade of iron ore in the concentrate (Figure 4).

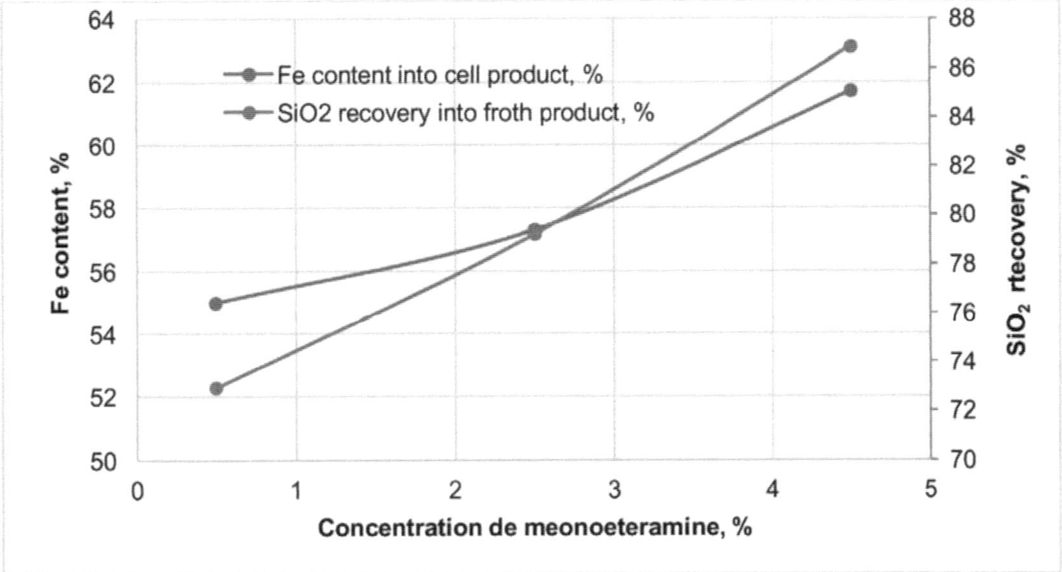

Figure 4 – Impact of monotheramine concentration on iron content in concentrate and silicon dioxide recovery into froth product

Based on the studies performed, regression models have been built for iron content in concentrate (**Y**$_1$) and for silicon dioxide recovery into froth product (**Y**$_2$) (Figures 5).
Two-factor interaction of impeller rotary speed and monoetheramine concentration results in a decrease of the parameter standard deviation at small values, and an increase at large values.

a) b)

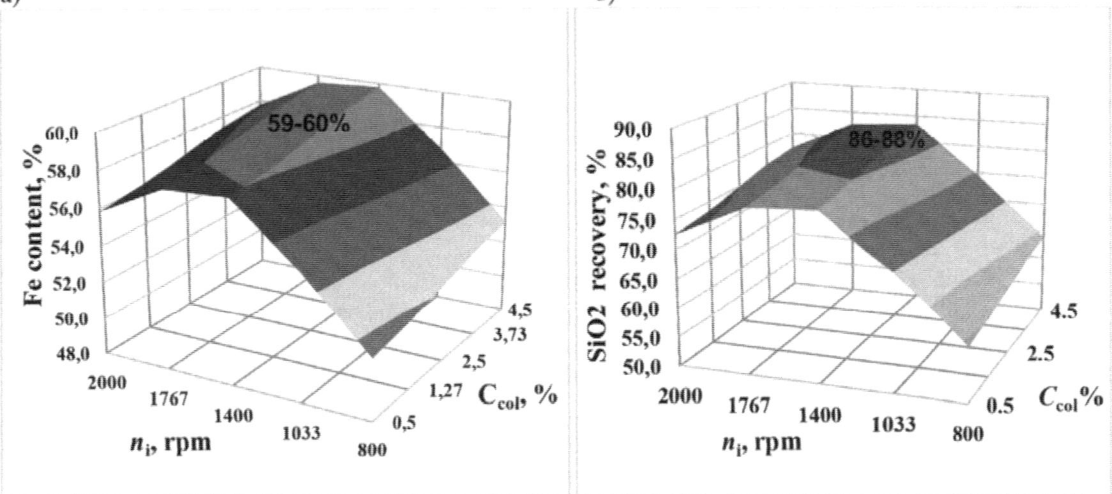

Figure 5 – Impact of impeller rotational speed (n_i) and monoetheramine concentration (C_{col}) on iron content in concentrate (a) and silicon dioxide recovery into froth product (b)

Therefore, the optimization of the flotation parameters, impeller rotation speed and monoetheramine concentration, provides maximum separation of iron and silicon dioxide.

5. Conclusions

After a large amount of research performed, several factors have been identified to improve the reverse cationic flotation of hematite ores:

- selective collectors based on modified fatty acids produce an efficient flotation of ferruginous amphiboles from fine-disseminated hematite ores and minimum iron losses;
- modified CMC used for hematite depression produces a concentrate iron grade over 68%;
- both impeller rotation speed and monoetheramine concentration influence the reverse cationic flotation of oxidized ferruginous quartzite.

6. References

1. Decision on grant of patent application. Claim № 2015136829/03 (056338). O. Y. Poperechnikova, E.N.Shumskaya
2. Froth Flotation. A Century of Innovation/Edited by M.C. Fuerstenau et al. Colorado, 2010, 891 p.
3. Eigeles, M.A. principles of non-sulfide minerals, Nedra, Moscow, 1964.
4. Foundations of theory and practice of minerallurgy, Wroclaw University of Technology, 2007.
5. Araujo, A.C., Viana, P.R.M., and Peres, A.E.C. (2005). Reagents in iron ores flotation. Miner. Eng. 18, 219–224.
6. Liu, Q., Zhang, Y., and Laskowski, J.S. (2000). The adsorption of polysaccharides onto mineral surfaces: an acid/base interaction. Int. J. Miner. Process. 60, pp. 229–245.
7. Montes-Sotomayor, S., Houot, R., and Kongolo, M. (1998). Flotation of silicated gangue iron ores: Mechanism and effect of starch. Miner. Eng. 11, pp. 71–76.
8. Filippov, L.O., Filippova, I., Severov, V.V. The use of collectors mixture of various molecular structure in the reverse cationic flotation of magnetite ore. Minerals Engineering, 23 (2) 91–98.
9. Flippov, L.O., Filippova, I., Severov, V.V. New technology for producing hematite concentrate from wastes generated in the processing of iron quartzites Metallurgist, 54 (5-6) 268-272.
10. Filippov, L.O., Severov, V.V., Filippova, I.V. (2013). Mechanism of starch adsorption on Fe–Mg–Al-bearing amphiboles. Int. J. Miner. Process. 123, pp. 120–128.
11. Severov V.V., **Filippov L.O.,** Filippova I.V. Relationship between cation distribution with electrochemical and flotation properties of calcic amphiboles. *Int J Mineral Processing* 147, 18–27.
12. Nanthakumar, B., Grimm, D., and Pawlik, M. (2009). Anionic flotation of high-iron phosphate ores – Control of process water chemistry and depression of iron minerals by starch and guar gum. Int. J. Miner. Process. 92, pp. 49–57.
13. Wen Qi, G., Parentich, A., Little, L., and Warren, L. (1992). Selective flotation of apatite from iron oxides. Int. J. Miner. Process. 34, pp. 83–102.
14. Filippov, L.O., Severov, V.V., and Filippova, I.V. (2014). An overview of the beneficiation of iron ores via reverse cationic flotation. Int. J. Miner. Process. 127, pp. 62–69.
15. Zedginidze, I.G. Experiment design for study of multicomponent systems. Moscow, Nauka,
16. Sheridan, M.S., Nagaraj, D.R., Fornasiero, D. and Ralston, J. (2002). The use of factorial experiment design to study the flotation properties of a copper ore with the N-allyl- O-alkyl thionocarbamate collector. Minerals Engineering, 15 (5), pp. 333-340.

GIS-based identification of active lineaments within the Krasnokamensk Area, Transbaikalia, Russia

V.A. Petrov[1], M. Lespinasse[2], S.A. Ustinov[1], C. Cialec[2]

[1]Institute of Geology of Ore Deposits, Petrography, Mineralogy and Geochemistry (IGEM) Russian Academy of Sciences, Staromonetny per. 35, Moscow, 119017 Russia

[2]UMR 7359, GeoRessources, Université de Lorrain-CNRS-CREGU, BP 70239 - 54506 Vandœuvre-les-Nancy, France

E-mail: vlad@igem.ru

Abstract. Lineament analysis was carried out using detailed digital elevation models (DEM) of the Krasnokamensk Area, southeastern Transbaikalia (Russia). The results of this research confirm the presence of already known faults, but also identify unknown fault zones. The primary focus was identifying small discontinuities and their relationship with extended fault zones. The developed technique allowed construction and identification of the active lineaments with their orientation of the compression and expansion axes in the horizontal plane, their direction of shear movement (right or left), and their geodynamic setting of formation (compression or stretching). The results of active faults identification and definition of their kinematics on digital elevation models were confirmed by measuring the velocities and directions of modern horizontal surface motions using a geodesic GPS, as well as identifying the principal stress axes directions of the modern stress field using modern-day earthquake data. The obtained results are deemed necessary for proper rational environmental management decisions.

1. Introduction

Lineament analysis is one of the most effective remote sensing methods for studying faults framework and deep structures of a geological region. This analysis is based on the identification of lineaments with diagnostic features. These lineaments are recognized by their: 1) rectilinear negative forms of relief, 2) rectilinear exposed slopes, 3) escarpments and ledges, 4) sequences of small streams straight segments crossing (in one direction) watersheds and valleys, 5) watersheds with straight axial lines, 6) areas with isoline thickening and bending. One of the most difficult problem with of the lineament analysis for regional tectonic structure features determination is the identification of lineaments of tectonic origin, thus fault structures. In addition, it is necessary to determine which of the lineaments associated with active faulting, thus reflecting faults in a modern tectonic stage of regional development. This task is not always easy. The possibility of identifying active lineaments (faults) and their kinematics using newly built digital elevation models and horizontal surface displacement analysis according to GPS observations is clearly shown in this work. This approach was carried out using various GIS software – ArcView 3.2, Easy Trace 7.99 (Free version), QuantumGIS 2.14.3, and Surfer 9.0, as well as new modules developed by the authors.

2. Objectives of the research study

The main objective of this research study is to identify the active fault system of the Argun area, which located near the town Krasnokamensk in southeastern Transbaikalia (Russia). This territory includes the Argun complex composed mainly of Archean-Proterozoic and Paleozoic granitoids. As a result of the Late Mesozoic tectonic and magmatic activity of the region, within the Kuytunsky and Streltsovsky volcano-tectonic structures (calderas), a unique system of faults was formed (Petrov et al., 2013). The Streltsovsky Ore Field (SOF) is located within the Streltsovsky caldera and comprises 19 molybdenum-uranium hydrothermal deposits. Most of the uranium, lead, zinc, gold, and tin occurrences of the Kuitunsky caldera area were discovered during the early stages of the exploration programs. For environmental planning purposes and prospectivity analysis for new ore deposits, it is necessary to identify the tectonic structural features of the area. This facilitated, geodynamic reconstructions, including estimation of the stressed-strained state of the rocks and the kinematics of movement along major faults. In order to meet these objectives, the authors used a lineament analysis technique in combination with structural data and GPS data observations.

3. Construction and analysis of detailed digital elevation models

Lineament analysis of different regions was carried out by a special technique developed by the authors on the basis of newly-created, detailed digital elevation models (DEM) (Ustinov and Petrov, 2016). Previously conducted studies (Ganas et al., 2005; Oguchi et al., 2005; Biggs et al., 2006) have shown that this technique is most effective in areas with highly dismembered relief.

The digital model of any geometrical (geographical) object is considered to be a specific form of basic data representation. The method of their structural description makes it possible "to calculate" (to recover) an object by interpolation, approximation or extrapolation (Novakovsky et al., 2003). The DRMs are special types of three-dimensional mathematical models representing both real and abstract geofields (surfaces). At the same time within a digital model, other indicators and characteristics, like atmospheric pressure, air temperature, precipitation, oil reservoir pressure, geophysical and geochemical fields, and concentration of pollutants, can act as "a surface relief" (Khromikh and Khromikh, 2007). It is necessary to distinguish digital models of the relief cartographical image (digital maps) from the actual digital relief models which in practice are often understood to be the digital models of heights created by using a limited set of source map data on the topography (X, Y, Z). DEMs creation and their recalculation from one type to another are based on the use of a mathematical algorithm. Its correct application depends not only on the developed model adequacy but also on the optimality of computer memory resources costs and calculation times.

The method of DEMs creation, using an irregular network of starting points, requires recovery (interpolation) of the surface and recalculation to regular network. Currently, there are many methods used to solve this problem. Among them – the interpolation based on Delaunay triangulation, the average weighted interpolation, kriging, etc. However, in all cases for calculating the elevation or value of a point, it is necessary to use some interpolation algorithms (i.e. the values received in starting points are matching true values precisely) or approximations (the values received in starting points are matching true values with some degree of accuracy).

As the actual material for creation of detailed DEMs in the research were topographic maps at a scale of 1:100000 (isolines conducted through 20 m), as well as the geological map at a scale of 1:200000, the map of anomalous magnetic field, the gravitational anomalies maps, and the tectonic setting map. Also used were satellite images and maps of a topographic relief received from satellite images analysis. All the available material was enough to provide the necessary detail of the DEMs at the scale of research with the possibility of using the created models for lineament analysis.

The initial stage of operation was vectorization of cartographic material. This task was carried out by means of the EasyTrace 7.99 Pro software (free version). EasyTrace is a software package for semi-automatic interactive vectorization of color and black-and-white bitmap images working within the Microsoft Windows operating system. Vectorizing lines on a black-and-white raster (image) is much faster and more convenient. However, it is not always possible because the initial materials for vectorization are color images in most cases. Therefore, the «binarization» tool in the EasyTrace program is provided. The binarization of a raster is intended for separation, from the color image, a subject black-and-white layer. For example, from a raster of the color topographic map, it is possible to select a two-color layer of isolines or rivers and save it in the form of a separate two-color (binary) raster. After this operation, the map objects of interest (isolines in this case) will be presented in one color and all other objects are painted in a different color. Also before vectorization, the raster of a topographical map can be modified in the graphical editor due to elimination of unnecessary elements, color correction, gain of definition, and visibility of the interesting elements. Such actions increase the accuracy of a binarization process.

Further vectorization (tracing) of initial topographic maps contour lines was also carried out. It is a rather long and monotonous process that takes place almost automatically therefore there is no sense to stop on it in detail. As the relief of the studied area is rather dismembered, this information is considered to be enough for objective digital relief model creation with adequate details for the lineament analysis. Besides, the considered vectorizing program allows connecting elevation values to each constructed vector contour and to appropriate attributes. As a result the constructed isolines were painted in various colors depending on values of heights.

DEMs of the studied area were constructed on the basis of two main mechanisms: 1) regular network of heights (GRID) and 2) irregular triangulation network (TIN). Both of these mechanisms of relief modeling are widely used in geographical and geological information systems and are supported by many types of GIS software. At the same time, each of these two mechanisms has both shortcomings and advantages, which need joint consideration. Model TIN is the model developed especially for relief description. During the creation of the TIN model, the irregular network of triangles corresponding to Delaunay triangulation is used. At the same time, flat areas are modeled by a small number of large triangles and areas of steep ledges where it is necessary to show in details all sides of a relief the surface is displayed by numerous small triangles. The GRID relief model provides splitting space into indivisible elements. At the same time the matrix of heights or regular network of elevation marks is formed. The regular network of heights is a lattice with equal rectangles or squares where tops are grid knots. The regular network of heights has received the name "GRID" because in calculating of similar model the Gridding method is used and files of such DRM have the GRD format (Khromikh and Khromikh, 2007).

It is more expedient to use the TIN model for the research of local objects – such as massifs of rocks. Since research in this study covered a large area, the GRID model was used (Figure 1). On the basis of the grid file of a GRID digital relief model, support maps were constructed in carrying out the lineament analysis.

Lineaments on DEMs were identified and analyzed by means of plugin module; created by the authors and integrated into GIS software. Initially, the module was developed as the tool for realization and partial automation of tasks of the microstructural analysis special technique (Ustinov, Petrov, 2015). After adding some user-friendly functions and tools, it became possible to use this module for spatial analysis of microstructures, and also most importantly of large linear zones corresponding to faults and extended linear segments of geological structures.

4. Results of the spatial lineament analysis

By using detailed digital relief models, lineaments within the investigated area have been identified and drawn. Identified lineaments are characterized by varying orientations with similar total lengths in most directions. So to understand the role of various structures within the tectonic regime of the area and to exclude lineaments of non-tectonic origin, it is necessary to carry out classification of the identified lineaments, based on length and orientation (Figure 2).

Small in accordance with the scale lineaments up to 2,000 m are characterized mainly by submeridional, sublatitudinal and less by NE orientations. Lineaments that are more than 2,000 m long form accurately defined systems, mainly with NW, sublatitudinal and less NEt and submeridional orientations. After lineaments identification and their ranging by length the technique proposed Petrov et al. (2010) was used.

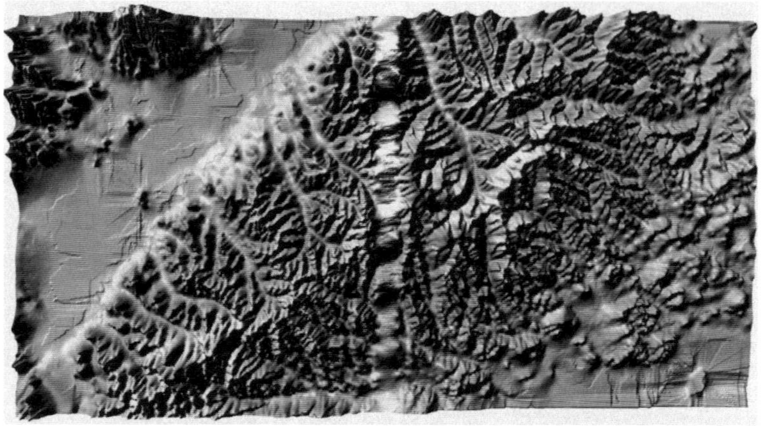

Figure 1. GRID relief model of the studied area

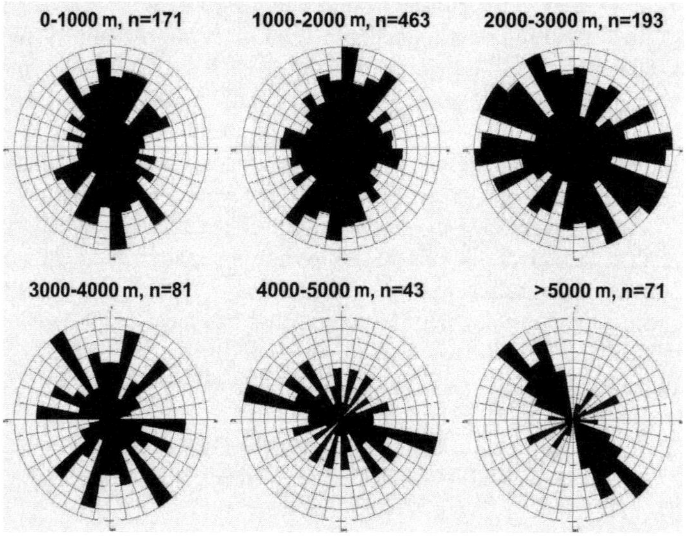

Figure 2. Rose diagrams of different lineament length orientation. n – quantity of the objects used for construction

Large-scale objects presumably faults are allocated among the lineaments and near them the orientations of smaller lineaments, conditionally called megacracks, are considered. These small lineaments correspond to feathering microfractures in the fault zone. Orientation of the feathering microfractures and faults, as well as the direction of horizontal compression and expansion axes which has caused shear movement in relation to the plane of a fault changes, is dependent on the geodynamic setting. Four possible options of relative orientation of the shear plane and feathering triad cracks are shown in Figure 3.

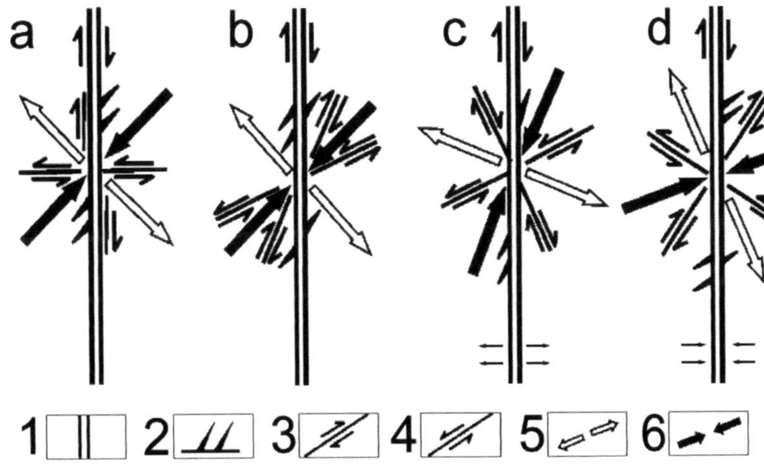

Figure 3. Options for stress-strained conditions at the shear angles 45° (a), <45° (b), the situation of additional extension (c) and compression (d). 1 – break; 2 – mode I cracks; 3, 4 – shears with right (3) and left (4) kinematics; 5, 6 – extension (5) and compression (6) axes orientations in the horizontal plane (after Gzovsky, 1975).

If the orientation of megacracks relative to each other and to their orientation with respect to estimated fault (lineament) corresponds to one of the options, then it is assumed that the lineament and megacracks are of tectonic origin. Concurrently, the orientation of compression and expansion axes in the horizontal plane, the direction of shear movement along the fault (right or left), and the geodynamic setting of fault formation (compression or extension) are reconstructed.

On the basis of the obtained results, the latest fault tectonics scheme included information on new structures of I, II and smaller ranks with the indication of faults kinematics and orientation of tension and compression axes was created (Figure 4).

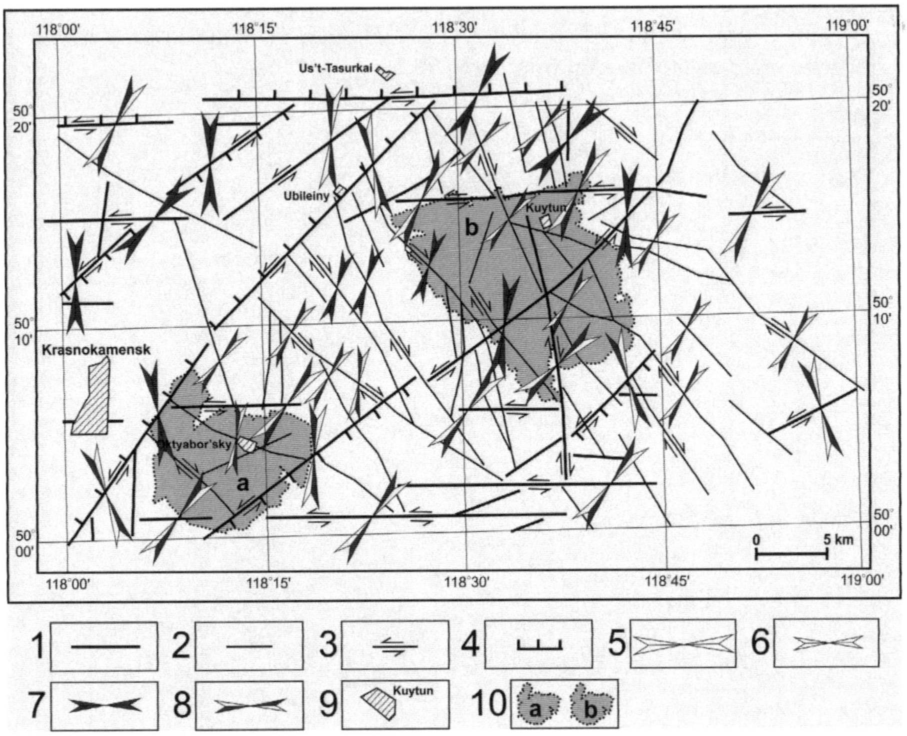

Figure 4. Scheme of the latest tectonics and neotectonic stress of the Streltsovsky and Kuytunsky calderas area. 1 – neotectonic faults of I rank; 2 – neotectonic faults of II and smaller ranks; 3 – shift movements; 4 – faults or uplifts (barbs directed to lowered wing); 5 – orientation of I rank faults compression axis in the horizontal plane; 6 – orientation of II and smaller ranks faults compression axis in the horizontal plane; 7 – geodynamic situation of three-axis tension; 8 – geodynamic situation of compression; 9 – settlements; 10 – contours of the calderas: a – Streltsovsky, b – Kuytunsky.

Within the considered territory the I rank faults are rather accurately expressed by relief. They are extensive (more than 5,000 m length) with predominant NE-SW and latitudinal orientations. The position of the I rank structures is also marked by a large number of feathering fractures formed in the zone of faults dynamic impact. These structures are presented by left lateral shifts with submeridional orientation of compression axes for NE-SW structures and NE orientation for latitudinal structures.

The II rank structures include all extended northwest-southeast faults, which are intensively breaking the Argun raising into numerous blocks. Within the considered study area there is a large amount of such faults. Within DEMs, there are signs that these faults displace and break into segments structures of other orientations that indicate a later time activation of the NW-SE structures. These faults are mostly right lateral strike-slips, which have been formed in the geodynamic setting of compression with (N)-NE orientation of the compression axis.

5. Results verification

The obtained results for the identification of active faults and the definition of their kinematics using digital relief models were confirmed by measured results of the velocities and directions measurements of modern horizontal movements monitored in the area of SOF by means of GPS geodesy. GPS measurements were carried out by specialists of the IZK SB RAS (Irkutsk) stations for permanent and temporary observations during 2010-2015. In addition, the controlling mechanism of earthquakes and orientations of the main tension compression and extension axes operating in the area were defined.

Conducted measurements of velocities and the directions of modern horizontal movements around the SOF by method of GPS geodesy have shown that the geodetic points located on surfaces of the Sukhoi Urulyunguy depression are displaced in a (E)-SE direction (Figure 5). It is determined that the velocities of horizontal movements vary from 20 to 25 mm/year.

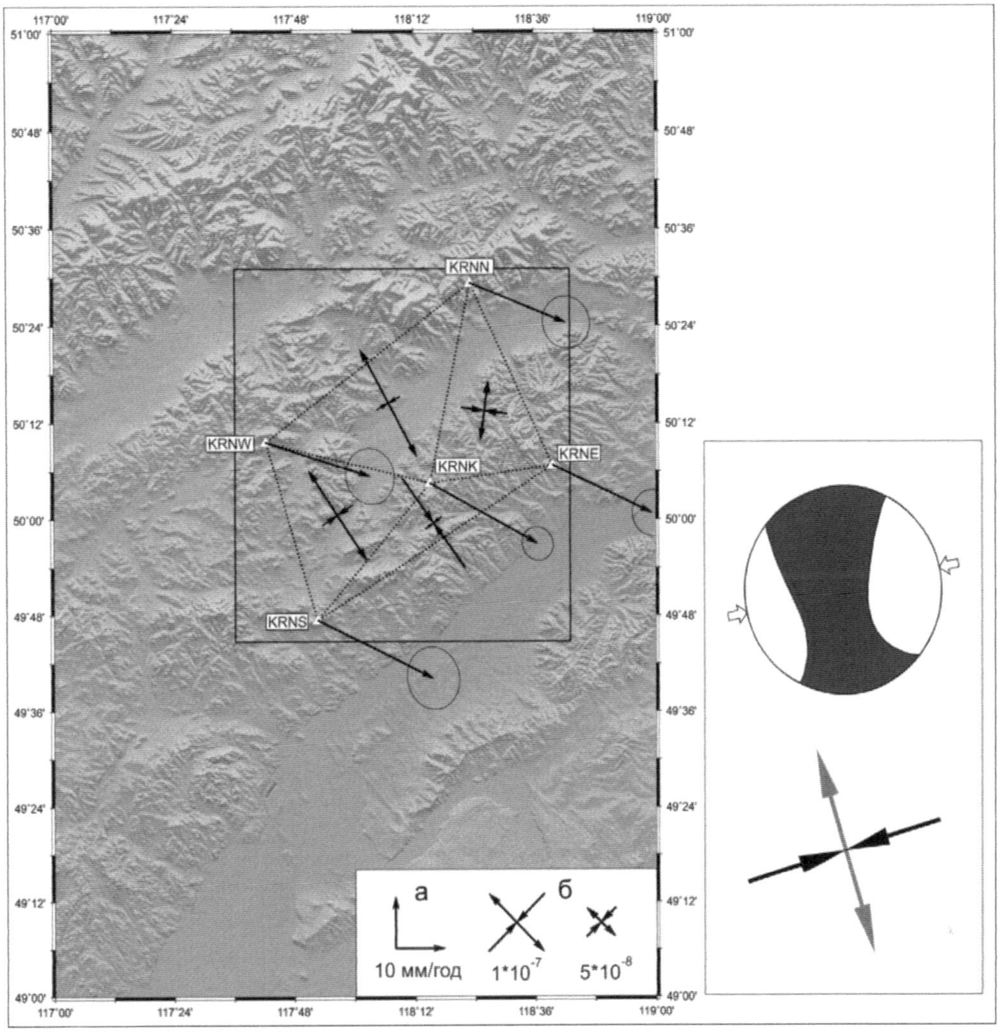

Figure 5. The current intensely deformed state of SE Transbaikalia. At the left – velocities of modern horizontal movements and relative deformations of geodetic network near the SOF: a – vectors of GPS station (KRNK, KRNW, KRNN, KRNE, KRNS) movement and velocities (mm/year) according to measurements in 2010-2015 in the basis of ITRF2008, in 95% confidential interval; b – velocities of the main horizontal deformations in a year-1 (the meeting arrows – shortening axes, the diverging arrows – lengthening axes). The square area has designated the area of the observation ground and tectonophysics modeling (an explanation in the text). On the right – the average mechanism of earthquakes (atop) and orientation (bottom) axes of the main stress of compression (the meeting arrows) and extension (the diverging arrows) in the area.

It is assumed that the geodynamic setting of horizontal extension takes place here. Herewith one triangle of the counted deformations has given the maximum reduction in the NW direction at very small lengthening in the NE direction that contradicts data for three other triangles. For the fourth triangle, it was found that the maximum lengthening is focused vertically. This corresponds to the geodynamic regime of horizontal reduction. Such significant differences in the stress state modes of adjacent crustal blocks can be explained with results for modern stress state tectonic reconstructions of different orogens. For example, in (Rebetsky et al., 2012) it is shown that in areas of ridge raisings the situation of horizontal compression or shear (transpression) prevails, and in deflections and lowlands the situation of horizontal extension or shear (transtension) prevails.

The studied area (according to the Set ..., 1999) is characterized by low seismicity. Therefore, for modern stress-strained state assessment of rock massifs by data about mechanisms of the earthquakes centers the extensive territory with coordinates 44°-54° of N latitude and 110°-124° of Et longitude is considered. In total on this territory according to various references solutions of focal mechanisms for seventeen events are known. The analysis of stereograms of focal mechanisms and their classification charts has shown that the most earthquakes are characterized by shear type of shift in the center and also there are several events of uplifting type and two earthquakes of faulted type. By data about mechanisms of the earthquakes centers the average mechanism and orientation of main tension compression and extension axes of the studied area were determined by the Yunga's (1990) method. The receiver results show that the average mechanism is close to the average shear type. Axes of compression and extension are nearly horizontal; however, the axis of compression is sub-latitudinal, and the axis of extension is sub-meridional (cf. Figure 5).

Using these observations for determination of the geodynamic setting in SE Transbaikalia the conclusion reached for the Baikal geodynamic polygon within which GPS network contained up to 20 stations (Sankov et al., 1999; 2005; Lukhnev et al., 2010) can be made. Probable influence on the horizontal movements is determined by rising and spreading of anomalous mantle substance in the Baikal rift. Although as shown by calculations (Lesne et al., 1998) modern strain can be explained by the influence of only one horizontal efforts from remote sources at plate boundaries. Important roles in formation of modern structures are played by erosion and sedimentation, the temperature field (for example, in connection with volcanic activity), fluid dynamics and other phenomena (Sherman, 2009).

On the basis of geostructural, tectonic, and geomorphological analysis of the study area, the lack of significant vertical block movements due to large raisings at the neotectonic stage is interpreted. Along with this interpretation, the question of the neotectonic stress field structure remains open so far. Statistically, most part of interpretations for the late Cenozoic by the Sim's (2011) method, points to extension in NW and compression in NE directions. Modern stress-strained state by data on mechanisms of the earthquakes centers is defined as NW-SE extension and NE-SW compression. These basic features are confirmed by the preliminary data analysis of GPS measurements.

6. Conclusions

By means of the lineament analysis carried out on the basis of digital relief model creation for the investigated area near the town Krasnokamensk in SE Transbaikali, linear elements such as lineaments were identified with a high degree of reliability. Among these, on the basis of a special technique, active lineaments were established and interpreted to reflect modern fault zones. Depending on the relative orientation of the large lineaments, which are presumably demarcating faults, and the small lineaments which are conditionally called megacracks, the tectonic setting and nature of these lineaments, orientation of compression and extension axes in the horizontal plane, the directions of shear displacements (right or left) and the geodynamic situation (compression or extension) of fault formation were reconstructed and established. The geodynamic setting of the recent fault tectonics, established by Petrov et al. (2010) was supplemented by new information on I, II and smaller rank structures with their kinematics and orientation of tension axes. The calculations of horizontal displacements according to GPS observations executed using triangulation and data on mechanisms of earthquakes have confirmed conclusions about the fault nature and kinematics of the revealed active lineaments.

Acknowledgements

This work was partly supported by program I.15P of Basic Researches of the Russian Academy of Sciences Presidium and by the ARCUS project. We would like to thank the collaboration of our colleagues: Rebetsky Yu.L and Sim L.A. from IFZ RAS (Moscow); Sankov V.A., Ashurkov S.V., Parfeevets A.V., Lukhnev A.V. from IZK SB RAS (Irkutsk); Rasskazov I.Yu. and Saksin B.G. from

IGD FEB RAS (Khabarovsk), as well as the authorities of ARMZ Uranium Holding Co., Sviatetsky V.S., Prosekin B.A., Shchukin S.I. and Tolstobrov V.A., for ongoing support of the research studies.

References

[1] Biggs et al. (2006) Fault identification for buried strike-slip earthquakes using InSAR: The 1994 and 2004 Al Hoceima, Morocco earthquakes Geophys. J. Int. 166, 1347–1362

[2] Ganas et al. (2005) DEM-based morphometry of range-front escarpments in Attica, central Greece, and its relation to fault slip rates. Geomorphology 65, 301–319

[3] Gzovsky M.V. Fundamentals of Tectonophysics. M.: Nauka, 1975. 536 pages (in Russian)

[4] Khromikh V.V., Khromikh O.V. Digital relief models: Tutorial. Tomsk: TML-Press publishing house, 2007. 178 pages (in Russian)

[5] Lesne O., Calais E., Deverchere J. Finite element modeling of crustal deformation in the Baikal rift zone: new insights into the active-passive debate. Tectonophysics. 1998. Vol. 289. P. 327-430.

[6] Lukhnev A.V., Sankov V.A., Miroshnichenko A.I. et al. Rotations and deformations of the Earth surface in the Baykalo-Mongolsky region according to GPS measurements. Geology and geophysics. 2010. Vol. 51. No. 7. P. 1006-1017 (in Russian)

[7] Novakovsky B.A., Prasolov S.V., Prasolova A.I. Digital relief models of real and abstract geofields. M.: Scientific world, 2003. 64 pages (in Russian).

[8] Oguchi et al. Identification of an active fault in the Japanese Alps from DEM-based hill shading. CSIS Discussion Paper #49.

[9] Petrov V.A., Sim L.A., Nasimov R.M., Schukin S.I. Fault Tectonics, Neotectonic Stresses, and Hidden Uranium Mineralization in the Area Adjacent to the Strel'tsovka Caldera. Geology of Ore Deposits. 2010. Vol. 52. No. 4. P. 279–288.

[10] Petrov V.A., Lespinasse M., Poluektov V.V., Cuney M., Nasimov R.M., Hammer J., Schukin S.I. Stress-time context of fault permeability at the Krasnokamensk Area, SE Transbaikalia. Journal of Physics: Conf. Series. 2013. 416.

[11] Rebetsky Yu.L., Sycheva N.A., Kuchay O.A., Tatevossian R.E. Development of inversion methods on fault slip data.Stressstateinorogenes of the central Asia. Tectonophysics. 2012. Vol. 581. P. 114-131.

[12] Sankov V.A., Levi K.G., Cale E., et al. Modern and Holocene horizontal movements at the Baikal geodynamic testing ground. Geology and geophysics. 1999. Vol. 40. No. 3. P. 422-430 (in Russian)

[13] Sankov V.A., Levi K.G., Lukhnev A.V., Miroshnichenko A.I. Modern movements of lithospheric blocks of Central Asia according to GPS geodesy. Topical issues of modern geodynamics of Central Asia. Under the editorship of K.G. Levi and S.I. Sherman. Novosibirsk: SB RAS, 2005. P. 165-179 (in Russian)

[14] Set of general seismic zoning maps of the Russian Federation territory (OSR-97). M.: IFZ RAS, 1999 (in Russian)

[15] Sherman S.I. Natural trigger mechanisms of metastable condition violation of the fault-block lithospheric environment in real time. Physics and technology problems of minerals development. Novosibirsk: Siberian Branch of the Russian Academy of Science publishing house, 2009. No. 5. Page 33-48. (in Russian)

[16] Sim L.A. Application of field methods of tectonic stress reconstruction by data about breaks for the solution of theoretical and practical tasks. Modern Tectonophysics: Methods and results. In 2 volumes. Vol. 2. M.: IFZ RAS, 2011. P. 156-171 (in Russian)

[17] Ustinov S.A., Petrov V.A. Use of GIS-technologies for the microstructural analysis in geology. Geoinformatica. 2015. No. 2. P. 33-46 (in Russian)

[18] Ustinov S.A., Petrov V.A. Use of detailed digital relief models for structural lineament analysis (on the example of the Urtuysky granite massif, SE Transbaikalia). Geoinformatica. 2016. No. 2. P. 51-60 (in Russian)

[19] Yunga S.L. Methods and results of seismotectonic deformations. M.: Nauka, 1990. 190 p. (in Russian)

3D geostatistical modelling for identifying sinkhole disaster potential zones around the Verkhnekamskoye potash deposit (Russia)

J J Royer[1], J Litaudon[1], L O Filippov[1], T. Lyubimova[2], N Maximovich[2]

[1]Université de Lorraine, Laboratoire GeoRessources, UMR7359 CNRS-CREGU, ENSG, 2 rue Doyen Marcel Roubault, 54505 Vandœuvre-lès-Nancy, France
[2]Perm tate University, Perm, Russia

Email: jean.jacques.royer@gmail.com

Abstract. This work results from a cooperative scientific program between the Perm State University (Russia) and the University of Lorraine (France). Its objectives are to integrate modern 3D geomodeling in order to improve sustainable mining extraction, especially for predicting and avoiding the formation of sinkholes disaster potential zones. Systematic exploration drill holes performed in the Verkhnekamskoye potash deposit (Perm region, Russia) have been used to build a comprehensive 3D model for better understanding the spatial repartition of the ore quality (geometallurgy). A precise modelling of the mineralized layers allows an estimation of the in-situ ore reserves after interpolating by kriging the potassium (K) and magnesium (Mg) contents at the node of a regular centred grid (over a million cells). Total resources in potassium vary according to the cut-off between 4.7Gt @ 16.1 % K_2O; 0.32 % $MgCl_2$ for a cut-off grade at 13.1% K_2O and 2.06 Gt @ 18.2 % K_2O; 0.32 % $MgCl_2$ at a cut-off of 16.5% K_2O. Most of reserves are located in the KPI, KPII and KPIII layers, the KPI being the richest, and KPIII the largest in terms of tonnage. A systematic study of the curvature calculated along the roof of the mineralized layers points out the location of potential main faults which play a major role in the formation of sinkhole during exploitation. A risk map is then derived from this attribute.

1. Introduction

Potassium is an alkaline element representing 2.6% of the crust's mass, and representing a crucial element for the agro-food industry which absorbs more than 90% of the world production. Potassium is used as a fertilizer for plants as it increases crop yield by 20%. Potassium stands for 15% of the world consumption of fertilizer, behind phosphate (25%) and nitrogen (60%). Due to the increase of the world demography, the FAO (Food and agriculture organization) predicts an increase in food demand of 70% by 2050, 80% of which being produced by better yields and cropping intensity. The slowdown in the arable lands expansion will stimulate further cropping intensity (multiple cropping and reduced fallow periods). Efficient and responsible use of mineral fertilizers will play a vital role in meeting future world demand for safe and nutritious food ([1], [2]). Consequently, potash production has to follow the increase of the world population for the future demand in fertilizer and for avoiding potential shortening in food production. Potassium resources are not regularly distributed worldwide as shown by the map of Figure 1. Canada, Belarus, and Russia have nearly 65% of all global reserves (Figure 2).

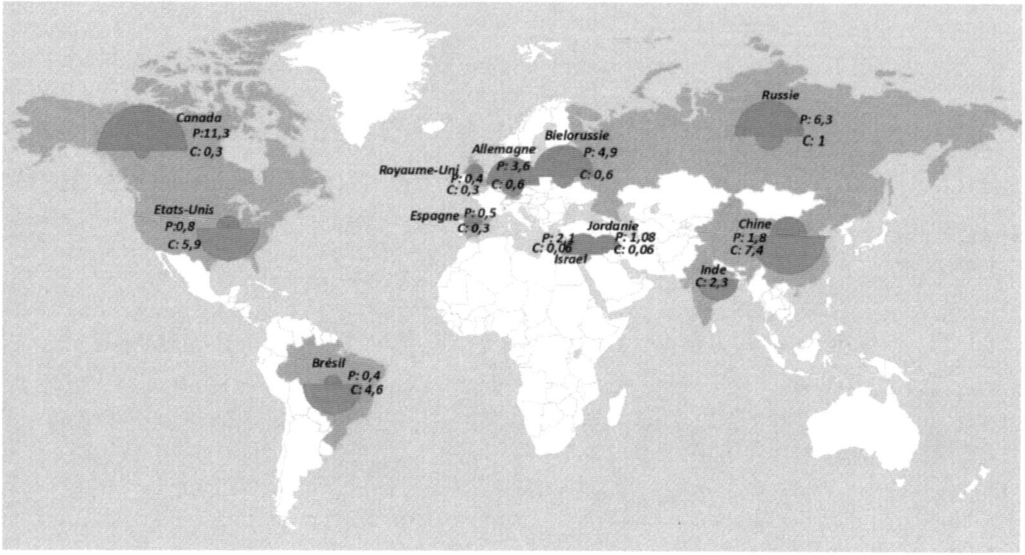

Figure 1. World potash reserves and production in Mt (Source: [1]).

Table 1 – Mineral resource estimations in sylvinite at Uralkali on 1st January 2015 (after [3])

	Category A			Category B			Category C_1			Category C_2			$A+B+C_1$		
Mine	**Ore (Mt)**	**K_2O (%)**	**K_2O (Mt)**	**Ore (Mt)**	**K_2O (%)**	**K_2O (Mt)**	**Ore (Mt)**	**K_2O (%)**	**K_2O (Mt)**	**Ore (Mt)**	**K_2O (%)**	**K_2O (Mt)**	**Ore (Mt)**	**K_2O (%)**	**K_2O (Mt)**
Berezniki 2	7.7	33.7	2.6	33.5	22.7	7.6	253.4	24.3	61.7	-	-	-	294.6	24.4	71.9
Berezniki 4	331.8	21.6	71.5	431.8	22.6	97.4	1,015.0	20.6	209.3	310.3	26.8	83.3	1,778.6	21.3	378.2
Ust-Yayvinsky	169.9	19.0	32.3	311.0	19.8	61.7	809.7	19.8	160.4	-	-	-	1,290.6	19.7	254.4
Solikamsk 1	115.6	17.8	20.6	14.7	15.4	2.2	86.0	16.3	14.0	-	-	-	216.3	17.0	36.8
Solikamsk 2	127.8	19.3	24.7	82.6	13.9	11.5	275.3	18.6	51.2	-	-	-	485.7	18.0	87.4
Solikamsk 3	103.8	17.5	18.1	64.6	18.9	12.2	1,239.6	17.4	216.0	-	-	-	1,408.0	17.5	246.3
Polovodovsky	-	-	-	694.1	16.7	1,15.8	2,386.6	17.4	415.2	260.8	15.3	39.8	3,080.7	17.2	531.0
Total	**857**	**20**	**170**	**1,632**	**19**	**308**	**6,066**	**18.5**	**1,128**	**571**	**21.5**	**123**	**8,555**	**19**	**1,606**

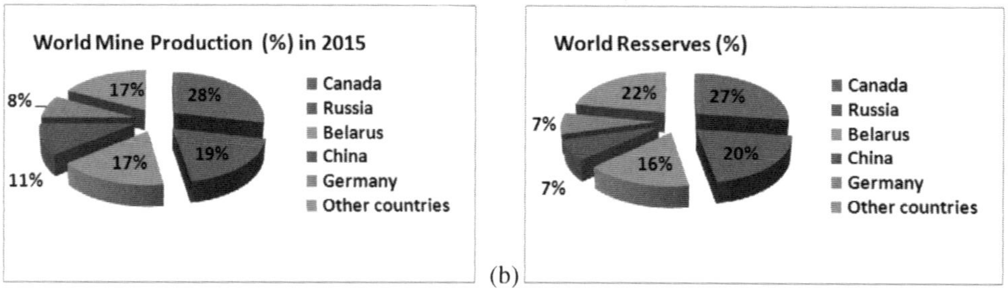

(a) (b)

Figure 2 : (a) Annual production (in %) and (b) global resources in potash (source: [4], [5], [6]).

The USGS currently estimated at 7 billion tons the recoverable global resources of potash[1] [4]. Canadian deposits are huge, thick, with high qualities and relatively easy to exploit, but lying at great depths

[1] About 200 years of estimated resources assuming current consumption.

(Saskatchewan> 900m; New Brunswick> 960m, Canada), and extraction costs tend to increase. 3D modeling can be very useful to estimate the recoverable resources by in-situ leaching and to anticipate future productions. The purpose of this study is to propose a model of the distribution of potash ore bodies, and to infer the areas at risk.

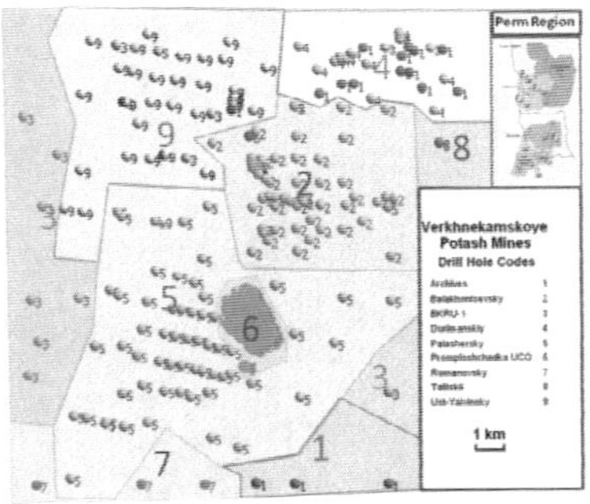

Figure 3. Drill location at Verkhnekamskoye deposit, Russia. Color indicates type of ores.

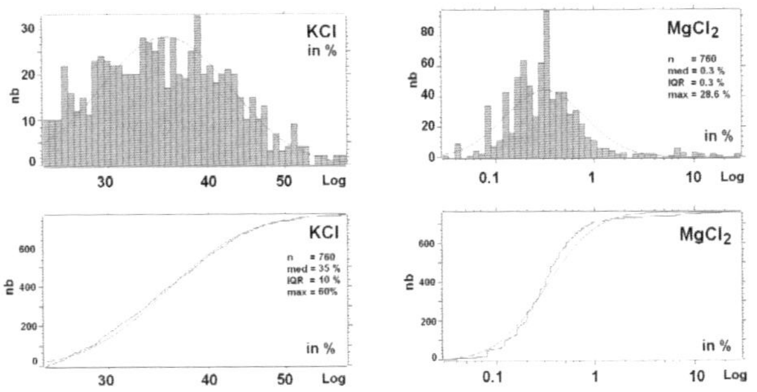

Figure 4. Histograms (top) and cumulative (bottom) distribution of the KCl and $MgCl_2$ grades showing a typical log-normal distribution.

2. Generality on Potash Deposits and available data

Primary potassium resources come from evaporation of sea water (evaporites). Potassium ores are mined from underground deposits, salt lakes and brines. Sylvite (KCl) and halite (NaCl) are the most abundant minerals, mixing the two forms is the ore commonly called sylvinite [7].

2.1. The Verkhnekamskoye Potash Deposit (Russia)

Exploited since the early XVth century, the oldest and biggest deposits of potash in the World are located in the Uralkali region (Russia). The ore reserves of the Verkhnekamskoye region potash mines are evaluated at 15.875 Gt (including proved categories ($A+B+C_1$, a resource classification used in former Russian countries), 56 Gt including the C_2 indicated one, [8]). These huge resources make Russia the third world producer of potash after Canada and Byelorussia. On the 1st January 2015, the total resources in sylvinite ($A+B+C_1$) of mines exploited by Uralkali (seven in total) were estimated at 4Gt @ 18.8% K_2O (estimated + inferred) presenting 1.388Gt of K_2O.

2.2. Use of IT techniques in geometallurgy

Advanced information technology (IT) methods are used for charactering the grade and quality spatial distribution of ores in stratiform deposits. These techniques allow better planning of operations, increased recovery, grade content at production, and better selection of blocks improving the efficiency of flotation. Another advantage of using 3D models is the ability to quantify the strains and stresses using backward restoration techniques [9], and thus ensure a better ground stability during underground operations, and optimization of in-situ leaching. Several dramatic collapses occurred in this mining area of the Verskmanskoy region (in 1986 [1]; in 2007, [10], [11]). These collapses were attributed to the formation of artificial karsts by water inflows intrusion, and had attracted the attention of the security agencies on better mining methods. 3D modelling of the Verkhnekamskoye deposit can help to better understand the geometry of geological structures, their connection with potential aquifers, and to provide the location of faults.

2.3. Available data for modelling

The industrial partner provided through the State University of Perm, an Excel™ database of 1,062 drill holes, 7,638 samples and 18 variables (including well number, type, mine, layer, drill, mine, sub-layer, X, Y, roof level, thickness, HO, KCl, $MgCl_2$, NaCl, $CaSO_4$, Comments, Br), used to build a 3D model, and to perform the geostatistical study of the deposit using the 3D gOcad modeller ([12], [13]). A stratigraphic column of the geological formation was reconstructed from borehole data.

3. 3D geomodeling and geostatistics

3.1. 3D modelling

The Excel™ drill holes database was imported in gOcad™ including the coordinates X,Y,Z, the layer label, and grades including KCL, $MgCl_2$, Br, $CaSO_4$, and NaCl. Tops of the A, A', A-A', B. KPI, KPII, and KPIII layers, were fit in the 3D space using a triangulated surface (TSurf)

(a)

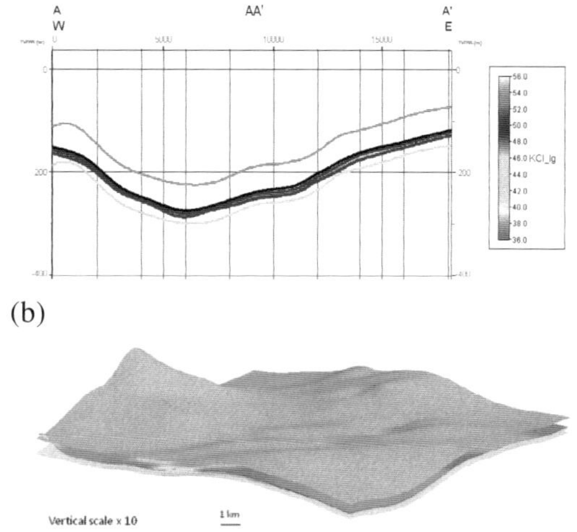

(b)

Figure 5. (a) EW Vertical cross-section showing the tops and bases of the A and KPIII mineralized layers; (b) 3D model of the potash mineralized zone. Top surfaces of the mineralized layers A-A', B, KPI, KPII, and KPIII.

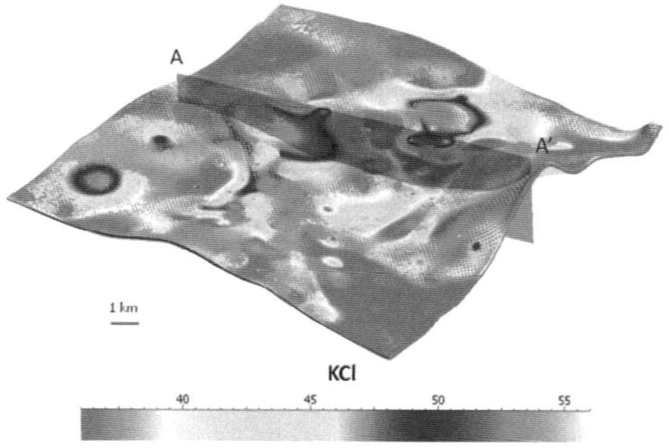

Figure 6. Stratigraphic grid adjusted to the mineralized layers A and KPIII. The color scale indicates the KCl content.

(one per layers) imposing layer thickness, or passing through drills constrains during DSI interpolation. The thicknesses were interpolated using DSI on the top surface for each layer, then used as constrains to re-interpolate the layers, avoiding self-cross-cutting among layers (Figure 5b). A vertical cross-section (Figure 5a) shows the result for layers A and KPIII.

3.2. Statistical study

A study performed on the KCl and $MgCl_2$ grades (mainly on layers A, A', B, KPI, KPII, and KPIII) and on the entire deposit, shows that grades have a classical log-normal distribution. The mean grades in KCl and $MgCl_2$, but also variability, are very high at the Verkhnekamskoye potash deposit, around $36.2_{\pm 6.9}$ %KCl ($\approx 22.7_{\pm 4.3}$ %K_2O) and $0.68_{\pm 2.08}$ %$MgCl_2$ (Figure 4). In details, two populations can be identified: (i) a *high* (within 10-50% KCl) grade minable one representing more than 60% concentrated in three homogeneous layers with low variability, and (ii) a *low* (0.1-8% KCl) grades; the same for $MgCl_2$ but with an inverse repartition compared to KCl, the lowest grades (0.01-5% $MgCl_2$) are the most common (70%) while the highest (5-30% $MgCl_2$), the least frequent. This shows that at the beginning of the evaporation cycle, the $MgCl_2$ concentration in the brine was not enough for forming carnallite (KCl, $MgCl_2$ ($6H_2O$)), and sylvinite precipitated while $MgCl_2$ concentrated in the brine. At the end of the cycle, the $MgCl_2$ concentration in the brine was sufficient to form carnallite in small proportions in the upper most layers. Similarly, the gypsite ($CaSO_4$) shows an inverse vertical zoning from top (0.1%) to bottom (2.4 but up to 8%$CaSO_4$), but with only log-normal distribution.

Correlations: Study of correlations between elements shows that they are reproducible from one layer to the others, such as $MgCl_2$ and OH. Thus, the abundance of $MgCl_2$ and OH would indicate the presence of carnallite (KCl, $MgCl_2$ ($6H_2O$)). However, this interpretation must be handled with care as OH can be included in clay minerals or in carbonates[2]. One can estimate the amount of carnallite from $MgCl_2$,

[2] In fact OH is the loss of ignition (L.O.I) and includes all the volatiles elements such as CO_2, F, H_2O, OH, ...

and calculate the residual amount of OH. Assuming that this residual corresponds to CO_2 contained in dolomite ($CaMg(CO_3)_2$) (observed in the deposit), one can estimated the average content in dolomite per layer, and observes a vertical increasing amount of dolomite from bottom (KPIII) to top (PT_B), the richest in dolomite [14].

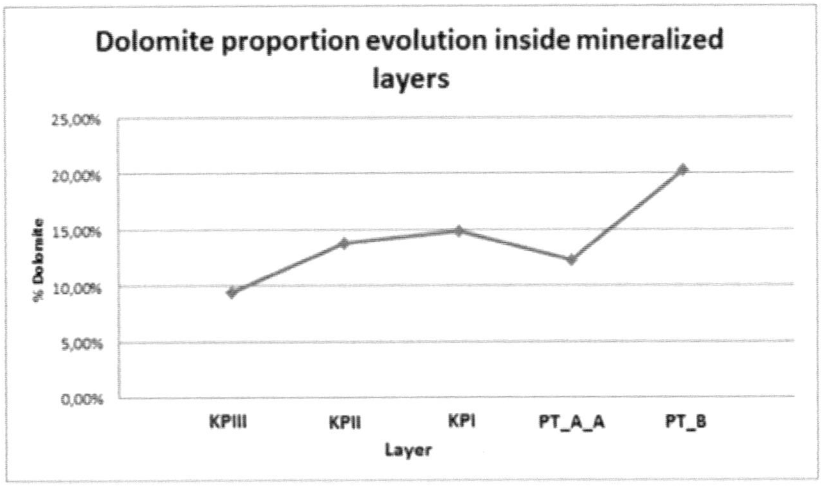

Figure 7. Indirect estimation per layer of the average content in dolomite calculated from the residual amount of OH not included in the carnallite.

3.3. Spatial variability

A regular stratigraphic grid (185×194×30 so in total over 1M cells) with elementary cells of about 100×100×2.5m was adjusted on the top and bottom of the mineralized zone in order to calculate the variograms (Figure 6). Variograms were calculated vertically and horizontally along the deformation of the layers (Figure 8a-b). Spatial variations of KCl and $MgCl_2$ point out the existence of an anisotropic structure in the vertical direction with a range of about 20m, and an isotropic structure along the layer with a range of about 1,500m. Variograms were fit to a spherical model with a nugget effect at the origin. Fitted parameters are reported in Figure 8c. They were used to estimate *in-situ* reserves by kriging. This shows that grades can be estimated vertically and horizontally at distances of about 20m and 1,500m, respectively. This spatial structure appears to be valid on the entire deposit and can be used to optimize the sample mesh of the drilling exploration campaign [15].

3.4. In-situ reserve estimations

In-situ reserves were estimated by kriging using the drill hole data and the 3D model after interpolating potassium (K) and magnesium (Mg) contents at the nodes of the regular centered stratigraphic grid over one million blocks (or cells). The estimation was then refined on a grid adjusted to the mineralized layers (Figure 9). Total resources in potassium of the mineralized zone vary depending on the cut-off grade. They were estimated between 4.7Gt @ 16.1% K_2O; $MgCl_2$ 0.32% for a cutoff grade exploitation of 13.1% K_2O, and 2.06 Gt @ 18.2% K_2O; $MgCl_2$ 0.32% for a cutoff grade of 16.5% K_2O, respectively. The most important resources are concentrated in the KPI layers KPII and KPIII, the KPI layer being the richest, while the KPIII layer is the largest in terms of tonnage.

4. Defining potential risk map using curvatures

Gaussian (K_G) and mean (K_H) curvatures were calculated on the topography and on each mineralized layer top surface and maps on Figure 11 and Figure 12. They were used to identify lineaments interpreted as possible fractures down cutting at depth the top of the mineralized layers, thus creating flow paths for possible fluid circulations from the surface. These lineaments are mainly oriented along a N30 direction, a regional tectonic direction which can be observed on the topography at a regional scale. These structures seem to continue at depth and can be identified very clearly on the map of curvatures calculated of the top of the KPIII mineralized layer, suggesting that the faults have a more or less vertical dip; no field observations are available to contradict this assumption. These maps were then used to define a risk map of possible collapses areas (Figure 12b).

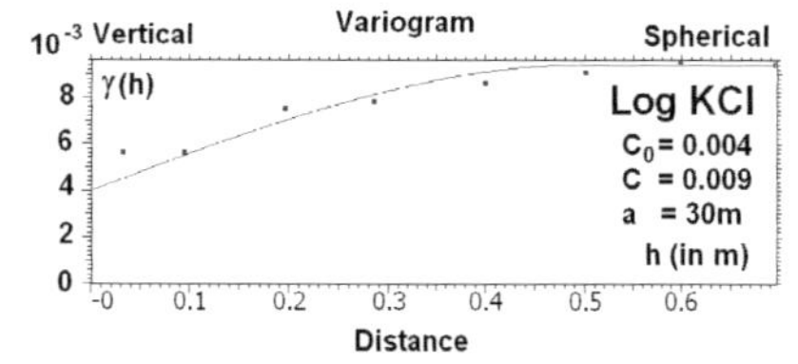

(a)

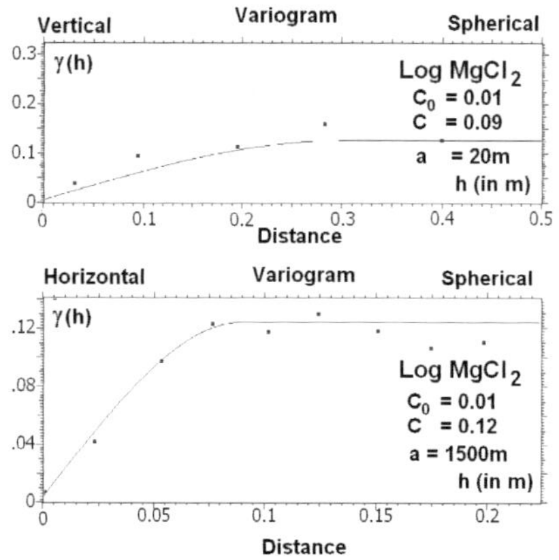

(b)

Region	Element	Variogram type	Range	C	Nugget
Layer 1	KCl	KCl	2 950 m	32	0
	$MgCl_2$	$Log_{10}MgCl_2$	2 300 m	0.44	0
Layer 2	KCl	KCl	2 600 m	14	0
	$MgCl_2$	$Log_{10}MgCl_2$	1 400 m	0.20	0.06
Layer 3	KCl	KCl	2 400 m	25	0
	$MgCl_2$	$Log_{10}MgCl_2$	2 000 m	0.24	0
PT_B	KCl	KCl	2 400 m	0.017	0
PT_A_A	KCl	KCl	0 m	0.019	0.019

(c)

Figure 8. Vertical (a) and horizontal (b) variograms calculated along the geological layers for KCl and $MgCl_2$ showing an anisotropic structure (see text); (c) parameters of the theoretical spherical variogram with nugget effect fitted to the experimental ones for both KCl and $MgCl_2$, and their logarithm.

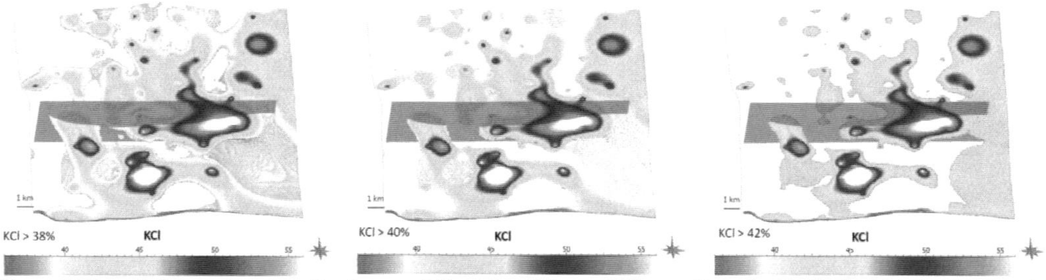

Figure 9. From left to right, mineralized zones defined for different KCl cut-off grades (i.e K_2O> 18.2%, 18.9% and 19.5%).

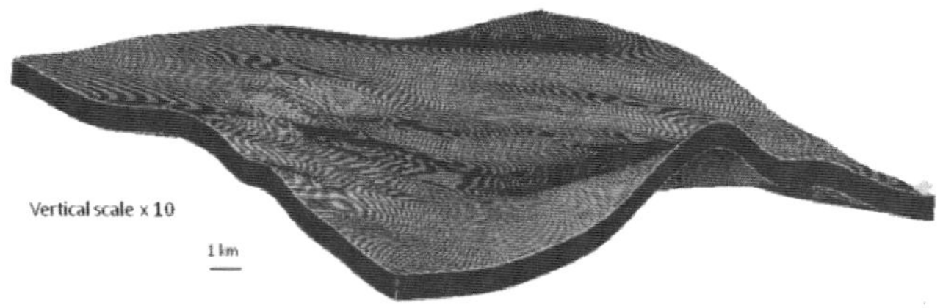

(a)

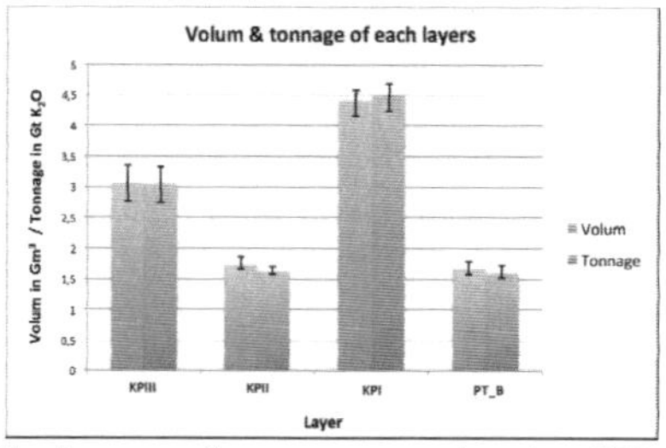

(b)

Figure 10. (a) Stratigraphic regular grid using to perform reserves estimation (185×194×30 cells of size 100×100×2.5 m); (b) In-situ volume and ore tonnage estimated per mineralized layer using a sequential Gaussian simulation (SGS) with uncertainty indicated by vertical bars.

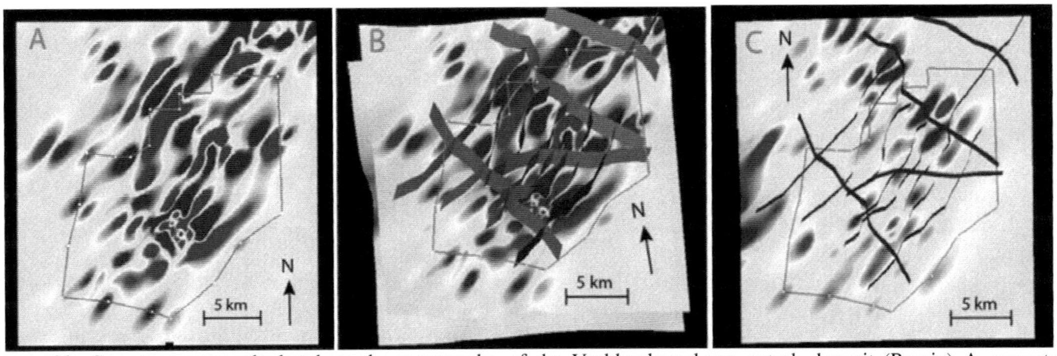

Figure 11. Curvature map calculated on the topography of the Verkhnekamskoye potash deposit (Russia) A: curvature calculated on the topography (without faults); B: same as A with identified faults; C: projection of the faults identified on the topography and extended downward on the mineralized KPIII layer.

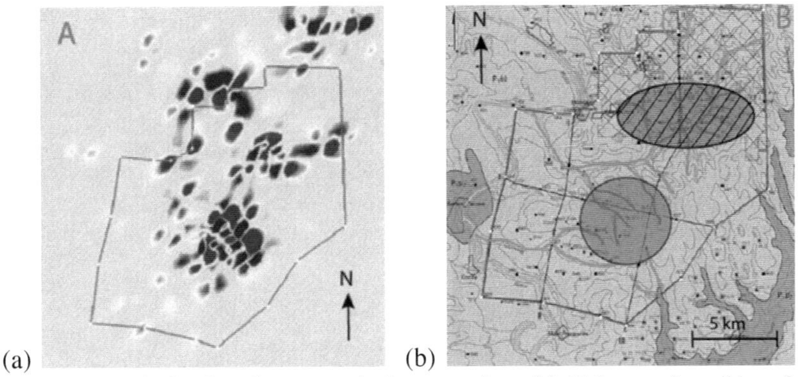

Figure 12. (a). Curvature map calculated on the topography for comparison; (b). Risk map of possible surface collapse areas due to deep dissolution of salt by fluid circulation through faults (in red higher potential risks, hashed smaller risks) superposed on the hydro-geological map with river network (in blue).

5. Conclusions

This work shows important results obtained both on the 3D modelling aspects of the Verkhnekamskoye (Russia) potash deposit, and on the in-situ reserves estimation. The reserve estimation based on the 3D model shows a vertical distribution of KCl and $MgCl_2$ grades compatible with actual knowledge on the formation of evaporites. Total in-situ reserves estimated by kriging vary according to the cut-off between 4.7Gt @ 16.1 % K_2O; 0.32 % $MgCl_2$ for a cut-off grade at 13.1% K_2O and 2.06 Gt @ 18.2 % K_2O; 0.32 % $MgCl_2$ at a cut-off of 16.5% K_2O. Most of reserves are located in the KPI, KPII and KPIII layers, the KPI being the richest, and KPIII the largest in terms of tonnage. Several other aspects were explored, in particular the impact of faulting identified from curvature map, on the hydrogeology system as a risk factor for water intrusion at depth provoking catastrophic dilution of the salt. The most direct consequences of this fault modelling is the defined of high probability risk map of possible collapses areas. As far as we know, this is the first comprehensive 3D model of the Verkhnekamskoye (Russia) potash deposit. However, this is a first attempt which can be further developed for hydro-geological, mechanical and environmental studies.

6. Acknowledgements

The work was supported by the Government of Perm Krai, Russia (grant C-26-004-07).
The authors would like to express their thanks for the support to ASGA and Lorraine-Russia Arcus Project.

References

[1] USGS, Mineral Commodity Summaries and Potash, January 2008.

[2] Potash Corp 2013. Online Annual report, 63p, http://www.potashcorp.com/media/POT_2015_AR_Full_Report.pdf
[3] Uralkali 2013. *Sustainable Leadership*, Intregrated Report & Accounts 2013, 178p.
[4] USGS, Jan 2016, http://minerals.usgs.gov/minerals/pubs/commodity/potash/mcs-2015-potas.pdf
[5] USGS, 2009, http://minerals.usgs.gov/minerals/pubs/commodity/phosphate_rock/mcs-2009-phosp.pdf
[6] USGS, 2017, https://minerals.usgs.gov/minerals/pubs/commodity/potash/mcs-2016-potas.pdf
[7] Garret D 1996. *Potash: Deposits, processing, properties and uses*. Chapman & Hall, London, UK.
[8] Kydryachov A I et al 2006. *Mineral and raw materials inventory of the Perm Region*, Koppektiv abtopov, Perm, 464p.
[9] Moretti I, Lepage F, Guiton M 2006, KINE3D: a new 3D restoration method based on a mixed approach linking geometry and geomechanics. *Oil & Gas Science and Technology, Rev. IFP*, **61**(2), 277-289.
[10] Andrejchuk A 2002. Collapse above the World's largest potash mine (Ural, Russia). *Int. J. Speleol,* **31**(1/4), 137-158.
[11] Ashrafiantar N, Hebel H P and Busch W 2011. *Monitoring of mining induced land subsidence - Differential SAR interferometer and persistent interferometer using TerraSAR-XData in comparison with Envisa Data*. 4th TerraSAR-X Science Team Meeting, 14-16 Feb., DLR, Oberpfaffenhofen, Germany, 7p
[12] gOcad Research Group - http://www.ring-team.org/
[13] Mira Geoscience - http://www.mirageoscience.com/
[14] Litaudon J 2015. *Geostatistic study analysis of the Verkhnekamskoye potash deposit*. Third year project in GeoRessources Laboratory, ENSG Nancy report, France, 27p.
[15] Royer J.J 2013. *Development of Scientific Bases of Mineral Resources High performance flotation & efficiency Improvement of ore enrichment of the K, Mg salts Verkhnekamskoye Deposit."3D modeling Part"*. 1st Annual Project Report, ENSG, UdL, Nancy, 17p.

Numerical simulation of the dynamics evolution of alluvial mining quarries

T P Lyubimova[1], A P Lepikhin[2], Ya N Parshakova[1]

[1] Institute of continuous media mechanics UB RAS, 614013, Academika Koroleva st, 1, Perm, Russian
[2] Mining Institute UB RAS, Sibirskaya, 78 –a, Perm, Russia

E-mail: lyubimovat@mail.ru

Abstract. Alluvial mining quarry (or placer mining) is one of the main techniques for extracting important building materials such as sand and gravels. Prediction of quarries detrimental effects on the hydraulic regimes of rivers, in particular on flow regimes, has been carried on in full details in 0, 1 and 2D problem formulations (in the latter case, a depth-averaging is applied). However, the prediction of the quarry behavior itself is unfeasible, though such information would be of paramount importance for estimating the adverse effect on the river bed. This work studies the dynamics evolution of alluvial mining quarries in the framework of two-dimensional formulation based on width-averaging. The Euler multiphase model, which allows simulating separately the behavior of several interacting phases, is implemented. The conducted numerical experiments show that the upstream part of the quarry is eroded more intensively than the downstream one, displacing the quarry up-stream. This effect was observed during numerous field case-studies.

1. Introduction

Exploitation of alluvial deposits of construction materials, principally sand and gravels, leads in general to the implantation of alluvial mining quarries. To minimize detrimental effects of such quarries, it is important to predict the impacts of extraction on the hydrological regime of water body. Traditionally, such predictions are made in the framework of hydrodynamic models using "zero"- one- or two-dimensional problem formulations. In [1] the results of calculations of the sand quarry behavior in the framework of a one-dimensional model are presented. The behavior of quarries of different sizes, differing both in length and width, is shown. The behavior of the quarry depending on the flow rate is investigated, the sanding of the quarry is found to be accelerated with the increase in flow rate, while the shape of the quarry varies with time in different ways for different flow rates. The paper [2] begins with an overview of the sand and gravel industry in general. It concludes with a discussion, by way of selected examples, of the broad range of potential impacts and describes some techniques that have been successfully used to prevent or limit those impacts. Where the river enters the pit immediately moves upstream, and the riverbed starts to be scoured [3]. In the latter case, a depth-averaging is applied. A more effective way to obtain the desired estimates is to perform computational hydrodynamic experiments in two-dimensional formulation based on a width-averaging. Such an approach is presented and implemented in this paper.

2. Mathematical model

The dynamics of alluvial mining quarry was modeled using the multiphase Euler model, which allows simulating separately the behavior of several interacting phases, which might be any of the three media – liquid, solid or gas phases in any combination.

The description of multiphase flows as interpenetrating continua includes the concept of phase volume fraction α_q. The phase volume is a space occupied by each phase. The laws of mass and momentum conservation are assumed to be satisfied for each phase independently. The volume V_q of phase q, is defined as $V_q = \int_V \alpha_q dV$, with $\sum_{q=1}^{n} \alpha_q = 1$.

The equation for the law of momentum conservation and continuity equation are solved independently for each phase.

The equation of continuity for phase q is written as [4]

$$\frac{\partial}{\partial t}\left(\alpha_q \rho_q\right) + \nabla \cdot \left(\alpha_q \rho_q \vec{v}_q\right) = \sum_{p=1}^{n}\left(\dot{m}_{pq} - \dot{m}_{qp}\right),$$

where $\vec{v}_q$ is the velocity of phase q, $\dot{m}_{pq}$ describes the mass transfer from phase p to phase q while $\dot{m}_{qp}$ describes the mass transfer in the opposite direction.

The law of conservation of momentum for a liquid phase ($q = l$) is written as

$$\frac{\partial}{\partial t}\left(\alpha_q \rho_q \vec{v}_q\right) + \nabla \cdot \left(\alpha_q \rho_q \vec{v}_q \vec{v}_q\right) = -\alpha_q \nabla p + \nabla \bar{\bar{\tau}}_q + \alpha_q \rho_q \vec{g} +$$

$$+\sum_{p=1}^{n}\left(\vec{R}_{qp} + \dot{m}_{pq}\vec{v}_{pq} - \dot{m}_{qp}\vec{v}_{qp}\right) + \left(\vec{F}_q + \vec{F}_{lift,q} + \vec{F}_{\mathrm{vm},q}\right)$$

where $\bar{\bar{\tau}}_q$ is the shear stress tensor of the phase q, $p = s$:

$$\bar{\bar{\tau}}_q = \alpha_q \mu_q \left(\nabla \vec{v}_q + \nabla \vec{v}_q^{\,T}\right) + \alpha_q\left(\lambda_q - \frac{2}{3}\mu_q\right)\nabla \cdot \vec{v}_q \bar{\bar{I}}$$

Here μ_q and λ_q are the shear and bulk viscosities of the phase q, $\vec{F}_q$ is the external force, $\vec{F}_{lift,q}$ is the buoyancy force, $\vec{F}_{\mathrm{vm},q}$ is the virtual mass force, $\vec{R}_{qp}$ is the force of interaction between the phases, p is common pressure for all phases, $\vec{v}_{pq}$ is the inter-phase velocity, which is defined as follows: if $\dot{m}_{pq} > 0$ (i.e. the mass of the phase q is transferred into the mass of phase p) then $\vec{v}_{pq} = \vec{v}_p$ and vice versa. Additionally, if $\dot{m}_{qp} > 0$, then, $\vec{v}_{qp} = \vec{v}_q$ and if $\dot{m}_{qp} < 0$, then $\vec{v}_{qp} = \vec{v}_p$.

The inter-phase force depends on the friction, pressure, particle packing and other factors. For this force the following conditions are verified: $\vec{R}_{qp} = -\vec{R}_{pq}$, $\vec{R}_{qq} = 0$,

$$\sum_{p=1}^{n}\vec{R}_{pq} = \sum_{p=1}^{n} K_{pq}\left(\vec{v}_p - \vec{v}_q\right)$$

Here, K_{pq} is the coefficient of momentum exchange between the phases.

For a solid phase the law of momentum conservation has the following form:

$$\frac{\partial}{\partial t}(\alpha_s\rho_s\vec{v}_s)+\nabla\cdot(\alpha_s\rho_s\vec{v}_s\vec{v}_s)=-\alpha_s\nabla p-\nabla p_s+\nabla\cdot\bar{\bar{\tau}}_q+\alpha_s\rho_s\vec{g}+$$

$$+\sum_{p=1}^{n}\left(\vec{R}_{ls}+\dot{m}_{ls}\vec{v}_{ls}-\dot{m}_{sl}\vec{v}_{sl}\right)+\left(\vec{F}_s+\vec{F}_{lift,s}+\vec{F}_{vm,s}\right)$$

Here p_s is the pressure of the solid phase.

The exchange coefficient between the liquid and the solid phases can be written as

$$K_{sl}=\frac{\alpha_s\rho_s f}{\tau_s}$$

where τ_s is "*the particle relaxation time* " (or *particle drag coefficient*), which is defined as $\tau_s=\rho_s d_s^{\,2}/(18\mu_l)$ and d_s is the diameter of particles in the solid phase, ρ_s the solid density, and μ_l the dynamic viscosity (assuming a Stokes flow and a Reynolds number less than unit).

The coefficient f can be derived from several models. In this study, we use the Gidaspow model of closely packed layer of particles [5]. Particle packing depends on their quantity and particle diameter. The calculation of maximum packing for a liquid phase with solid inclusions is presented in [6].

The stress tensor includes the shear and bulk viscosities of the solid phase, which are related to the particle momentum due to particle motions and their collisions. The shear viscosity consists of the collision-induced part, the kinetic part and the part associated with additional friction:

$$\mu_s=\mu_{s,col}+\mu_{s,kin}+\mu_{s,fr}$$

To describe the turbulence pulsations of velocity, we use the disperse $k-\varepsilon$ model of turbulence. The random motion of the solid phase in the liquid phase turbulent flow is the major contribution to the energy change in the system. The contribution of inter-particle collisions is negligible.

The equations for the turbulent kinetic energy k and the rate of its dissipation ε are given as

$$\frac{\partial}{\partial t}(\alpha_l\rho_l k_l)+\nabla\cdot(\alpha_l\rho_l k_l\vec{U}_l)=\nabla\cdot\left(\alpha_l\frac{\mu_{t,l}}{\sigma_k}\nabla k_l\right)+\alpha_l G_{k,l}-\alpha_l\rho_l\varepsilon_l+\alpha_l\rho_l\Pi_{k_l}$$

$$\frac{\partial}{\partial t}(\alpha_l\rho_l\varepsilon_l)+\nabla\cdot(\alpha_l\rho_l\vec{U}_l\varepsilon_l)=\nabla\cdot\left(\alpha_l\frac{\mu_{t,l}}{\sigma_\varepsilon}\nabla\varepsilon_l\right)+\alpha_l\frac{\varepsilon_l}{k}(C_{1\varepsilon}G_k-C_{2\varepsilon}\rho_l\varepsilon_l)+\alpha_l\rho_l\Pi_{\varepsilon_l}$$

Here $G_{k,l}=\mu_t S^2$ is the generation of the turbulent kinetic energy due to the mean velocity gradient, where $S=\sqrt{2S_{ij}S_{ij}}$ is the norm of the tensor of the mean strain rate, $S_{ij}=\frac{1}{2}\left(\frac{\partial v_j}{\partial x_i}+\frac{\partial v_i}{\partial x_j}\right)$; Π_{k_l} defines the influence of the solid phase on the liquid phase, $C_{1\varepsilon}, C_{2\varepsilon}, \sigma_k, \sigma_\varepsilon$ are constants.

The problem was solved using unsteady approach. The calculations were carried out in two-dimensional formulation based on the averaging in the horizontal direction perpendicular to the flow of the river. At initial time, the quarry was assumed to be rectangular (Fig.1) or with the inclined lateral walls (Fig.2).

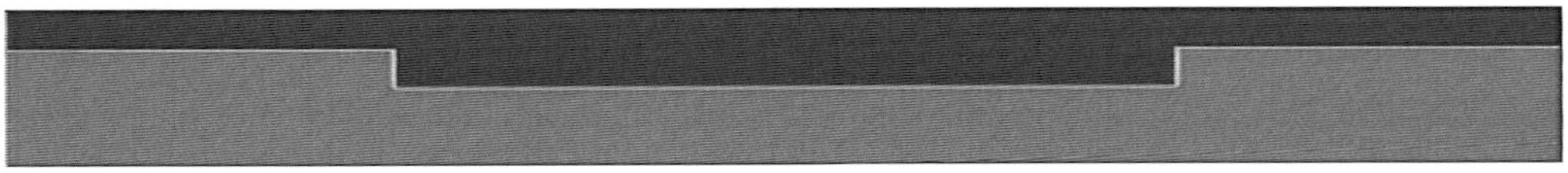

(a)

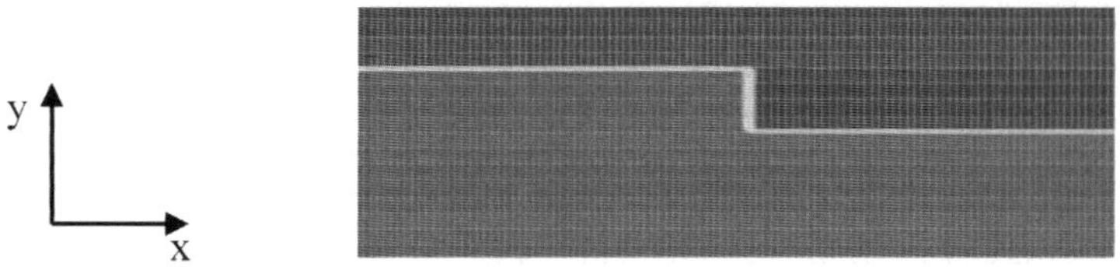

(b)

Fig. 1. Computational domain at initial time with water (in blue), and sand (in red). (a) the entire computational domain, (b) the fragment of the grid covering the studied domain

The no-slip condition was imposed at the lower boundary of the studied domain, modeling the river bottom. The upper boundary of the studied domain, corresponding to the free liquid surface, was considered as nondeformable. It was supposed to obey the vanishing condition of the normal velocity component and tangential stresses. On the lateral boundaries of the domain, the condition of zero velocity derivative with respect to the normal was imposed.

A constant flow velocity $\vec{V}$ with one nonzero component was assumed at the input of the studied domain.

$$x=0: \quad v_{xq}=V, \; v_{yq}=0$$

At the exit of the studied domain, the derivatives with respect to x of all functions were assumed vanishing (equal to zero) as the boundary conditions.

As initial conditions, the velocity of the basic flow is imposed equal to the velocity at the input of the studied domain: $\vec{v}_q=\vec{V}$.

3. Mathematical model

The calculations were performed in a 200x14 m domain, with horizontal dimension of 200 m and vertical extension of 14 m. The length of the quarry was 100 m and the depth was 6 m. The velocity of the base flow was taken to be equal to 0.3 m/s. The size of the particles forming the bottom and walls of the quarry were about d = 150 μm, the particle density was $\rho_s = 2.65\cdot10^3$ kg/m^3. The calculations were made up to 400,000 seconds (about 4.63 days).

The numerical simulations showed that a vortex appears near the bottom of the quarry (Fig. 2). During simulations, it appears that this vortex is moving in such a way that it causes the erosion of the front (up-stream) edge of the quarry more intensively than the downstream one.

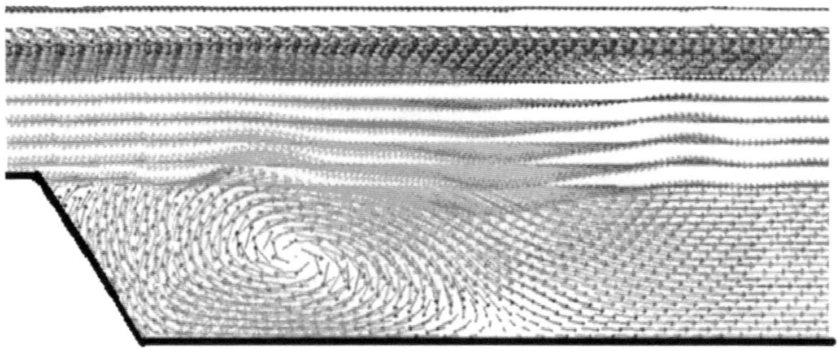

Fig. 2. Velocity vector field in the quarry

For a sand shear viscosity of the order of $10^5\ kg/m\,s$, the sanding of the quarry occurs slowly (Fig.3). The quarry erosion moves upstream and is accompanied by the accumulation of the sediments inside the quarry. With decreasing levels and increasing slopes one can observe the erosion of the river bottom, which propagates upstream (against the base stream). In the downstream part of the quarry, the growth of the slopes of the water surface and flow velocities is observed, which, as a rule, increases the size of the bottom ridges. The depth of the riverbed penetration is proportional to the quarry length. The area occupied by the quarry is gradually covered by sediments intensively delivered from the first (upstream) region.

(a)

(b)

Fig.3. Dynamics of the quarry erosion. (a) – 9,000 s (2.5 h), (b) – 400,000 s (4.63 days).

Figure 4 illustrates the regimes with uniform sanding of the quarry in the case where the flow velocity is higher than the non-eroding velocity. One can see the erosion of the bed bottom in the upstream region adjacent to the quarry and gradual sanding of the quarry itself.

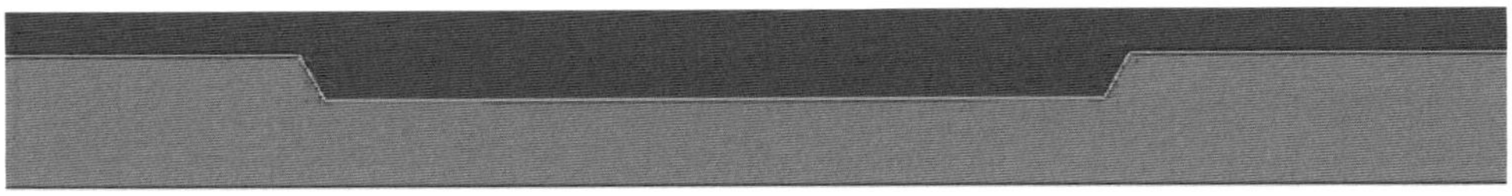

(a)

(b)

Fig. 4. Dynamics of erosion of quarry (a) – 5,000 s (1.39 h), (b) – ~10^6 s (11.574 days).

With increasing distance from the quarry, the sediment flow rate gradually increases along the length extension of the river. For length quarries of 50, 100 and 200 m, the time of sanding of the quarry was found to be equal to 6, 13 and 27 months, respectively, for the average size of the river bed particles equal to 0.15 mm and an average velocity of the base flow equal to 0.3 m/s.

4. Conclusions

The calculations showed that, at the bottom of the quarry, the water stream motion generates a vortex in such direction that the upstream wall of the quarry is eroded more intensively than the downstream wall. Filling of the quarry with sediments occurs due to a subsequent downward displacement of the upstream wall of the quarry. The suspended sediments can partially settle out in the downstream part of the quarry, which leads to gradual sanding of the quarry accompanied by a change of the flow depth in the flow regions upstream and downstream from the quarry at a distance exceeding its length. In the downstream region, the overall erosion of the river bed takes place. The zone of maximal erosion of the bottom is adjacent to the downstream wall of the quarry and spreads over this wall. This effect was recorded in multiple field observations.

Acknowledgements
The study was supported by Russian Foundation for Basic Research (grant N 16-41-590005).

References

[1] Bovolin V, Ponce V M 2008 *Internal report, San Diego St. Univ.* 34
[2] Langer W H 2003 U.S. Geological Survey Open-File Report 02-153
[3] Collins B, Dunne T 1990 *Special Publication, California Division of Mines and Geology, Sacramento* **98** 29
[4] Anderson T B, Jackson R 1967 *I & EC Fundam.*, 6:527-534
[5] Gidaspow D, Bezburuah R, Ding J 1992 *Hydrodynamics of Circulating Fluidized Beds, Kinetic Theory Approach.* In Fluidization VII, Proceedings of the 7th Engineering Foundation Conference on Fluidization 75
[6] Fedors R F, Landell R F 1979 *Powder Technology* 23:225-231

IOP Conf. Series: Journal of Physics: Conf. Series **879** (2017) 012020 doi:10.1088/1742-6596/879/1/012020

Identifying potential disaster zones around the Verkhnekam-skoye potash deposit (Russia) using advanced information technology (IT)

J J Royer[1], L O Filippov[1,2]

[1]Université de Lorraine, Laboaratoire GeoRessources, CNRS / CREGU, ENSG, 2 rue du Doyen marcel Roubault, BP F-54505 Vandœuvre-lès-Nancy, France.
[2]National University of science and technology MISIS, 4 Leninsky pr., 117049 Moscow, Russia

Email: jean.jacques.royer@gmail.com

Abstract. This work aims at improving the exploitation of the K, Mg, salts ore of the Verkhnekamskoye deposit using advanced information technology (IT) such as 3D geostatistical modeling techniques together with high performance flotation. It is expected to provide a more profitable exploitation of the actual deposit avoiding the formation of dramatic sinkholes by a better knowledge of the deposit. The GeoChron modelling method for sedimentary formations (Mallet, 2014) was used to improve the knowledge of the Verkhne-kamskoye potash deposit, Perm region, Russia. After a short introduction on the modern theory of mathematical modelling applied to mineral resources exploitation and geology, new results are presented on the sedimentary architecture of the ore deposit. They enlighten the structural geology and the fault orientations, a key point for avoiding catastrophic water inflows recharging zone during exploitation. These results are important for avoiding catastrophic sinkholes during exploitation.

1. Introduction

Potassium is used in over 90% as fertilizers. This commodity is crucial for contemporary food industry. FAO[1] forecasts an increase by 70% to 80% of food demand in 2050 on expectations of increase in world population [1]. The efficient and responsible use of mineral fertilizers will play a vital role to meet the challenges of tomorrow's agriculture and future global demand for food [5]. World reserves of potash are reported in Table 1 showing that potassium resources are not regularly distributed worldwide. Canada, Belarus, and Russia have nearly 65% of all global reserves. Canadian deposits, the most important, are huge, thick, with high qualities and relatively easy to exploit. However, most of these lie at great depths (Saskatchewan > 900m; New Brunswick > 960m, Canada), and extraction costs tend to increase. They become difficult to exploit by traditional extractive methods, but technical solutions exist such as in-situ leaching (Solvay process) and could be an alternative where deposits are irregular in shape.

[1]Food and Agriculture Organization.

Table 1 – Production and potassium global resources (source: [21], [20])

Country	Mine Production[1]		Reserves	
	2014	2015[4]	Recoverable ore	K_2O Equivalent
Canada	11,000	11,000	4,200,000	1,000,000
Belarus	6,290	6,500	3,300,000	750,000
Russia	7,380	7,400	2,800,000	600,000
Jordan[3]	1,260	1,250	NA[2]	270,000
Israel[3]	1,770	1,800	NA	270,000
China	4,400	4,200	NA	210,000
Germany	3,000	3,000	NA	150,000
Chile	1,200	1,200	NA	150,000
United States	850	770	1,500,000	120,000
Other countries	50	50	250,000	90,000
United Kingdom	610	610	NA	70,000
Spain	715	700	NA	20,000
Brazil	311	311	300,000	13,000
World total (rounded)	38,800	38,800	NA	3,700,000

[1] Data are rounded to avoid disclosing company proprietary
[2] NA = Not available; [3] Total reserves in the Dead Sea are arbitrarily divided equally between Israel and Jordan; [4]Estimated.

Advanced 3D modelling can be very useful to estimate the recoverable resources by in-situ leaching and to anticipate future productions. The objectives of this study are to propose a model for studying the distribution of potash in ore deposits, and to infer areas at risk of surface collapses due to underground mining exploitation ([2], [3]).

2. The GeoChron theory

2.1. Introduction

The modern GeoChron approach tries to reproduce the sedimentary mechanisms which have led to the successions of the sedimentary formations [11]. One of the original ideas of GeoChron is to turn the complex 4D space-time into a simplified 3D one by mapping time on depth using a judicious choice of a curvilinear coordinate system. This model allows introducing *attributes* like *acceleration* or *sedimentary curvature* that can be directly connected to basin formation parameters such as *diffusion* or *advection sedimentation rate* [17]. It provides a framework for studying deformation in structural geology [13] and mineral resources potential study ([7], [4], [18]). It can be applied with benefit to the Verkhmekamskoye (Russia) potash deposits in order to better understand the sedimentary architecture of the geological formations.

2.2. What is GeoChron?

GeoChron is a mathematical model based on a transformation which relates the *present time Euclidean coordinates system* (in which geological observations are made), or Geological space (G-space), to a new depositional *time curvilinear coordinate system*, or Geo-Chronological space ($\overline{G}$-space). It can be seen as a time-stratigraphic model built in a depositional curvilinear coordinate system (u, v, t) superposed onto the geological present time coordinate system (x, y, z). The basic idea of the time stratigraphic model was initially suggested by [22] for 2D vertical cross-sections; the mathematical formulation of this concept was extended to the 3D space by [10].

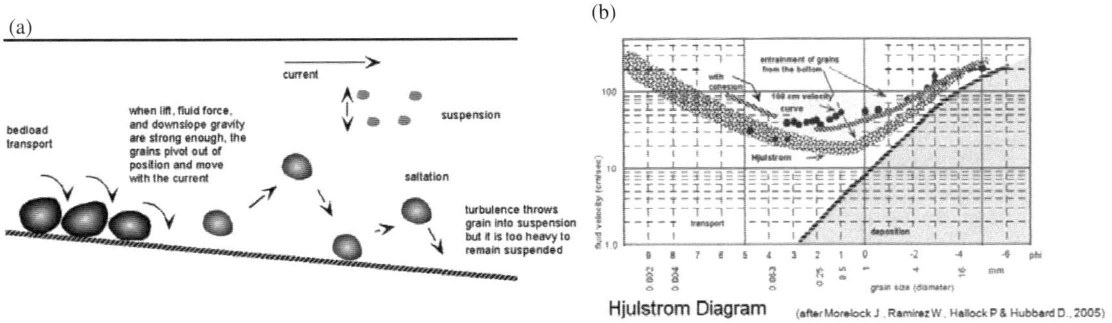

Figure 1. (a) Sediments transported under the combined effect of waves, currents, tides, gravity, rivers, evaporation, and tectonic subsidence. (b) Transport and deposition of sediments are functions of the grain size and the stream velocity or current velocity).

2.3. Intrinsic Sedimentary Attributes

Several attributes based on the lateral variation of the sedimentation rate can be introduced for characterizing the depositional environment, such as: (i) the *Einstein's curvature*, and (ii) the *sedimentary potential* ([11], p. 140). Some basic concepts are discussed: (i) the *vertical sedimentation rate* V_Φ corresponds to the amount of sediments (thickness) deposited per unit time ($m.s^{-1}$) defined by:

$$V_\Phi = \frac{1}{1-\Phi}\frac{1}{\|\boldsymbol{grad}\ t\|} \tag{1}$$

where t is time and Φ porosity. The sedimentation rate can vary or not laterally (Figure 1b). The sedimentation potential φ_ϕ is defined as the logarithm of the sedimentation rate $\varphi_\phi = \log V_\phi$.By definition, it is an *intrinsic*[2] property; (ii) the *lateral variation* of the sedimentation rate along the horizons (or sedimentary acceleration $\boldsymbol{A}$) is the gradient of the sedimentary potential calculated along the horizons (or isochrones surface). It is defined by: $\boldsymbol{A} = \boldsymbol{grad}_H\ \varphi_\phi$. This is a vector tangent to the horizons. The reader can consult the book by [11] for further information on this subject.

2.4. Einstein's curvature

A result of the Riemannian geometry shows that the volume variation of an elementary infinitesimal sphere submitted to deformations is proportional to the Einstein's curvature κ_E, an intrinsic property of the space (independent of the chosen coordinate system). In GeoChron, κ_E characterizes the lateral variations of the sedimentation rate ([11]; [17]). It is defined by:

$$\kappa_E = (\Delta_H V_\Phi)/V_\Phi \tag{2}$$

where Δ_H is the Laplacian calculated along a depositional horizon. It characterizes the relative variation of the layer thicknesses along the horizon. Mallet ([11], p. 142) shows that the Einstein's curvature (or sedimentary expansion) is related to the sedimentary acceleration by:

$$\kappa_E = div(\boldsymbol{A}) + \|\boldsymbol{A}\|^2 \tag{3}$$

where *div* is the divergence operator, $\boldsymbol{A}$ the sedimentary acceleration, a vector field measuring the variation of the sedimentary rate along the depositional horizons. The sedimentary acceleration vector is: (i) *tangent* to the horizons; (ii) *null* for constant sedimentation rate in the vicinity of the study location; (iii) oriented along the *increasing* layer thickness directions; (iv) the larger the module of the acceleration, the faster layer thickens in the direction of $\boldsymbol{A}$.

[2] e.g. independent on the coordinate system.

2.5. Sedimentary Models

Sedimentary basins result from the transportation of fluvial and/or marine sediment through waves, currents, gravity, tidal effects, rivers input, subsidence and tectonic events [12]. Models describing the redistribution of fluvial sediments during the formation of basins or the time-dependent evolution of landforms are usually based on two fundamental equations: (i) the *conservation* of sediments (continuity or mass balance) and (ii) the *sediment flux* evolution as a function of waves, currents, and gravity [6]. In the literature the conservation is written as (neglecting the porosity):

$$\partial_t Z = div\,(\boldsymbol{q}_s) + U \tag{4}$$

where $\partial_t Z$ is the derivative against time t, $div = [\partial u, \partial v]$ the 2D-divergence along the sedimentary horizon H, Z is the relative height of the column of sediment vertically stacked in the G-space (or equivalently, the relative altitude of the sea bottom counted from the sea level), $\boldsymbol{q}_s$ sedimentary flux, U accounts for combined effects of subsidence and compaction. Several models have been suggested for describing the sediment transport through the flux $\boldsymbol{q}_s$. One of the most used is the *diffusion-advection* model:

$$\partial_t Z = div(\,\boldsymbol{D}\,.\,\mathbf{grad}_H\, Z) + \boldsymbol{k}\,.\,\mathbf{grad}_H\, Z + U \tag{5}$$

where $\boldsymbol{D}$ is the diffusivity tensor (or dispersion) of sediments, $\boldsymbol{k}$ is the transport velocity of the sediments describing the advection term, the velocity is assumed to be conservative with null divergence ($div\,\boldsymbol{k} = 0$).

2.6. Sedimentary Models in GeoChron

The above classical approaches can be transposed directly into the GeoChron formalism noticing that the sedimentation rate plus the subsidence term $V_\phi + U$ can be expressed in function of the sedimentary rate according to ([11], p.35, Eq. 1.129; [17]):

$$\partial_t Z = V_\Phi + U = div\,\boldsymbol{q}_s + U \tag{6}$$

Deriving against time t, it comes (the subsidence term U is assumed to be global at the basin and time considered scales, it is locally constant by intervals, with a null derivative against time):

$$\partial_t V_\Phi = div\,(\partial_t\, \boldsymbol{q}_s) \tag{7}$$

where $\partial_t\, \boldsymbol{q}_s$ is the variation of the sedimentary flux as a function of time t at location (u, v), a quantity similar to the acceleration. If the flux is stationary, its derivative against time is null $\partial_t\, \boldsymbol{q}_s = \boldsymbol{0}$. This is not a realistic case as the sedimentary flux generally varies as a function of time (under Milankovitch cycles). Thus, the sedimentary acceleration $\partial_t\, V_\Phi$ captures the combined effects of the sedimentation, subsidence, compaction, loading and vertical uplifts of the basin as a function of time. Reporting Eq. (7) into Eq. (5) in the case of an isotropic diffusion coefficient d, it comes [17]:

$$(\partial_t V_\Phi)\,/\,(\Delta_H(V_\Phi)) = d + \boldsymbol{k}_d\,.\,(\mathbf{grad}\, V_\Phi)/(\Delta_H(V_\Phi)) \tag{8}$$

where d is the isotropic diffusion coefficient, $\boldsymbol{k}_d$ is the transport velocity of the sediments describing the advection term, Δ_H the Laplacian along the depositional surface. Introducing the sedimentary acceleration, Eq. (7) can be simplified into [17]:

$$\partial_t \ln V_\Phi = \kappa_E\, d + \boldsymbol{k}_d\,.\,\overline{\boldsymbol{A}} \tag{9}$$

where d is the isotropic diffusion coefficient, $\boldsymbol{k}_d$ is the transport velocity of the sediments describing advection term, $\overline{\boldsymbol{A}}$ the sedimentary acceleration, κ_E the Einstein's curvature.

2.7. Discussion

The time variation of the sedimentation rate $\partial_t \ln V_\Phi$ in eq. (9) is a relative dimensionless quantity that measures the vertical relative changes in the sedimentation rates V_Φ along isochrones of deposition in the geographical (or geological) space. The above relationship shows that this amount (which is easily calculated) is connected to the physical parameters describing the sedimentation (diffusion rate d,

Einstein's curvature $\varkappa_E$) plus an advection term described by a velocity $\boldsymbol{k}_d$ measuring the lateral sediment transport, multiplied by the acceleration vector $\overline{\boldsymbol{A}}$, a local measurement of the lateral variation of sediment rates along deposition surfaces. Several simple cases can be considered: (i) the sedimentation rate is *constant*, V_Φ = cst, the curvature $\varkappa_E = 0$ is null because the lateral variation of the

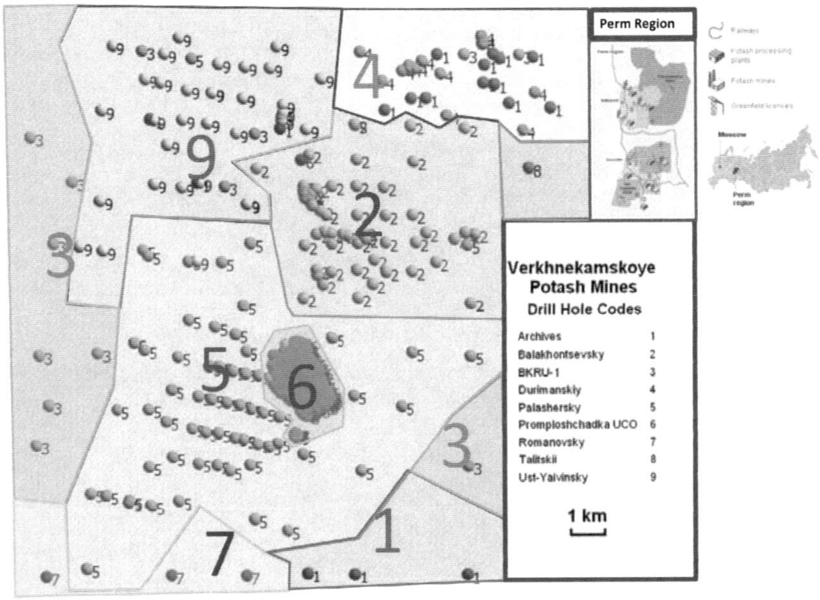

Figure 2. Drill location at Verkhnekamskoye deposit, Russia. Colors indicate type of ores.

sediment rate is zero, ***grad*** $V_\Phi = \boldsymbol{0}$, only vertical sediment inputs are accounted for (no lateral input); (ii) the sedimentation rate is an *harmonic* function, $\Delta_H(V_\Phi) = 0$, Eq. (9) simplifies into $\partial_t \ln V_\Phi = \boldsymbol{k}_d \cdot \overline{\boldsymbol{A}}$ for which lateral sediment incomes due to the advection are taken into account; (iii) the general case (not an harmonic function) where diffusion and advection are considered. The above concepts has been developed and usually applied to clastic sediments; these concepts can also be extended to evaporitic sediments, the sediment rate being the proportion of precipitated minerals. This method has been applied to the case of the Verkhnekamskoye deposit.

3. Case study: the Verkhnekamskoye ore deposit

3.1. Geological settings

The Verkhnekamskoye deposit is a multi-layered fromation of potassium and potassium-magnesium salts; it is elongated along a North South (NS) direction and located within the Cherdynsky, Solikamsky and Usolsky districts of the Perm region, Ural, Russia. The NS extension of the deposit is about 140 km for a maximum width of about 42 km, making this 3,750 km^2 area, one of the biggest potassium salts producer in the World. Identified tectonic structures of sub EW strike, referred as the Durinsky and Borovitsky troughs, divide the sedimentary formation into three parts: the Northern, Central and Southern. The Verkhnekamskoye deposit is confined to the central part of the Solikamsk depression in the Predural edge trough. The Solikamsk depression, as part of the Predural trough, is bounded from the west by the Russian plate of the East European Platform, and from the east by the West Ural Folding Zone. Its western boundary is limited by the Krasnoufimsky deep fault, and the eastern one by the West Ural regional fault. The Palashersky and Balakhontsevsky areas host the Lower Permian halogen deposits made of saline rocks which contain potassium and potassium-magnesium formations, interbedded by terrigenous formations (lower sub-thickness of the salt-marl layer), and on top, the over-salt layers comprising the upper sub-layer of the salt-marl, carbonate and various clastic layers. Quaternary deposits complete the stratigraphic section.

3.2. Methodology

Sedimentary attributes in §2 have been calculated on the Verkhnekamskoye deposit [9] using Gocad/SKUA scripts ([8], [14]) on a stratigraphic grid of 185×194×30 cells of size 100×100×2.5m

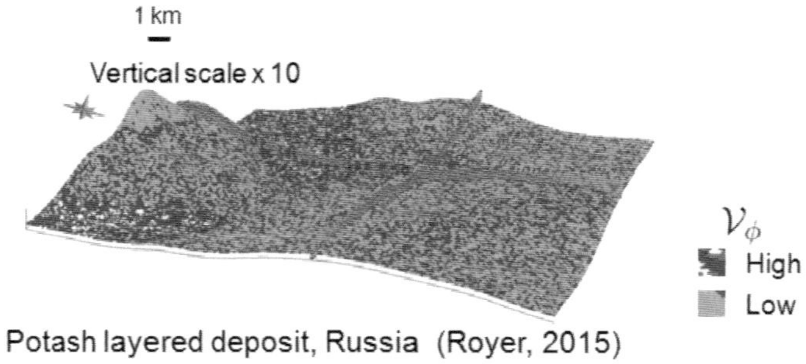

Figure 3. Sedimentation rate V_ϕ calculated on the stratigraphic grid. Highs (red) correspond to strong thicknesses with a high sedimentation rate V_ϕ, indicating the thickest layers [17].

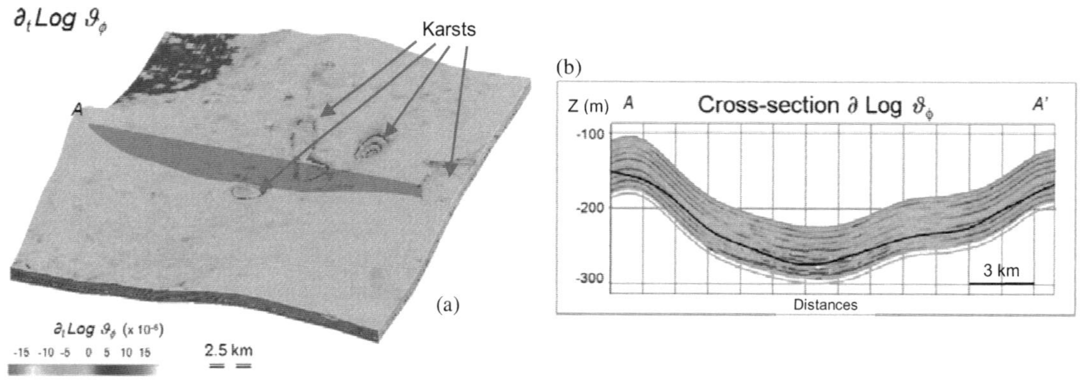

Figure 4. (a) Derivative against time of the potential (∂_t ln V_Φ), painted on an isochronous surface (left): the sedimentation rate (V_Φ) is almost constant everywhere along a layer, except in the upper left corner, and around the collapsed zones (karsts); (b) it varies however in function of the depositional time (or depth) as shown on the vertical cross-section (right), in particular at the center of the basin, and clearly delimits the sedimentary layers.

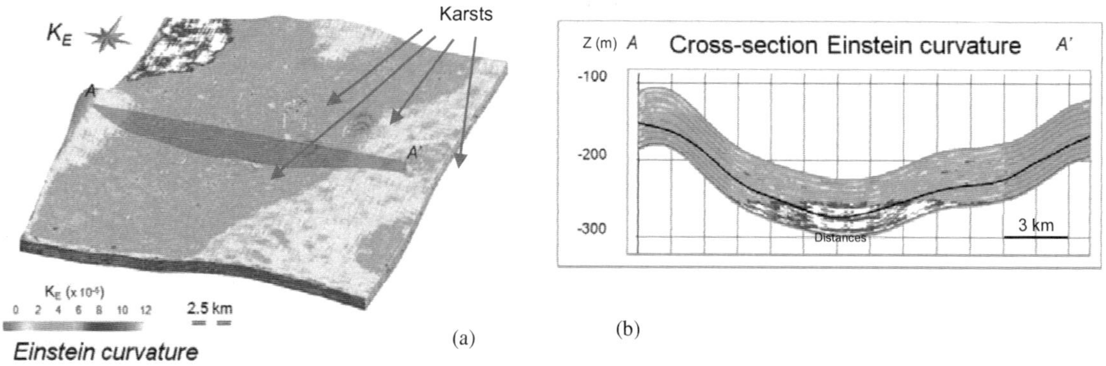

Figure 5. (a) The Einstein curvature emphasizes a NE oriented structure to the East of the studied area, and strong values in the NW upper left corner; (b) Vertical cross-section showing a strong change in curvature values in the middle of the basin.

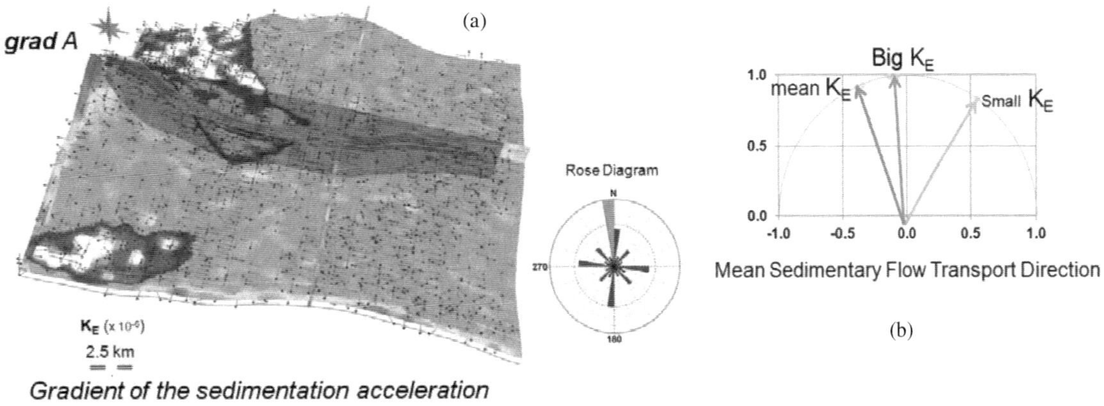

Figure 6. (a) Orientation statistical diagram (rose diagram, right) of the sedimentary acceleration gradient (arrows) (left) points out a main orientation along the NNW directions, with small contributions in EW and NS directions. These directions seem related to the main on-filling sedimentation supply directions during the formation of the basin; (b) Isotropic diffusion coefficient and velocity vector estimated by multi-variable linear regression on three classes of curvatures, strong (purple), medium (green) and small (yellow). Directions of infillings are given on the orientation diagram for each class.

(a total of more than one million (1,000,000) cells) of the 3D model built by [16] (Figure 2). The stratigraphic grid was adjusted between the top and the base of the mineralized zones to calculate sedimentary attributes along the geological layers (Figure 3).

3.3. *Temporal variation of the sedimentation rate $\partial_t \ln V_\phi$*

The variation over time of the sedimentation rate $\partial_t \ln V_\Phi$, a dimensionless relative quantity, measures the relative vertical variations on the sedimentation rate V_ϕ along the isochrones of deposition. It was calculated in 3D throughout the stratigraphic grid (Figure 4a). Represented in vertical section, it is often zero but sometimes varies brutally and emphasizes remarkably the mineralized boundary layers (Figure 4b). Painted on isochronous surfaces, this attribute is almost constant to zero everywhere along a sub-layer showing that the sedimentation rate is almost constant everywhere except in the NW left corner, and around the collapse zones (karsts). In the case of Verkhnekamskoye potash deposit, the derivative against time of the sedimentation rate $\partial_t \ln V_\Phi$ is a sedimentary attribute well suited to identify potential karsts areas and water infiltration, and can help to define risk areas.

3.4. *Einstein's Curvature*

The Einstein's curvature κ_E characterizes the lateral variations of the sedimentation rate and is related to the sedimentary acceleration vector $\boldsymbol{A}$. It was calculated on the entire mineralized zone (Figure 5a, and Figure 6a). Along the layer, the Einstein's curvature structures the mineralized zone in three areas: (i) *strong* values (indicating strong variations in the sedimentation rate) in the NW corner at the same locations where the acceleration was variable; (ii) a *central* area with average values oriented NE with a NW branch at the top left of the map; (iii) a zone with *low* values close to zero on topographic highs and emphasizing a NE direction to the east of the map. The karsts in this area are clearly outlined as zero-curvature contours. Figure 6a represents the curvature of the deepest layers (KPIII) and shows strong variations in the NS corner of the study area. NE directions appear much less clear at depth. In vertical cross-sections (section AA ') (Figure 5a,), the κ_E curvature (i) highlights the limits of the mineralized layers, (ii) shows tenuous fluctuations along the layers showing that the rate of sedimentation is not constant as one can think reading the maps of the acceleration, (iii) identifies areas with high variation of the curvature at the bottom centre of the basin. Thus, the curvature κ_E, like the acceleration, can be used to identify the layers boundaries and areas with karsts risks. Its resolution is better than that of the acceleration to identify weak relative variations.

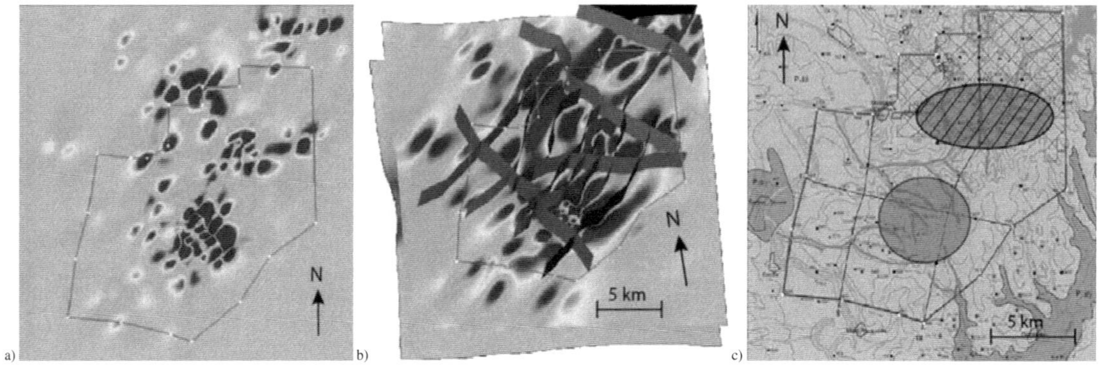

Figure 7. (a) Gaussian curvatures calculated on the topography for identifying faults; (b) Identified faulted structures extended at depth and reported on the mean curvatures map of the salt top layer showing a good fitting; (c) Surface map of the Verkhnekamskoye deposit (red line) with rivers and pools, and potential disaster zones (orange) identified as downward infiltrations of water along faulted zones which dissolve salts at depth.

3.5. Gradient sedimentary acceleration

The sedimentary gradient vector, or sedimentary acceleration, $\boldsymbol{A} = \boldsymbol{grad}\ V_{\Phi}$ was calculated at each point of the stratigraphic grid, and its orientation has been materialized as arrows along a stratigraphic horizon (Figure 6a). At first glance, the directions seem to be randomly distributed along the horizons. However, the directions diagram (rose graph) shows a preferential NNW orientation, with contribution from both conjugated directions EW and NS. These directions seem related to on-filling sediment supply directions.

3.6. Diffusion coefficient and transport velocity vector

The isotropic diffusion coefficients *d* and the transport velocity vector $\boldsymbol{k}_d$ defined by the formulation of Eq. (8) were estimated using multivariable linear regression on the three classes of curvatures, *strong* (purple), *medium* (green) and *small* (yellow) (Figure 6b) at the nodes of the stratigraphic grid over the study domain. The isotropic diffusion values *d* are relatively similar and vary between $0.67_{\pm.06}$ (high values) and $0.54_{\pm.02}$ (small to median values).
The transport velocity vector $\boldsymbol{k}_d$ and its direction change according to the curvature κ_E, and its module is in the same order of magnitude as those of the isotropic diffusion coefficient *d*; its orientation is given in the stereogram of Figure 6b for each of the classes. The direction of $\boldsymbol{k}_d$ is oriented towards NE for low curvature values, along the NNW direction for the mean values, and along NS for high values. The transport velocity vector materializes the direction and flux of the incoming sediment flow during sedimentation.

3.7. Identifying potential faults and disaster zones

In order to identify possible faulted lineaments [15], the Gaussian (κ_G) and mean (κ_H) curvatures maps have been calculated both on the topographic surface and at the top of the salts layers (Figure 7a & b). The lineaments identified on the topography map (Figure 7a) seem to extend at depth as shown by the curvature maps calculated on the top of each salt layers (Figure 7b); thus, they were interpreted as sub-vertical faults that cross-cut the mineralized salt layers. They are mainly oriented along a NE direction (N45), while their conjugates are approximately orthogonal along a NW one (N135). These two directions are visible on the rose diagram of Figure 6a, indicating that these faulted structures were originally active at the early stages during the basin in-filling, and are probably of deep crustal origin. These faults are weak structures that may open potential favorable drains for surface water infiltrations, provoking the dissolution of salts at depth, and therefore causing disaster at the surface by the development of sinkholes.

4. Conclusions

Several significant scientific results were obtained, both on the 3D modelling aspects of the Verkhnekamskoye potash deposit (Russia) and on the sedimentary characteristics of the deposit. The curvature of the horizon was used to define areas at risk of water invasion, and to establish a risk map. Several sedimentary attributes were calculated (sedimentation rate, acceleration, curvature, isotropic diffusion coefficient, sediment transport velocity). These parameters emphasize remarkably well the layers boundaries, the karsts areas, and the sedimentary architecture of formations with the main directions of the sediment income, the position of lineaments and fractures. However, several aspects remain to be explored. In particular, the role of water and hydrogeology in the deposit as a risk factor of liquid intrusion during the mining operation. This could be a research subject for the coming years. This study provides a 3D model of the mineralization of the Verkhnekamskoye region, which to our knowledge has never been done before.

5. Acknowledgements

The authors would like to express theirs thanks to the Lorraine Region for the financial support of Arcus Project and to the Gocad consortium.
The work was partially supported by the Government of Perm Krai, Russia (grant C-26-004-07).

References

[1] Alexandratos N, and Bruinsma J, 2012. World agriculture towards 2030/2050: the 2012 revision. ESA Working paper No. 12-03. Rome, FAO, 154p.
[2] Andrejchuk A. (2002) - Collapse above the World's largest potash mine (Ural, Russia). *Int. J. Speleol.*, **31** (1/4), 137-158
[3] Ashrafiantar N, Hebel H P and Busch W 2011. *Monitoring of mining induced land subsidence - Differential SAR interferometer and persistent interferometer using TerraSAR-XData in comparison with Envisa Data.* 4th TerraSAR-X Science Team Meeting, 14-16 Feb., DLR, Oberpfaffenhofen, Germany, 7p
[4] Camara N., Royer J.J. (2015) - 3D Modeling and Mineral Resources Estimation of the Kofi Birrimian Gold deposit, Zone C, Mali, 13th Biennial SGA meeting, Nancy, France, 1-4.
[5] FAO, 2008 : ftp://ftp.fao.org/docrep/fao/003/Y1997E/FRA%202000%20Main%20report.pdf
[6] Fagherazzi, S. and Overeem, I.(2007) - Models of Deltaic and Inner Continental Shelf Landform Evolution. *Annu. Rev. Earth Planet. Sci.*, **35**, 685-715.
[7] Fuentes L. (2012) – Characterizing and modeling phosphate rocks of the La Culebra area of the La Linda Los Bancos deposit (Venezuela) in order to evaluate phosphate resources. Msc University de Lorraine, ENSG, 94p.
[8] gOcad Research Group - http://www.gocad.org/w4/
[9] Kydryachov A I et al 2006. *Mineral and raw materials inventory of the Perm Region*, Perm, 464p.
[10] Mallet, J. L. (2004) – Space-Time Mathematical Framework for Sedimentary Geology. *Mathematical Geology*, **36**(1), 1-32.
[11] Mallet, J. L. (2014) – *Elements of Mathematical Sedimentary Geology: the GeoChron Model*. EAGE, European Ass, of Geoscientists & Engineers, ISBN 978-90-73834-81-1, 374p.
[12] Martinez, P. A. and Harbaugh, J. W. (1993) - *Simulating Nearshore Environments*. Computer Methods in the Geosciences, Vol. 12, Pergamon Press, Oxford.
[13] Mejia-Herrera P., Kakurina M., Royer J.J. (2015) Predictors for Off-fault Au Assessments from 3D Restoration – Mount Pleasant Area, Western Australia - 13th Biennial SGA meeting, Nancy, France, 1-4.
[14] Mira Geoscience - http://www.mirageoscience.com/
[15] Roberts A. (2001) – Curvature attributes and their application to 3D interpreted horizons. *First Break*, **19**(2), 85-100.
[16] Royer J.J. (2015a) - Développement des Connaissances Scientifiques en Flottation Hautes Performances

utilisées pour Exploiter les Ressources Minérales. Amélioration des Procédés d'Enrichissement du Minerai de la Mine de K-Mg de Verkhmekamskoye (Russie) - "Modélisation 3D et Optimisation de la Flottation". Programme de Coopération Scientifique avec l'Université d'État de Perm, Russie. Rapport interne, ENSG, UL, Nancy, 43p.

[17] Royer J.J. (2015b) - *Some GeoChron Unrevealed Sides: Linking the GeoChron Einstein Curvature with Classical Sedimentary Depositional Models* . 35th gOcad Meeting, Nancy, 1-12.

[18] Tungyshbayeva Z, Royer J.J., and Zhautikov T.M. 2015. *3D Modeling and Resources Estimation of a Gold Deposit, Zhungarie Province, Kazakhstan*. 13th Biennial SGA meeting, Nancy, France, 1-4.

[19] Uralkali 2013. *Sustainable Leadership*, Intregrated Report & Accounts 2013, 178p.

[20] USGS, Jan 2016, http://minerals.usgs.gov/minerals/pubs/commodity/potash/mcs-2015-potas.pdf

[21] USGS, 2009, http://minerals.usgs.gov/minerals/pubs/commodity/phosphate_rock/mcs-2009-phosp.pdf

[22] Wheeler, H. E. (1958) -Time-Stratigraphy. *Bulletin of the American Association of Petroleum Geologists*, **42**(5), 1047-1063.

Coal-oil gold agglomeration assisted flotation to recover gold from refractory ore

A. Otsuki[1], C. Yue[2]
[1] Université de Lorraine, GeoRessources UMR 7359 CNRS, ENSG, 2 rue du Doyen Marcel Roubault, TSA 70605, 54518, Vandoeuvre-lès-Nancy, France
[2] WA School of Mines, Curtin University, Kalgoorlie, Western Australia, Australia

E-mail: akira.otsuki@univ-lorraine.fr

Abstract
This study aimed to investigate the applicability of coal-oil gold agglomeration (CGA) assisted flotation to recover gold from a refractory ore. The ore with the grade of 2-5 g/t was tested with the CGA-flotation process in six different size fractions from 38 to 300 μm using different collector types and dosages. In addition, the flotation without CGA was performed under the same condition for comparison. The results showed that the higher gold grade and recovery were achieved by applying the CGA-flotation, compared with the flotation without CGA. More than 20-60 times grade increase from the head grade was obtained with CGA-flotation. The elemental analysis of gold and sulphur explained their relationship with gold recovery. The results well indicated the applicability of CGA to upgrade the refractory gold ore.

1. Introduction

Gold is always highly demanded in jewelry as well as in high-tech industries and medical applications due to its unique physical and chemical properties. This high demand constantly leads to the development of a new gold recovery process [1] with economically viable and eco-friendly technologies. The process can be determined mainly by the ore mineralogy. In addition, the project scale, location and life of the mine site, site-specific environmental condition, and cost should be considered to optimize the process [2].

Hydro- and pyro-metallurgical processes are major approaches to recover gold, and have been extensively used. The pyro-metallurgical process is cost and energy intensive, and also produces toxic gas (e.g. sulphur dioxide). The hydrometallurgical process is, however, more predictable and easily controlled [1]. Its conventional method applied to refractory gold ores, on the other hand, is cyanidation which poses environmental concerns. Chemicals used in the process would contaminate not only the plants but also animals near the site which could lead devastating effects to local residents [3]. Thus, less cost intensive and more environmentally-friendly pre-concentration process is highly demanded to reduce the dependency on hydro- and pyro-metallurgical processes.

Several gold recovery processes were proven to efficiently beneficiate a gold ore on an economical profitable and environment friendly way [4]. Coal-oil gold agglomeration (CGA) process is one of these alternatives. CGA was developed by British Petroleum (BP) research team in the early eighties [5] and is based on the natural hydrophobicity of coal and gold compared with most gangue materials [6]. It allows a selective recovery of hydrophobic gold particles attaching and penetrating into the pre-formed coal-oil agglomerates from aqueous slurry. The coal-oil agglomerates form in a coal slurry and oil under intensive agitation which enhances the homogeneous distribution of oil in fine spherical droplets and thus their collisions with coal particles. Further, the oil droplets spreading on the coal surfaces form the "oil bridges" which combine the coal particles to agglomerates (Fig.1 a) due to the interfacial tension of the oil and capillary attraction of the oil bridges between particles. Once coal-oil agglomerates form, they are added into gold-bearing ore slurry under intensive agitation; the gold particles thus collide, contact and attach with the coal-oil agglomerates, and eventually penetrate into the agglomerates (Fig.1 b) [4]. Then, the CGA agglomerates can be recovered by flotation or other methods (e.g. screening), dewatering to reject water and coal, and burning remained agglomerates to extract gold are suggested [6].

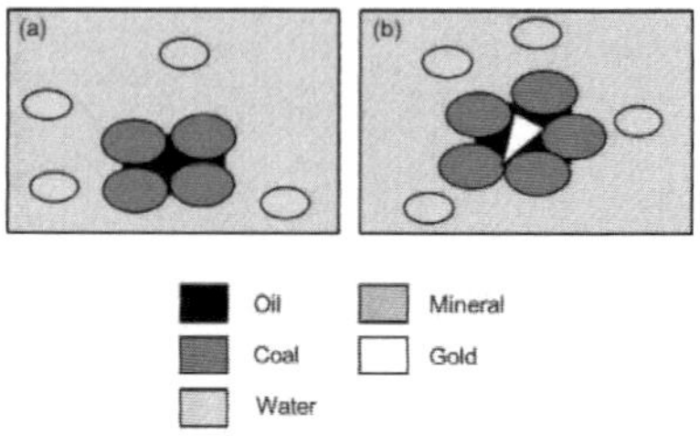

Figure 1. Schematic diagram of coal-oil gold agglomeration steps (after ref. 4). (a) Coal-oil aggregate formation, followed by (b) entrapment of gold/gold bearing particle into aggregate.

One of the significant progresses in relation to the CGA method was the application of froth flotation as the agglomerate recovery method. Moses and Peterson (2000) first implemented this combination and showed its scale-up capability over the other methods [3]. After this investigation, the effectiveness of this CGA-flotation combination was reported; but their investigations were limited to free gold [6-8]. In this investigation, on the other hand, we tested the CGA-flotation, in order to selectively upgrade refractory gold ore.

In most metal processing operations, tailings disposal is one of the biggest concerns in terms of economic and environmental impact. Gold processing plants generate the tailings containing cyanide and cyanide complexes. On the other hand, the tailings produced in CGA process contain only very small amounts of less harmful contaminants (e.g. coal fines, oil, collectors, lime). In addition, the tank storage should be much less than it is required for a conventional cyanidation plant, since the residence time needed for CGA is low [4].

The previous studies were, however, focused on the application of CGA to process artificial and natural ore samples having only free or native gold particles [3, 6-8]. Only available report on CGA application to refractory ore showed the contradictory results [4]. One reported the poor gold recovery (0-10 % for sulphide type flotation concentrate) while the other reported the successful recovery (93.3% for arsenopyrite-pyrite-quartz type ore with the head grade of 6.1 g/t; 95.5% for limonite-pyrite-quartz type ore with the head grade of 2.4 g/t; 79.7 % for pyrite-quartz type ore with the head grade of 1.5 g/t). Therefore, it is worth to make more effort to investigate the feasibility of CGA to upgrade refractory gold ore. Also, there have been only a limited numbers of researches performed to identify the effect of ore particle size and flotation collectors on gold recovery. Wu et al.[9, 10] investigated the effect of collector type on flotation kinetics; but they used a synthetic binary mixture composed of gold particle and quartz sand with collectors of large carbon chain length difference (i.e. C=6 vs. C=10). All the other CGA research investigated the agglomeration parameters (e.g. oil/ore ratio, agglomerate/ore ratio, coal/oil ratio) [e.g. 7, 11] and operational parameters (e.g. stirring rate, aeration rate, impeller speed) [e.g. 3, 8, 12].

Within these contexts, in this study, refractory gold ore was tested with the combination of CGA and froth flotation. Batch tests were performed in different flotation conditions (i.e. collector type and concentration) to evaluate the flotation behavior of the gold ore in different size fractions. This study revealed the relationship between the experimental parameters (e.g. particle size, flotation collector) and gold recovery in order to discuss the applicability of CGA to refractory gold ore processing. For comparison, the performance of two processes, the CGA-flotation and the flotation (without CGA), was also investigated.

2. Materials and methods

Reagents and CGA-flotation procedure stated in this work followed the procedure reported by Sen et al. [6] with necessary modifications.

2.1 Materials

The following reagents were used for flotation: 50 or 100 g/t of potassium amyl xanthate (PAX, $C_6H_{12}OS_2K$) (ORICA Ltd.), and 50 g/t of sodium isobutyl xanthate (SIBX, $C_5H_{10}OS_2Na$) (ORICA Ltd.) and copper sulphate ($CuSO_4 \cdot 5H_2O$) (Chem-Supply Ltd.). The following reagents were used for agglomerate formation: 400 g/t olive oil (100% pure, MORO), 1200 g/t diesel oil (TITAN ULTRALUBE Ltd.), 200 g/t amyl alcohol (AJAX CHEMICALS SYDNEY AUSTRALIA Ltd.), and 200 g/t methyl isobutyl carbinol (MIBC, $C_6H_{14}O$) (ORICA Ltd.). The reagent used for gold analysis was aqua regia prepared with hydrochloric acid (32% by weight, Chem-Supply Ltd.) and nitric acid (92% by weight, Chem-Supply Ltd.) at the ratio of 2:1 by volume.

Refractory gold ore from a Western Australian mine was used as a sample. XRD analysis identified the following major minerals: quartz, sheet silicates (e.g. muscovite, kaolinite), dolomite, and pyrite. The ore was ground by a laboratory rod mill (ESSA, 520490) and sieved to obtain six different size groups (from 38 to 300 μm). Table 1 shows the gold and sulphur contents in each size fraction analysed by Inductively Coupled Plasma Optical Emission Spectrometry (ICP-OES, VARIAN) and sulphur analyser (LECO), respectively. The products of CGA-flotation and flotation (without CGA) were also analyzed by the same methods. The results show the average values of 2 analyses of each size fraction/product. Mass of each size group tested in CGA-flotation and flotation was approximately 250 g for each flotation batch test after splitting for a representative sample. At least 2 flotation tests at each condition were performed separately and their average results were reported in this paper. Oxidized coal from Bowen Basin, Queensland, Australia was ground by laboratory rod mill and the -150 μm fraction was used for agglomerates. The results of duplicated ash analysis showed that the ash content of the coal used in this investigation was 16.7%.

Table 1 Gold and sulphur content of ore sample in different size fractions.

Size (μm)	**Gold (g/t)**	**Sulphur (%)**
38-53	4.75	0.64
53-75	3.21	0.53
75-106	3.58	0.46
106-150	3.04	0.30
150-212	3.67	0.25
212-300	1.54	0.21

2.2 Methods & Procedures

2.2.1 Coal-oil gold agglomeration (CGA)-flotation

Preparation of oil emulsions: olive oil, diesel oil, amyl alcohol and MIBC with the ratio of 2:6:1:1 by volume were added into a two-baffled emulsifying cell (Sunbeam) and agitated at 800 rpm for 5 min to produce an oil emulsion. The dosage of agglomeration reagents used for each batch test is mentioned in the section 2.1, following the values reported by Sen et al. [6].

Pre-agglomerate formation and coal-oil-gold agglomeration flotation: pre-agglomerate that coal particles agglomerated with oil emulsion was then formed before adding the agglomerate into the gold ore pulp. A mixture of 15 g of coal particles (coal/oil ratio was kept as 30:1) and oil-water emulsion were added to the emulsifying cell and agitated at 800 rpm for 5 min. In a 3.2 L flotation cell (Agitair model LA-500 R, Baker process technology), flotation collector (i.e. PAX or SIBX) was added to gold ore pulp, which contains 250 g of one size range sample produced in the manner stated in the section 2.1, and agitated at 850 rpm for 10 min. Pre-agglomerated coal particles-oil-water suspension was then added to the pulp and agitated at 1150-1250 rpm for another 5 min. Stirring speed was decreased to 850 rpm and pulp was conditioned for 5 min to form agglomerates with gold bearing minerals. Finally the produced coal-oil-gold agglomerates were recovered by flotation. The air valve was turned on and the froth was collected as a concentrate for 2 min. The CGA-flotation process was performed with six different size fractions of refractory ore to reveal the relationship between the particle size and gold recovery. To test the effect of collectors on the performance of CGA-flotation process, different types and dosages of collectors were used stated in the section 2. The natural pH of ore pulp was 8.5 and not modified during the flotation. Each series of tests were performed twice and their average values were reported in this paper.

2.2.2 Refractory gold ore flotation without CGA

To compare the CGA-flotation with the flotation (without CGA), a series of batch flotation tests with gold ore in six different size fractions was also conducted using the same flotation reagents to recover gold. Gold ore sample prepared by following the procedure in the section 2.1 Materials was added to the laboratory flotation cell with tap water to produce a gold ore pulp. The pulp was agitated at 850 rpm with the air valve off. Then 50 g/t of copper sulphate was added into the pulp and conditioning for 5 min. 50 g/t of PAX and 200 g/t of MIBC frother were then added into the pulp for conditioning another 5 min. The pulp pH was maintained at natural pH 8.5. Finally the concentrate was collected for 2 min after the air valve was turned on.

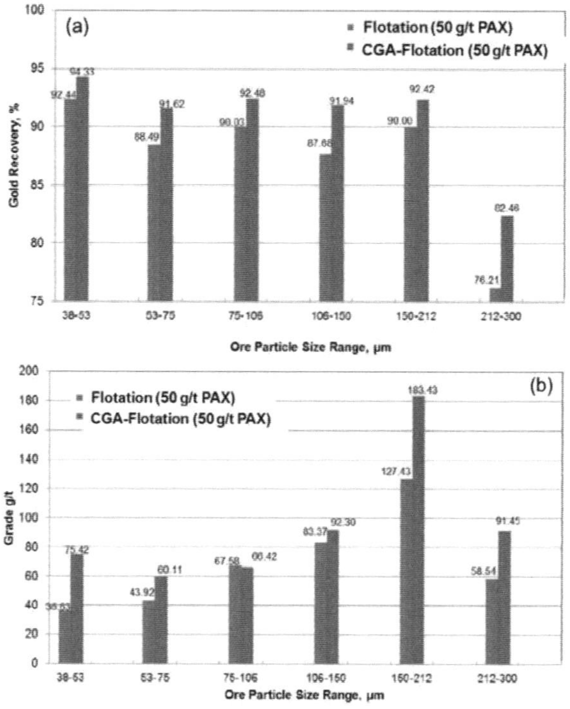

Figure 2. (a) Gold recovery and (b) gold grade in the concentrate of flotation and CGA-flotation.

3. Results and discussion

3.1 Comparison between CGA process and the conventional flotation process

The performance of two processes, CGA-flotation and the flotation without CGA, on refractory gold ore, was compared. These batch tests were conducted under comparable conditions except CGA process. The pre-formed coal-oil agglomerates were added into the gold bearing pulp before flotation and flotation was used to separate the coal-oil-gold agglomerates from the pulp. Figure 2-a shows that the gold recoveries in all six different size fractions were higher with CGA-flotation, compared with the flotation (without CGA). The gold recovery increased in the range between 2 and 6 % by using CGA-flotation. The largest difference was observed in the coarsest fraction (i.e. 212-300 μm) while higher than 90% recovery was achieved in all the other fractions (i.e. 38-212 μm) with CGA-fotation. This good recovery in the wide size range agreed with the point Moses and Peterson made, i.e. gold grains in 1-500 μm can be recovered by CGA in equal ease [3]. In the case of CGA-flotation applied in this study, the wide size range of good recovery can be explained by the following two reasons: (1) for fine grain gold, it can form agglomerate to become suitable size (i.e. 20-200 μm) to be captured by air bubbles; (2) for coarse grain gold, it can form agglomerate until the upper limit of flotation with high hydrophobicity (i.e. 500 μm for coal particles [13]). Slight increment in recovery could be due to the less selectivity of coal-oil agglomerates to gold-bearing sulphide minerals that have less gold exposure in the ore than to free gold particles. The gold recovery can increase more significantly if CGA process is further optimized for processing refractory gold ore.

Figure 2-b indicates that the gold grades of concentrates in all six different size fractions are generally higher with CGA-flotation compared to the ones by using the flotation without CGA. As the recovery increased with finer size fractions, the gold grades in those fractions were relatively low compared to that of coarse size fractions. However, CGA-flotation obtained twice much grade compared with the flotation in the fine fraction of 38 to 53 μm. In the coarsest size fraction (212-300 μm), the gold grade was concentrated 60 times more than head grade of 1.5 g/t.

As a summary, the higher gold recovery and grade were achieved by applying CGA process prior to froth flotation compared to the flotation without agglomeration. The grade-recovery cures in Fig. 3 clearly captured the enhancement of gold recovery by CGA. The results proved that CGA process is applicable to upgrade a refractory gold ore. The further discussion on gold grade and recovery with two different methods will be provided with the results shown in Table 2 indicating the correlation between the recoveries of gold and sulphur that is the major element associated with gold in refractory gold ores.

Table 2 shows the correlation between the gold and sulphur recoveries in flotation concentrate in different size fractions by (a) flotation and (b) CGA-flotation. The general observed trend was that the increase in sulphur recovery strongly correlates with gold recovery increase, and the recovery values of gold and sulphur is similar in fine size fractions (i.e. 38-106 μm). For example, in the 38-53 μm fraction with CGA-flotation, the gold recovery was 94.3% while the sulphur recovery was >92.2%. It means that in fine size fractions most of the gold particle was associated with sulphide minerals (e.g. pyrite). This agreed with the previous reports on gold associations [e.g. 14]. Also, generally higher sulphur recovery by CGA-flotation than flotation led that higher gold recovery by CGA-flotation. The similarity between sulphur and gold recoveries decreases with increasing the particle size, i.e. higher similarity in fine size fraction while less similarity in coarse size fraction. In the coarsest fraction (i.e. 212-300 μm) showed the largest difference between the sulphur and gold recoveries (e.g. 57.1% S vs. 82.5% Au in CGA-flotation). It indicates that more recovered gold was associated with sulphide minerals in fine size fractions while less recovered gold was associated with sulphide minerals and more recovered gold was associated with other host minerals (e.g. quartz) in coarse size fractions.

3.2 Effect of collector type on the CGA-flotation

The effect of collector type on the performance of the CGA-flotation was investigated. Two sets of batch flotation tests were carried out under same conditions except for the collector types, i.e. 50 g/t of PAX (C=6) or 50 g/t of SIBX (C=5). SIBX was selected as it is one of the most common collectors used in refractory gold ore flotation [15]. The comparison indicated that the gold recoveries between two series had little difference (Fig. 4-b) with different collectors which did not significantly enhance the physic-chemical properties of mineral particles by changing their wettability. In relatively coarse particle size fraction (+150-300 μm), the gold recoveries were slightly higher with SIBX. This can be due to the higher selectivity of a shorter chain length collector, SIBX in order to recover the gold bearing mineral particles with air bubbles. On the other hand, in general, the different collectors from PAX and SIBX had little effect on the CGA-flotation performance. This agreed with the froth flotation results of refractory gold ore reported by Bulatovic [16] that there was no significant change in the gold recovery with different collectors (i.e. 82% with 50 g/t of sodium ethyl xanthate ($C_3H_5NaOS_2$; C=3) while 88% with 50 g/t of PAX (C=6).

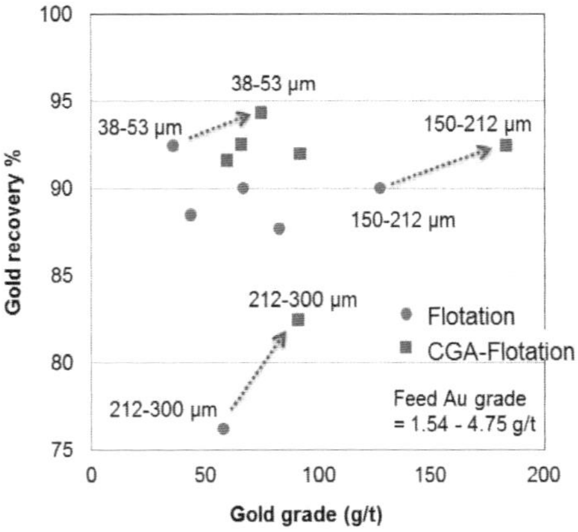

Figure 3. Grade-recovery curves of two different gold recovery processes.

Table 2 Gold and sulphur recoveries in flotation concentrate in different size fractions with (a) conventional flotation and (b) CGA-flotation. Note that the recoveries not having the exact value (e.g. >92.2%) are due to the small amount of sulphur content available in flotation tail below the detection limit of sulphur analysis for determination of sulphur recovery.

Size (μm)	Flotation		CGA-flotation	
	Gold recovery (%)	Sulphur recovery (%)	Gold recovery (%)	Sulphur recovery (%)
38-53	92.4	89.1	94.3	>92.2
53-75	88.5	86.8	91.6	>90.6
75-106	90.0	87.0	92.5	>89.1
106-150	87.7	80.0	91.9	>83.3
150-212	90.0	>80.0	92.4	>80.0
212-300	76.2	66.7	82.5	57.1

Wu et al. [9] investigated the gold adhesion to agglomerates, by performing CGA-flotation with two different xanthate collectors of different chain lengths and synthetic ore (i.e. gold + quartz). After 5 min, 50% of the gold particles were recovered using decanthioxanthate (DX; C=10) collector. With shorter chain-length amyl xanthate collector (PAX; C=6), the gold flotation rate was slower and achieved only 30% gold recovery after 5 min [9]. In froth flotation general, difference in the length of the non-polar part of the molecule results in the different collecting ability of target minerals due to their difference in wettability. Wu et al. [9] proposed that, in CGA-flotation system, a collector with a longer chain-length was able to immerse into the oil film on the agglomerate surface more in depth and rigid. Additionally the gold particle with longer chain-length collector molecules would have a higher hydrophobicity, leading to a greater hydrophobic interaction force between the gold particles and the agglomerates.

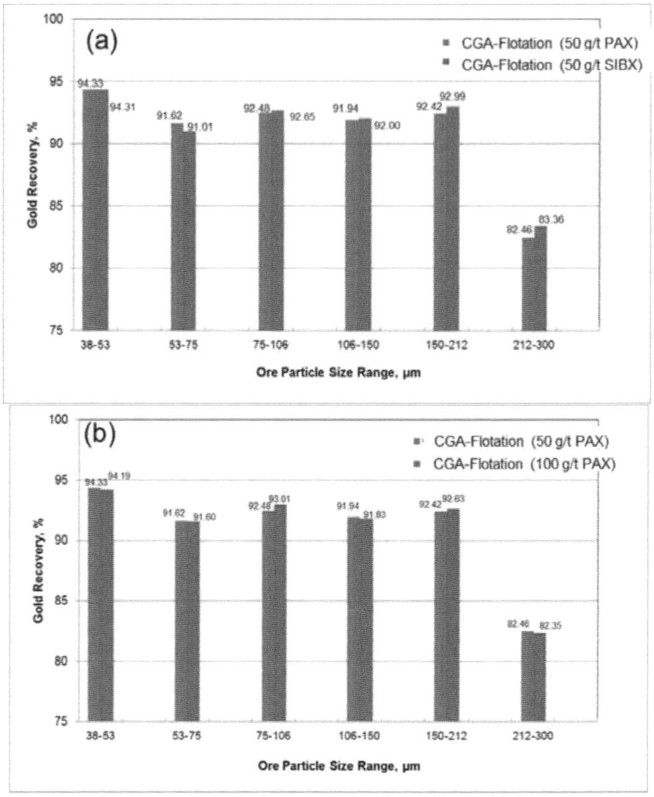

Figure 4. Gold recoveries in different size fractions with different (a) collectors (PAX, SIBX) and (b) collector dosages (50, 100 g/t of PAX).

Their study was performed with a synthetic binary ore composed of gold particles and quartz sand. It means that their gold particles were fully liberated from quartz and no other gangue minerals co-existed. In the case of refractory gold ore composed of many different minerals and limited degree of gold particle liberation due to their fine dissemination (e.g. solid solution), the use of longer chain collector may distract the selective collector attachment passively leading the low gold recovery. Also, the price of the collector can be the strong decision indicator. As a matter of fact, PAX is the most common collector being used in commercial refractory gold ore processing plant [15].

3.3 Effect of collector dosage on CGA-flotation

The effect of collector dosage on gold recovery with CGA-flotation process was investigated. Two sets of batch tests were conducted under similar conditions except for the collector dosage used in the flotation, i.e. 50 and 100 g/t of PAX. Figure 4-b shows that doubling the collector had almost no effect on the gold recovery in all six different size fractions. This may result from the high hydrophobicity of coal-oil-gold agglomerates even with 50 g/t. Since coal-oil-gold agglomerates were readily floatable after the agglomeration, higher collector addition in the flotation cell did not change the hydrophobicity of gold bearing particles; thus, no significant increase in the gold recovery was achieved. In flotation of refractory ore, adding more collectors would decrease the gold recovery because some sulphide minerals without gold particle were hydrophobized and floated as a concentrate. This reverse effect was not observed (Fig.4-b). The results suggest that changes in flotation conditions (i.e. collector type and concentration) did not significantly affect the flotation performance in this investigation.

4. Conclusions

The present study aimed to investigate the applicability of CGA to a refractory gold ore. Refractory gold ore was beneficiated with the combination of CGA and froth flotation in comparison with the flotation (without CGA) to improve the gold recovery under the same condition. The higher gold grade and recovery were achieved with the CGA-flotation tests than conventional flotation. Higher than 20-60 times grade increase from the head grade was obtained by CGA in coarse fractions. The differences in gold recovery by two methods were explained by examining the relationship between gold and sulphur recoveries in different size fractions. In addition, the effect of type and dose of collectors were investigated. It was found that there was no difference under different (a) collector type (i.e. PAX vs. SIBX) and (b) dosage (50 g/t vs. 100 g/t), explained by (a) fine dissemination of gold particle and (b) high enough hydrophobicity given by even small collector dosage. The results indicated the applicability of CGA to upgrade a refractory gold ore. Application of CGA to the cleaning stage of flotation circuit would be recommended due to its high grade product after CGA.

References

[1] Syed, S. Recovery of gold from secondary sources—A review. Hydrometallurgy **115-116,** 30-51 (2012).
[2] Lunt, D., Weeks, T.: Chapter 3 Process flowsheet selection. In: Adams, M.D. (Ed.) Developments in Mineral Processing 15**,** 73-96 (2005).
[3] Moses, L.B., Petersen, F.W. Minerals Engineering **13**, 255-264 (2000).
[4] Akcil, A., Wu, X. Q., Aksay, E.K.: Coal-Gold Agglomeration: An alternative separation process in gold recovery. Separation & Purification Reviews 38**,** 173-201 (2009).
[5] BP Australia Limited. Recovery of metal values from mineral ores by incorporation in coal-oil agglomerates. Australia Int. Patent Pct/au 84/00071, 1984.
[6] Sen, S., Seyrankaya, A., Cilingier, Y. Minerals Engineering **18,** 1086-1092 (2005).
[7] Sen, S., Ipekoglu, U., Cilingir, Y. Separation Science and Technology **45**, 610-618 (2010).
[8] Sen, S., Akar, G., Cilingir, Y., Malayoglu, U., Tanriverdi, M., Ipekoglu, U. Separation Science and Technology **48**, 1585-1890 (2013).
[9] Wu, X.Q., Gochin, R.J., Monhemius, A.J. Minerals Engineering. **17**, 33-38 (2004).
[10] Wu, X.Q., Gochin, R.J., Monhemius, A.J. Int. J. Miner. Process. **74**, 327-336 (2004).
[11] Calvez, J.P.S., Kim, M.J., Wong, P.L.M., Tran, T. Minerals Engineering **11**, 803-812 (1998).
[12] Kotze, W., Peterson, F.W. The South African Institute of Mining and Metallurgy **100**, 57-62 (2000).
[13] Parekh, B.K.: III / Coal: flotation. In: Poole, C.F., Cooke, M., Wilson, I.D. (eds.) Encyclopedia of Separation Science, pp.2490-2496, Academic Press, London (2000).
[14] Vaughan, J.P., Kyin, A. Mineralogical Magazine **68**, 255-277 (2004).
[15] O'Connor, C.T., Dunne, R.C. Minerals Engineering **4**, 1057-1069 (1991).
[16] Bulatovic, S.M. Minerals Engineering **10**, 895-908 (1997).

Surfactant effect on interaction of rising bubble and particle in a liquid subjected to vibrations

D.V. Lyubimov[1], L.S. Klimenko[2], T.P. Lyubimova[1,2], L.O. Filippov[3]

[1]Perm State University, Perm, Russia
[2]Institute of Continuous Media Mechanics, Perm, Russia
[3]Universite de Lorraine, Nancy, France

Email : lubimova@psu.ru

Abstract. The paper investigates the surfactant effect on the interaction of solid particle and gas bubble in a liquid subjected to vibrations. Surfactant transport between the bubble surface and the surrounding liquid is limited by the adsorption-desorption process. The particle is subjected to the Stokes, Basset and buoyancy forces, and average force related to the inhomogeneity of the pulsational field. The problem is solved numerically. It is found that in the presence of surfactant, the impact of vibrations on the particle-bubble collision is weaker than in the absence of surfactant. Thus, higher vibrations intensity is needed for increasing the collision efficiency, together with a higher consumption in energy.

1. Introduction

The effect of surfactant on particle-bubble interactions is the subject of interest in many studies (see, for example, [1], [2]). As well as direct measurements of the particle-bubble interaction [3], there are several theoretical investigations together with their comparison with experimental data [4]. The majority of these studies are focused on particle-bubble attachment, although for fine particles the most important sub-process is collision. The interest to this problem is justified by its applications to many technological and physical processes, especially flotation.

In flotation theory, the bubble–particle interaction can hardly be modeled without taking into account for the hydrodynamic approach. However, existing theories of the flotation process are either of empirical nature [5], [6] or do not account for the action of the external fields [7], [8].

A theoretical model for bubble-particle collision in a liquid under external vibrations was developed in [9]. The ambient liquid was assumed to be incompressible and submitted to gravitational forces. Analysis of the different components of the pulsational field scattered on the bubble allows the estimation of their contributions to the radiation force acting on the particle. Trajectories of the particle motion near a pulsating bubble were calculated. The evaluation of the efficiency of the flotation process exposed to an ultrasound action is then discussed.

In this study, the previously developed model is extended by taking into account the presence of surfactants. Analyzing the different components of the pulsational field scattered on the bubble surface in presence of surfactants, it is possible to calculate the average vibrational force acting on the particle. These results are compared with the cases where the surfactants are absent.

2. Problem statement

We study the behavior of a single solid particle near a gas bubble (Fig. 1). The system is subjected to the gravity field and to a vibrational field. The vibrations are considered as harmonic with amplitude b and frequency ω. We assume that the dissipative processes caused by the viscosity can be neglected as the investigated frequency is high. Thus, the viscosity of the fluid around the bubble and the particles is null (case of inviscid flow). Moreover the amplitude of the vibrations b is considered to be small compared to the bubble radius, this allows to linearize the governing equations. Additionally, we assume that the applied vibrations are such that the liquid can be considered as incompressible. To increase the collision efficiency between the bubble and the particle surface, a soluble surfactant is added. The transport of surfactant between the bubble surface and the surrounding liquid is limited by an adsorption-desorption process.

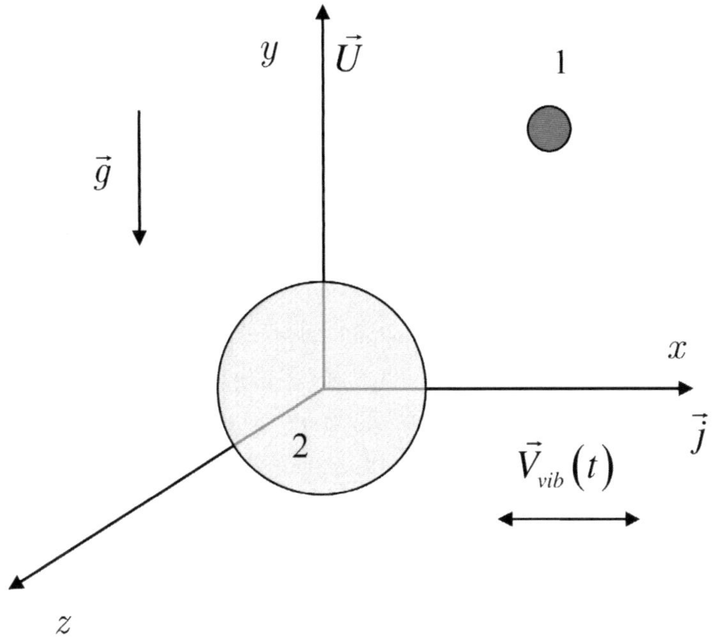

Figure 1 Problem configuration, 1 – solid particle, 2 – gas bubble

For studying the behavior of the system, it is convenient to decompose the motion into a fast (pulsational) and slow (average) components. Let us begin with the bubble pulsational motion under vibrations.

3. Pulsational velocity of ambient liquid

We assume that the particle radius is much smaller than the bubble radius. It allows ignoring the effect of the particles on the pulsational flow scattered on the bubble. Under the effect of the vibration of frequency ω, the total liquid flow around the bubble is composed of a *vibration field* in the absence of the bubble and a *pulsational field* scattered by the bubble and associated with its radial (or *monopole* [10]) and translational motion

$$\boldsymbol{V} = \boldsymbol{v}_{\mathrm{p}} + \boldsymbol{v}_{\mathrm{m}} + \boldsymbol{v}_{\mathrm{tr}}, \quad (1)$$

where the uniform pulsating field of amplitude V_{vib} can be written as

$$\boldsymbol{v}_{\mathrm{p}} = \mathrm{Re}(\boldsymbol{V}_{\mathrm{vib}} \mathrm{e}^{i\omega t}) \quad (2)$$

Monopole bubble motion. We define the liquid flow v_{m} associated with the monopole (it does not depend on the angle), the bubble motion induced by the external pressure part is independent of the coordinates. The pressure at a long distance from the bubble can be represented in the form

$$p = \mathrm{Re}\left(q\mathrm{e}^{\mathrm{i}\omega t}\right) \tag{3}$$

with q is the pressure amplitude, Re is the real part, t is the time, and i is the imaginary unit. As liquid is assumed incompressible, only the radial component must be account for

$$v_{\mathrm{m}} = \mathrm{Re}\left(\frac{c_1}{r^2}\mathrm{e}^{\mathrm{i}\omega t}\right) \tag{4}$$

where c_1 is an unknown constant, and r is the spherical coordinate. Then, according to the linearized Euler equation ($\rho\partial v_m/\partial t = -\partial p/\partial r$), the pressure in the surrounding liquid is given by

$$p = \mathrm{Re}\left\{\left(i\omega\rho\frac{c_1}{r} + q\right)\mathrm{e}^{\mathrm{i}\omega t}\right\} \tag{5}$$

where ρ is the fluid density. On the bubble surface a kinematic condition and a normal stresses balance condition in the presence of surfactants are imposed:

$$\frac{\partial\zeta}{\partial t} = v_{\mathrm{m}},$$
$$p - p' - \frac{2}{R_{\mathrm{b}}}\gamma'\Gamma = 2\frac{\gamma}{R_{\mathrm{b}}}\zeta, \tag{6}$$

where $p' = -3\rho_{\mathrm{g}}c^2\zeta/R_{\mathrm{b}}$ is the gas pressure inside the bubble where ρ_g is the gas density and c is the speed of sound, R_{b} is the bubble radius in equilibrium, γ is the surface tension coefficient and its derivative $\gamma' = \partial\gamma/\partial\Gamma$, where Γ is the surface concentration in surfactant, and ζ is the bubble surface deviation (i.e. the position of the bubble surface at time t determined from it position when the bubble is at equilibrium).

We assume that the bubble remains spherical and that the thickness of the film formed by a surfactant is sufficiently small; these assumptions allow writing the equation of mass conservation for the surfactant as [11], [12]

$$\frac{\mathrm{d}\Gamma}{\mathrm{d}t} + 2\Gamma_0\frac{v_{\mathrm{m}}}{R_{\mathrm{b}}} = 0 \tag{7}$$

where Γ_0 is the surface concentration of the surfactant at equilibrium, and Γ at time t.

The surface deviation can be obtained from Eq. (6) by substitution Eq. (4) and Eq. (5) and taking into account Eq. (7). Indeed, from Eq. (7) $\omega\Gamma = 2i\Gamma_0 c_1 R_b^{-3}$, thus the boundary conditions Eq. (6) reads

$$i\omega\zeta = c_1 R_b^{-2},$$
$$\frac{i\omega\rho}{R_b}c_1 + q + 3\zeta\frac{\rho_{\mathrm{g}}c^2}{R_b} - \frac{4i\gamma'\Gamma}{\omega R_b^4}c_1 = 2\frac{\gamma}{R_{\mathrm{b}}}\zeta,$$

which gives the amplitude of surface deviation as

$$\zeta = \frac{q}{\omega^2\rho R_{\mathrm{b}} - c^2\frac{3\rho_{\mathrm{g}}}{R_{\mathrm{b}}} + \frac{2\gamma}{R_{\mathrm{b}}^2} - \frac{4\Gamma_0}{R_{\mathrm{b}}^2}\gamma'} \tag{8}$$

Taking into account that the natural frequency of the bubble $\omega_0^2 = \left(3\rho_{\mathrm{g}}c^2R_{\mathrm{b}} - 2\gamma\right)/\rho R_{\mathrm{b}}^3$ and $\gamma' < 0$, the surface deviation can be rewritten in the following form

$$\zeta = \frac{qR_{\mathrm{b}}^{2}}{\rho R_{\mathrm{b}}^{3}\left(\omega^{2}-\omega_{0}^{2}\right)+4\Gamma_{0}\left|\gamma'\right|} \tag{9}$$

Thus, the velocity and pressure in the surrounding liquid caused by a monopole bubble motion is

$$\mathrm{v}_{\mathrm{m}} = \mathrm{Re}\left(\frac{\mathrm{i}\omega\zeta R_{\mathrm{b}}^{2}}{r^{2}}\mathrm{e}^{\mathrm{i}\omega t}\right) = \mathrm{Re}\left(\frac{\mathrm{i}\omega qR_{\mathrm{b}}^{4}}{\rho R_{\mathrm{b}}^{3}\left(\omega^{2}-\omega_{0}^{2}\right)+4\Gamma_{0}\left|\gamma'\right|}\frac{1}{r^{2}}\mathrm{e}^{\mathrm{i}\omega t}\right) \tag{10}$$

$$p = \mathrm{Re}\left(\frac{-\omega^{2}qR_{\mathrm{b}}^{4}\rho}{\rho R_{\mathrm{b}}^{3}\left(\omega^{2}-\omega_{0}^{2}\right)+4\Gamma_{0}\left|\gamma'\right|}\frac{1}{r}\mathrm{e}^{\mathrm{i}\omega t}+q\mathrm{e}^{\mathrm{i}\omega t}\right) \tag{11}$$

The velocity and pressure depend on the unknown liquid pressure q far from the bubble, the far-field pressure is found from the linearized Euler equation in the form

$$q\mathrm{e}^{i\omega t} = -\mathrm{i}\omega\rho\left(\boldsymbol{V}_{\mathrm{vib}}\cdot\boldsymbol{r}_{\mathrm{b}}\right)$$

where r_b is the radius-vector of the bubble center which is counted from the pressure nodes, and $\boldsymbol{V}_{vib}$ the fluid velocity due to the far field oscillating pressure. The coordinates of the nodes are determined by the type of generation of oscillations in the liquid.

Finally, the velocity of the surrounding liquid due to the monopole bubble motion is obtained as

$$\boldsymbol{v}_{\mathrm{m}} = \mathrm{Re}\left(\frac{\omega^{2}\rho R_{\mathrm{b}}^{4}}{\rho R_{\mathrm{b}}^{3}\left(\omega^{2}-\omega_{0}^{2}\right)+4\Gamma_{0}\left|\gamma'\right|}\left(\boldsymbol{V}_{\mathrm{vib}}\cdot\boldsymbol{r}_{\mathrm{b}}\right)\frac{\boldsymbol{r}}{r^{3}}\mathrm{e}^{\mathrm{i}\omega t}\right) \tag{12}$$

Translational bubble motion. Let us proceed to the determination of the liquid flow induced by the translational oscillations of the bubble. The Euler and the continuity equations for incompressible liquid in a reference frame moving with the bubble center read

$$\frac{\partial \boldsymbol{v}_{\mathrm{tr}}}{\partial t} = -\frac{\nabla p}{\rho}-\frac{\mathrm{d}\boldsymbol{U}_{\mathrm{b}}}{\mathrm{d}t}, \tag{13}$$

$$\mathrm{div}\,\boldsymbol{v} = 0, \tag{14}$$

where $\boldsymbol{U}_{\mathrm{b}}(t)$ is the velocity of the bubble center relative to the laboratory reference frame.

Far from the bubble the velocity is given by $\boldsymbol{v} = b\omega\cos(\omega t)\boldsymbol{j}-\boldsymbol{U}_{\mathrm{b}}$, where $\boldsymbol{j}$ is a unit vector in the vibration direction.

The kinematic condition and the condition of stress balance at the bubble surface under the assumption of negligible thickness of the surfactant film, are (see, for example[11], [12]):

$$\frac{\partial\zeta}{\partial t} = \boldsymbol{v}_{\mathrm{tr}}\cdot\frac{\boldsymbol{r}}{r},$$
$$p-\frac{2}{R_{\mathrm{b}}}\gamma'\Gamma = \frac{\gamma}{R_{\mathrm{b}}^{2}}\left(\nabla_{\vartheta}^{2}+2\right)\zeta, \tag{15}$$

where $\nabla_{\vartheta}^{2} = \frac{1}{\sin\vartheta}\frac{\partial}{\partial\vartheta}\left(\sin\vartheta\frac{\partial}{\partial\vartheta}\right)$ (we assume that the velocity field has axial symmetry), and ϑ is the angle counted from the vibration direction. To have the problem closed, the mass conservation equation for the surfactant is needed

$$\frac{\mathrm{d}\Gamma}{\mathrm{d}t}+\Gamma_{0}\frac{1}{R_{\mathrm{b}}}\left(\frac{\partial v_{\vartheta}}{\partial\vartheta}+v_{\vartheta}\cot\vartheta\right)+2\Gamma_{0}\frac{v_{r}}{R_{\mathrm{b}}} = 0 \tag{16}$$

It is convenient to express the velocity in terms of flow potential φ so that the continuity equation has the form of a Laplace's equation whose solution can be written as [13]

$$\varphi = \mathrm{Re}\left\{\left(b\omega r+\frac{c_{2}}{r^{2}}\right)\cos\vartheta\,\mathrm{e}^{\mathrm{i}\omega t}-\boldsymbol{U}_{\mathrm{b}}\cdot\boldsymbol{r}\right\}, \tag{17}$$

$$p=\mathrm{Re}\left\{-\mathrm{i}\rho\omega\left(b\omega r+\frac{c_2}{r^2}\right)\cos\vartheta \mathrm{e}^{\mathrm{i}\omega t}\right\}.$$

where c_2 is an unknown constant, and $\boldsymbol{U}_b$ is the velocity of the bubble center.

From the kinematic condition, the surface deviation can be thus written as $\zeta=\mathrm{Re}\{c_3\cos\vartheta \mathrm{e}^{\mathrm{i}\omega t}\}$. However as the bubble is immobile in the reference frame attached to the bubble center, the translational oscillations of the bubble as a whole is null so $c_3=0$. Thus, the bubble surface deviation vanishes. It follows that the ambient liquid velocity due to the bubble translation motion takes the form

$$\boldsymbol{v}_{\mathrm{tr}}=\frac{\rho b\omega^3 R_{\mathrm{b}}^3}{\rho R_{\mathrm{b}}^3\omega^2+12\Gamma_0|\gamma'|}\mathrm{Re}\left\{\left(-2+2\frac{R_{\mathrm{b}}{}^3}{r^3}\right)\cos\vartheta \mathrm{e}^{\mathrm{i}\omega t}\boldsymbol{e}_r+\left(2+\frac{R_{\mathrm{b}}{}^3}{r^3}\right)\sin\vartheta \mathrm{e}^{\mathrm{i}\omega t}\boldsymbol{e}_\vartheta\right\}. \quad (18)$$

Notice that the velocities and coordinates rules for converting these quantity from the local reference frame attached to the bubble center into the fix laboratory reference frame are:

$$\boldsymbol{V}'=\boldsymbol{V}+\boldsymbol{U}_{\mathrm{b}},\ \boldsymbol{r}'=\boldsymbol{r}+\boldsymbol{R}, \quad (19)$$

where the velocity $\boldsymbol{U}_b$ of the bubble center reads

$$\boldsymbol{U}_{\mathrm{b}}=3b\omega\frac{\rho R_{\mathrm{b}}^3\omega^2-4\Gamma_0|\gamma'|}{\rho R_{\mathrm{b}}^3\omega^2+12\Gamma_0|\gamma'|}\boldsymbol{j}\cos\omega t, \quad (20)$$

The coordinate are found by integrating Eq.(20) over time t (assuming $|\gamma'|$ and R_b constant)

$$\boldsymbol{R}=3b\frac{\rho R_{\mathrm{b}}^3\omega^2-4\Gamma_0|\gamma'|}{\rho R_{\mathrm{b}}^3\omega^2+12\Gamma_0|\gamma'|}\boldsymbol{j}\sin\omega t, \quad (21)$$

Thus, the velocity in the laboratory reference frame can be obtained from the converting rules Eq.(19)

$$\boldsymbol{V}'(\boldsymbol{r}')=\boldsymbol{V}(\boldsymbol{r}'-\boldsymbol{R})+U_{\mathrm{b}}\boldsymbol{j}=\boldsymbol{V}(\boldsymbol{r}')+U_{\mathrm{b}}\boldsymbol{j}-\nabla\boldsymbol{V}(\boldsymbol{r}')\cdot\boldsymbol{j}R \quad (22)$$

Neglecting the nonlinear terms ($\nabla\boldsymbol{V}(\boldsymbol{r}')\cdot\boldsymbol{j}R$ which is small under the small amplitude of vibrations assumptions, and omitting the primes in the liquid translation velocity in the laboratory reference frame, it comes:

$$\boldsymbol{V}_{tr}=\frac{\rho b\omega^3 R_{\mathrm{b}}^3}{\rho R_{\mathrm{b}}^3\omega^2+12\Gamma_0|\gamma'|}\left(3\frac{R_{\mathrm{b}}{}^3}{r^5}(\boldsymbol{j}\cdot\boldsymbol{r})\boldsymbol{r}+\left(1-\frac{R_{\mathrm{b}}{}^3}{r^3}\right)\boldsymbol{j}\right)\cos\omega t. \quad (23)$$

Finally, the liquid pulsational velocity around the gas bubble under external vibrations has the following form

$$\boldsymbol{V}=\frac{\rho\omega^2 R_{\mathrm{b}}^4}{\rho R_{\mathrm{b}}^3\left(\omega^2-\omega_0^2\right)+4\Gamma_0|\gamma'|}(\boldsymbol{V}_{\mathrm{vib}}\cdot\boldsymbol{r}_{\mathrm{b}})\frac{\boldsymbol{r}}{r^3}+\frac{\rho\omega^2 R_{\mathrm{b}}^3}{\rho R_{\mathrm{b}}^3\omega^2+12\Gamma_0|\gamma'|}\left((\boldsymbol{r}\cdot\boldsymbol{V}_{\mathrm{vib}})\frac{3R^3}{r^4}\frac{\boldsymbol{r}}{r}+\boldsymbol{V}_{\mathrm{vib}}-\boldsymbol{V}_{\mathrm{vib}}\frac{R^3}{r^3}\right), \quad (24)$$

which is in a good agreement with the results obtained in the absence of surfactant [7].

4. Trajectories of particle

Let us consider the average particle motion. The particle is subjected to gravity $\boldsymbol{g}$, buoyancy, Stockes, Basset forces and average force caused by the vibrations. The equation of the particle motion reads

$$\left(\rho_{\mathrm{s}}+\frac{1}{2}\rho\right)\frac{\mathrm{d}^2\boldsymbol{r}}{\mathrm{d}t^2}=\boldsymbol{F}_{\mathrm{vib}}-\frac{9}{2R_{\mathrm{p}}{}^2}\eta\left(\frac{d\boldsymbol{r}}{dt}-\boldsymbol{U}\right)-\frac{9\rho}{2R_{\mathrm{p}}}\sqrt{\frac{\nu}{\pi}}\int_0^t\frac{\mathrm{d}}{\mathrm{d}t_1}\left(\frac{d\boldsymbol{r}}{dt_1}-\boldsymbol{U}\right)\frac{\mathrm{d}t_1}{\sqrt{t-t_1}}+(\rho_{\mathrm{s}}-\rho)\boldsymbol{g}$$

where ρ_{s} is the particle density, ρ,η are density and dynamic viscosity of the surrounding liquid, and the $\boldsymbol{U}$ is the velocity field induced by the rising bubble, which according to [13] has the form

$$\boldsymbol{U} = \frac{uR_b}{2r}\left[\frac{\boldsymbol{r}}{r^2}(\boldsymbol{k}\cdot\boldsymbol{r}) + \boldsymbol{k}\right] - u\boldsymbol{k}$$

where $u = gR_b^2/3\nu$ is the bubble rising velocity, $\boldsymbol{k}$ is the direction of its motion.

For incompressible fluid, the average vibrational force acting on the particle from the oscillating liquid, is given in [14]. In the laboratory reference frame after averaging over the vibration period, the expression for the average vibrational force per unit volume was obtained in the form:

$$\boldsymbol{F}_{\text{vib}} = \frac{3}{4}\rho\frac{\rho_s - \rho}{\rho_s + \frac{1}{2}\rho}\nabla\overline{V^2}.$$

Here $\boldsymbol{V}$ is the pulsational velocity of the liquid, described in the previous section (Eq. (24)).

The equation of particle motion was solved numerically using the Euler method. For each time step, the Basset force was computed from the previous time steps using a Caputo-type fractional derivative algorithm for solving partial differential equations [15]. The calculations show that in the presence of surfactant, the effect of vibrations on the particle-bubble interaction is lower than in the absence of surfactant which means that to increase the interaction efficiency, higher vibrations intensity is needed, and thus more energy.

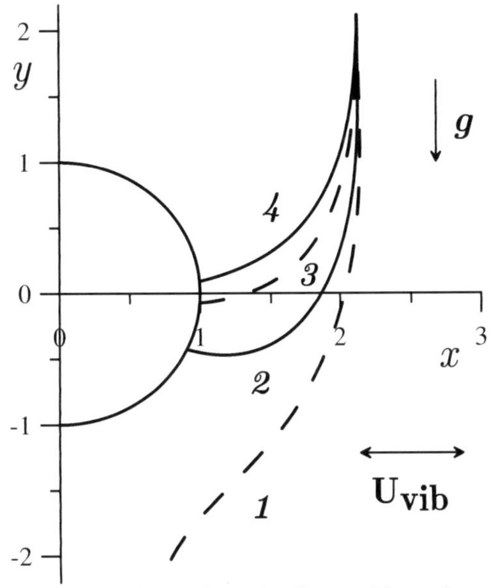

Figure 2 Trajectories of the particle motion, dot lines with surfactants, solid lines without surfactants, lines 1,2 – $\rho_s/\rho = 1.2$, lines 3, 4 – $\rho_s/\rho = 2$.

5. Conclusions

The solid particle behavior near a rising gas bubble in an incompressible liquid with surfactants was investigated under a gravity field and an external vibrational field. The surfactant transfer between the bubble surface and the surrounding liquid is limited by the adsorption-desorption process. The particle is subjected to the Stokes, Basset and buoyancy forces, and the average force due to the inhomogeneity of the pulsational field. The surfactants impact the pulsational bubble motion and decrease the bubble radius variations. It changes the radiation force value acting on the particle.

The problem is solved numerically using the Euler method; the trajectories of the particle motion have been calculated. In the presence of surfactant, it has been shown that vibrations decrease the particle-bubble collision efficiency compared to the case without surfactant; thus, to increase the collision efficiency a higher vibrations intensity is required, leading to larger energy consumptions.

It should be noted, that the surfactants are widely used in the flotation process and their effects on other flotation sub-processes (attachment-detachment) are important. In future we plan to extend this analysis to other sub-processes of bubble-particle interaction.

The work was supported by the Russian Science Foundation (grant 14-21-0090).

References

[1] Preuss M & Butt H J 1998. Direct measurement of particle-bubble interactions in aqueous electrolyte: Dependence on surfactant. *Langmuir*, **14**(12), 3164-3174.

[2] Fielden, Matthew L, Robert A, Fielden M L, Hayes R A, & Ralston J 1996. Surface and capillary forces affecting air bubble-particle interactions in aqueous electrolyte. *Langmuir*, **12**(15), 3721-3727.

[3] Gillies G, Kappl M, & Butt H J 2005. Direct measurements of particle-bubble interactions. *Advances in colloid and interface science*, **114**, 165-172.

[4] Nespolo S A, Chan D Y, Grieser F, Hartley P G, & Stevens G W 2003. Forces between a rigid probe particle and a liquid interface: Comparison between experiment and theory. *Langmuir*, **19**(6), 2124-2133.

[5] Dobby G S & Finch J A 1987. Particle size dependence in flotation derived from a fundamental model of the capture process. *Intl J. Miner. Process*. **21**, 241–260.

[6] Gaudin A M 1957. *Flotation*, McGraw-Hill, 573p.

[7] Nguyen A V 1999. Hydrodynamics of liquid flows around air bubbles in flotation: a review. *Intl J.Miner. Process.* **56**, 165-205.

[8] Phan C M, Nguyen A V, Miller J D, Evans G M & Jameson J G 2003. Investigations of bubble–particle interactions. *Intl J. Miner. Process*. **72**, 239-254.

[9] Lyubimov D V, Klimenko L S, Lyubimova T P, & Filippov L O 2016. The interaction of a rising bubble and a particle in oscillating fluid. *Journal of Fluid Mechanics*, **807**, 205-220.

[10] Leighton T G 1994. *The acoustic bubble*, Academic Press, London, 613.

[11] Lyubimov D V, Konovalov V V, Lyubimova T P, & Egry I 2011. Small amplitude shape oscillations of a spherical liquid drop with surface viscosity. *Journal of Fluid Mechanics*, **677**, 204-217.

[12] Levich V G 1962. *Physicochemical Hydrodynamics*, Prentice Hall, p. 573.

[13] Landau L D & Lifshitz E M 1987. *Fluid Mechanics* – Course of Theoretical Physics. 473

[14] Lyubimov D V, Lyubimova T P & Cherepanov A A 2002. Behavior of a drop (bubble) in a non-uniform pulsating flow. *Adv. Space Res*. **29**, 667–672.

[15] Butterworth-Heinemann, Diethelm K, Ford N J, Freed A D & Luchko Y 2005. Algorithms for the fractional, calculus: A selection of numerical methods. *Comput. Methods Appl. Mech. Eng.,* **194**, 743–773.

Improvement of ore recovery efficiency in a flotation column cell using ultra-sonic enhanced bubbles.

L O Filippov[1,2]**, J J Royer**[1]**, I V Filippova**[1,2]

[1]Université de Lorraine, GeoRessources, ENSG, 2 rue Doyen Marcel Roubault, BP10162, F-54505, Vandœuvre-lès-Nancy, France.
[2]National University of science and technology MISIS, 4 Leninsky pr., 117049 Moscow, Russia

Email: lev-filippov@univ-lorraine.fr

Abstract. The ore process flotation technique is enhanced by using external ultra-sonic waves. Compared to the classical flotation method, the application of ultrasounds to flotation fluids generates micro-bubbles by hydrodynamic cavitation. Flotation performances increase was modelled as a result of increased probabilities of the particle-bubble attachment and reduced detachment probability under sonication. A simplified analytical Navier-Stokes model is used to predict the effect of ultrasonic waves on bubble behavior. If the theory is verified by experimentation, it predicts that the ultrasonic waves would create cavitation micro-bubbles, smaller than the flotation bubble added by the gas sparger. This effect leads to increasing the number of small bubbles in the liquid which promote particle-bubble attachment through coalescence between bubbles and micro-bubbles. The decrease in the radius of the flotation bubbles under external vibration forces has an additional effect by enhancing the bubble-particle collision. Preliminary results performed on a potash ore seem to confirm the theory.

1. Introduction

Froth flotation is considered as the most widely used effective process in environmental engineering for mineral separation [1], wastewater treatment [2], removal hydrocarbon from refinery waters [3], and oil saturated beach sands [4]. Its efficiency comes from the selective ability of separating hydrophobic from hydrophilic materials ([5], [6]). Surfaces of particles are rendered hydrophobic by adding to the aqueous slurry a surfactant (or reagents such as xanthate salts) known as collector. The physical principle of froth flotation is based on the differences in the ability of air bubbles to selectively adhere to specific mineral surfaces in particle/water slurry in a flotation cell (Figure 1) to create the bubble-particle aggregates which move to the surface of slurry and form a stable froth phase. The removal of this froth with hydrophobic particles allows a spatial separation from remaining hydrophilic particles in the cell.

Recent works have investigated froth flotation enhanced by ultrasonic stimulation ([7], [8], [9]). Power ultrasonic waves applied to liquid may reduce the pressure sharply below the saturated vapor pressure, causing the dissolved air to separate out as micro-bubbles, a phenomenon known as *hydrodynamic cavitation* ([10], [11], [12], [13]). Under ultrasonic oscillation, the shape and geometry of stimulated pre-existing bubbles are modified increasing the attachment probabilities of mineral particles in the froth [9]. Dynamics of bubbles immersed into a non-compressible Newtonian liquid under oscillating pressure fields have been extensively studied these past decades for engineering processes ([14], [15], [16]) such as micro cavitation [10], micro-bubbles as contrast agent in clinical science [17], ultrasonic enhancement on fine particles flotation [8], and measurement of bubble solids mass loading in cell flotation [18]. If ultrasonic treatments are commonly used to enhance mineral processing flotation recovery [19], little attention on theoretical models has been paid to predict the capture, adhesion, and release probabilities and overall recovery of forced oscillating bubbles. Further interests have been motivated nowadays by the importance of such improvement in a number of applications ranging from cleaning fine coal particles, separating mineral ores, processing waste mining, cleaning waste industrial water or in environmental issues. On the other hand, a number of works have been done on predicting the capture, adhesion and release probability of particles on a single bubble in flotation cells, and a vast literature is now available to describe the particle behavior at the bubble scale ([20], [21], [22], [23]). In addition to these theoretical works at the micro-scale, models for predicting the overall recovery of flotation cells are available since more of a decade. The simplest models rely on first order kinetics type of law ([24], [25]). They both involved cell flotation parameter such as injected gas rate, column diameter and length, percentage of pulp but also collection efficiency through the particle-bubble capture probability.

In this work, we approximate the theoretical oscillating bubble radius time dependent curves by a trigonometric polynomial using results obtained previously ([15], [16]). These approximations are used to predict the particle-bubble ratio against time in order to derive the bubble capture probability as a function of time. This bubble diameter time-dependent function is then use to evaluate the enhancement in flotation recovery due to the application of ultrasonic external fields. It is shown that the size fluctuations of bubbles under ultrasonic wave increase the bubble-particle collision probability, and hence the efficiency of the flotation process.

2. Flotation column in a nutshell

A froth flotation cell consists typically of a *collection zone* (or *pulp zone*) located below the feed particle point, and a *cleaning zone*, (or *froth zone*), above the particle feed point [26] (Figure 2).

The *collection zone* is fed back-stream by the pulp pushed by the bubble swarm created by a gas sparger at the bottom of the flotation cell. Hydrophobic particles collide with, attach to bubbles and rise to the froth zone. The other remaining particles not attached to the bubbles fall at the bottom of the cell. In the *cleaning zone*, a downward liquid flow prevents hydraulic entrainment of fine particles into the concentrate [26].

2.1. Key parameters of a flotation cell

Several parameters govern the functioning of a flotation cell including the: (i) *type of floated mineral*; (ii) *superficial air rate* J_g, measured as the volumetric air flow rate per unit cross-section (typical values between 1 and 3 cm.s^{-1}) which flows minerals of interest to the froth zone; (iii) *bubble size* which is generally function of the air flow rate generated by the sparger at the bottom of the column [5], *collection rate* which is optimum for bubble size of about 1-2mm; (iv) *gas holdup* which can affect significantly the particle residence time and collection of minerals. It depends on gas rate, bubble size, slurry rate, slurry density, and bubble loading; (v) *wash water rate* J_w which increases the froth stability, allows a deep froth layer to develop, and prevents from hydraulic entrainment of non-desirable minerals; (vi) *froth depth* which is around 0.5 to 1.5 m. A shallow froth depth is used against hydraulic entrainment, while a deep froth is common to select hydrophobic minerals [27]; and (vii) *column height*. Yianatos [26] gives the following typical values for superficial gas rate: 1-3 cm.s^{-1}, superficial wash water rate: 0.3-0.5 cm.s^{-1}, superficial bias rate: 0.1-0.2 cm.s^{-1}, froth depth: 100 cm, average bubble size: 0.05-0.2 cm, and height/diameter ratio: 10/1. Others parameters are also important and characterize the flotation process including: the *overall* (k) and *collection zone* (k_c) flotation *rate constants*, the *floatability* factor F, the *collection efficiency* E_k, the *Peclet number* P_e ($P_e = (v_p L)/(2D)$), where v_p is the superficial velocity of solids, L the length of the collection zone, and D the dispersion coefficient of the solid phase in the column.

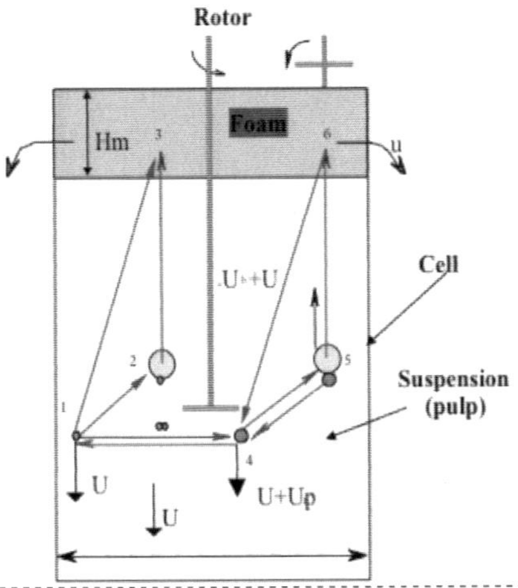

Figure 1. Schematic of flotation sub-processes in a froth flotation machine. 1, 4 - particle with no inertia (1) and with inertia forces (4); 2, 5- Particles attached to bubbles: 3, 6 – particles in the froth. Arrows show the particles transfer between different states: 1-4 aggregation; 1-2 and 4-5 – bubble-particle attachment; 2-3 and 5-6 – transport of particle; 5-2 - particle detachment; 1-3 and 6-4 – entrainment and particle drop-back

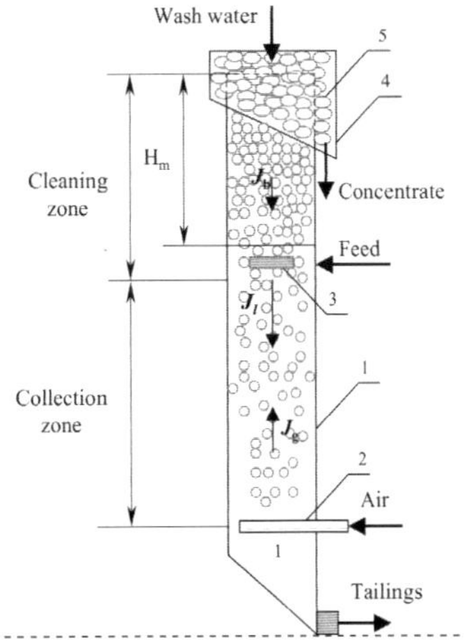

Figure 2. Diagram of a typical column flotation: 1 - column body, 2 – bubble sparger, 3 - feed distributor, 4 - froth launder, 5 - froth. H_m: froth depth; J_b: superficial bias velocity of the wash water; J_l: superficial feed velocity; J_g: superficial gas velocity.

2.2. Flotation kinetics

The principle of the froth flotation is based on the hydrophobic property of some mineral particles. After colliding with a bubble, a hydrophobic particle has the probability P_a to attach to the bubble and to be transported to the froth zone (Figure 3). Based on the sub-process presentation of a flotation process, the probability P of a given particle being collected to the froth zone is given by:

$$P = P_c P_a (1 - P_d) \qquad (1)$$

where P_c is the probability of collision with a bubble, P_a the probability of attachment, and P_d the probability of detachment from the bubble. Several expressions have been proposed to calculate P_c, P_a and P_d depending on the hydrodynamic conditions [9] (Figure 3). The collision angle θ_c depends weakly on the particle size, but strongly on the particle density and the bubble Reynolds number R_{eb}. Flotation is generally modeled as a first-order rate process with respect to the number of particles and the number of bubbles in the collision-attachment step, and a detachment process with respect to the number of bubble-particle aggregates [28].

2.3. Ultrasonic enhancements of the flotation process

Compared to classical flotation, ultrasonic stimulated flotation provides many advantages: it acts directly on the pulp by creating a large amount of cavitation tiny bubbles in the collection zone, with diameter from microns down to nano sizes [29], [30]. From energy consideration, it has been shown that micro cavitation bubbles occur preferentially at the surface of hydrophobic particles by rupture of the liquid-solid interface, making hydrodynamic cavitation a selective process with positive effect on flotation efficiency [12]. Nano bubbles created by hydrodynamic cavitation have the following effects [29]: (i) favor particles aggregation by tiny bubble bridging increasing the collision probability with air bubbles of the enlarged particle aggregate; (ii) increase the particle-bubble attachment probability

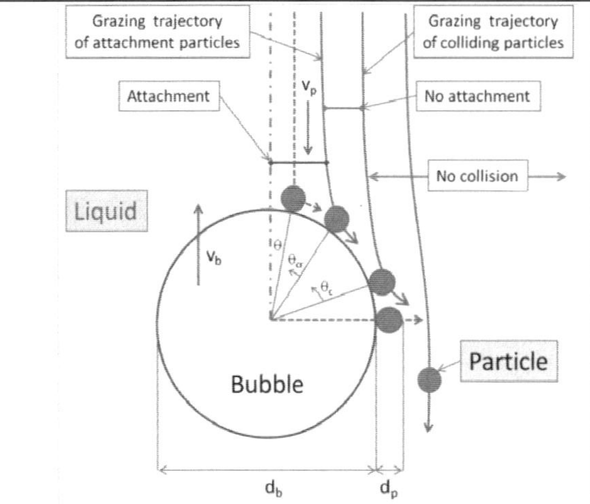

Figure 3. Schematic view of bubble-particle collision and attachment in flotation:

v_p = down-stream particle velocity; v_b = up-stream bubble velocity; d_p = particle diameter; d_b = bubble diameter; θ_c = collision (or critical) angle depends on the particle density and the bubble Reynolds number; beyond this angle, liquid pushes particles away and the particle cannot be attached to the bubble surface (after [21])

through coalescence of flotation bubble with tiny bubble frosted on particle surfaces; (iii) increase the contact angle and attachment force between particles and bubbles thanks to tiny bubbles frosted on particle surfaces; (iv) clean the particles surfaces through collapse of tiny bubbles by removing slime coatings, oxidation films [8], [30]; (v) decrease the reagent consumption [8], [30]. Negative effects on the functioning of the flotation cell can be also noticed when using ultrasound treatment [30] including (i) *high energy* consumption limiting its use in industry; (ii) temperature increase of both bubbles and liquid; (iii) increase of the reagent solubility, and (iv) induced destabilization of the mineral aggregates by splitting effects.

3. Governing equations for bubble evolution under ultrasonic waves

Hydrodynamics of single bubble immersed in a liquid subjected to ultrasonic waves can be described by the Navier-Stokes equations applied to the gas inside the bubble, initially at rest, and to the incompressible liquid adjacent to the bubble wall. The Navier-Stokes equations can be solved *analytically* in spherical coordinates [15] or *numerically* using a radial formulation [16].

3.1. Solving Navier-Stokes equations

A bubble initially at rest with mean radius R_0 is immersed in an unbounded incompressible liquid submitted to external pressure driven periodically. Neglecting gravitational forces, the Navier-Stokes equations (or mass and momentum conservation equations) describes the flow field in the incompressible liquid with density ρ surrounding the bubble [16]:

$$\nabla . \mathbf{u} = 0; \quad \rho \left(\frac{\mathrm{Du}}{\mathrm{Dt}}\right) = -\nabla P + \nabla . \boldsymbol{\tau} \tag{2}$$

where $Df/Dt = \partial f/\partial t + \mathbf{u} . \nabla f$ is the particle derivative of function *f*, *t* the time, ***u*** the velocity field, $\boldsymbol{\tau}$ the deviatoric part of the total stress tensor $\boldsymbol{\pi} = -\mathbf{P}(I) + \boldsymbol{\tau}$, and P the pressure of the liquid. The shape of the bubble-liquid interface is described by the kinematic equation:

$$\frac{D\mathbf{R}_b}{Dt} = \mathbf{u}|_{r=R_b} \tag{3}$$

where $\mathbf{R}_\mathrm{b}(\theta, t) = R_b(\theta, t)\, \mathbf{e}_\mathbf{r}$ is the position vector. The above set of equations can be solved either using (i) an *analytical solution* in spherical coordinates assuming an ideal gas in the bubble [14].

3.2. Analytical solution to the Navier-Stokes equations for a spherical bubble

Assuming bubbles being spherical, Eq. (3) can be simplified into ∂ (r^2 $u(r,t)$) = 0 using spherical coordinates, with solution for the radial velocity:

$$u(r,t) = \frac{1}{r^2}\Phi(t) \tag{4}$$

where $\Phi(t)$ is a function of time determined by the boundary conditions. Assuming no mass transfer across the bubble boundary, and using a local coordinate system attached to the bubble, the wall velocity is equal to the change rate of its radius $u(R_b,t) = \dot{R}_b$ (where $\dot{R}_b$ is the derivation of R_b against time t), thus $\Phi(t) = R_b^2(t)\dot{R}_b$, and consequently Eq.(4) becomes:

$$u(r,t) = \left(\frac{R_b(t)}{r}\right)^2 \dot{R}_b \tag{5}$$

Neglecting the dynamic viscosity of the liquid, and substituting the radial component of the velocity, the radial component of the Navier-Stokes equation applicable to the liquid can be written as:

$$\frac{1}{\rho}\frac{\partial P_l}{\partial r} + \frac{\partial u}{\partial t} + u(r,t)\frac{\partial u}{\partial r} = 0 \quad -\frac{1}{\rho}\frac{\partial P_l}{\partial r} = \frac{R_b^2\ddot{R}_b + 2R_b\dot{R}_b^2}{r^2} - \frac{2R_b^4\dot{R}_b^2}{r^5}$$
$$\Rightarrow -\frac{1}{\rho}\int_{P_l(R_b,t)}^{P_\infty(t)} dP = \int_{R_b}^{\infty}\left(\frac{R_b^2\ddot{R}_b + 2R_b\dot{R}_b^2}{r^2} - \frac{2R_b^4\dot{R}_b^2}{r^5}\right)dr \Rightarrow \frac{P_l(R_b,t) - P_\infty}{\rho} = R_b\ddot{R}_b + \frac{3}{2}\dot{R}_b^2 \tag{6}$$

where $P_l(R_b,t)$ is the liquid pressure on the bubble wall which is equal to the sum of all forces acting on the wall bubble: (i) the *gas pressure* in the bubble P_g, (ii) the *viscous stress* σ_{rr} of the liquid; and (iii) the *superficial tension* τ. It comes $P_l = \sigma_{rr} + \tau + P_g$. The radial component of the *viscous stress* σ_{rr} acting on the bubble wall depends on the liquid cinematic viscosity μ [10]:

$$\sigma_{rr}(r,t) = 2\mu\frac{\partial u}{\partial r} = -4\mu\frac{\dot{R}_b}{R_b(t)} \tag{7}$$

The tension pressure τ (or Laplace's pressure) acting on the surface bubble is given by the Laplace's formula $\tau(t) = -\frac{2\sigma}{R_b(t)}$ where σ is the *surface tension* and $1/R_b$ the bubble curvature. Assuming an ideal gas polytropic transformation for the gas trapped in the bubble, the gas pressure is thus equal to [9]:

$$P_g(t) = P_{g_0}\left(\frac{R_{b_0}}{R_b(t)}\right)^{3m} \tag{8}$$

with P_{g_0} the initial gas pressure in the bubble at rest. At $t = 0$, $P_l(R_b,0) = P_0$ and $\dot{R}_b = 0$, the initial pressure in the bubble becomes $P_{g_0} = \frac{2\sigma}{R_{b_0}} + P_0$. Thus, the total pressure at the bubble boundary P_l is given by combining Eqs. (7), and (8):

$$P_l(R_b,t) = \left(\frac{2\sigma}{R_{b_0}} + P_0\right)\left(\frac{R_{b_0}}{R_b(t)}\right)^{3m} - 4\mu\frac{\dot{R}_b}{R_b(t)} - \frac{2\sigma}{R_b(t)} \tag{9}$$

Other gas models (including ideal gas, Soave-Redlich-Kwong, and Peng-Robinson) can be used to predict the evolution of the internal bubble pressure (see [9]).

3.3. Numerical solution to the Navier-Stokes equations for complex bubbles

Assuming adiabatic conditions for the gas in the bubble, and accounting for the elasticity conditions represented by the Deborah number $D_e = \lambda\omega$ (with λ the polymer relaxation time), several authors including [15], and [16] gives a *numerical solution* to the bubble set of the Navier-Stokes equations:

Model (I) Kim and Kwak [15] $\frac{R_b(\omega t)}{R_{b_0}} = \frac{R}{R_0}[1 + \varrho\sin(\omega t + \omega\tau_0)\exp\{\sin(\omega t + \omega\tau_0)\}]$

Model (II) Foteinopoulou and Laso [12] $\frac{R_b(\omega t)}{R_{b_0}} = \frac{R}{R_0}(D_e)[1 + \varrho(D_e)\cos\{\omega t + \omega\tau_0(D_e)\}]$ (10)

where the radius equilibrium bubble R_{b0} = 500-700 μm, $\omega = 2\pi f$ angular frequency with f the ultrasonic frequency (around 28.5 kHz), external pressure at P_a = 0.142 MPa, the normalized amplitude $R/R_0 = 1$, $\varrho = 0.623$, $\omega\tau_0 = -0.374$, $k = 1.114$ and τ_0 is a delay term describing the time necessary for a bubble to change its shape under a compressive wave stress. In the model (II), the coefficients $R/R_0(D_e)$, $\rho(D_e)$ and $\omega\tau_0(D_e)$ are functions of the elasticity conditions represented by the Deborah number D_e. These two expressions can be generalized into [9]:

$$\frac{R_b(\omega t)}{R_{b0}} = \frac{R}{R_0} f_{(\varrho,\nu,\mu,\omega\tau_0)}(\omega t)$$

$$f_{(\varrho,\nu,\mu,\omega\tau_0)}(\omega t) = 1 + \varrho \cos\left(\omega t + \omega\tau_0 + \nu\frac{\pi}{2}\right)\{1 - \mu[1 - \exp\{\sin(\omega t + \omega\tau_0)\}]\} \quad (11)$$

where the coefficients (R/R_0, ϱ, $\omega\tau_0$, $\nu = \mu = 1$) can be constant as in model (I), or function of D_e ($R/R_0(D_e)$, $\varrho(D_e)$, $\omega\tau_0(D_e)$, $\nu = \mu = 0$) as in model (II). The tuning parameter is always observed to be less than 1: $\varrho < 1$ (value $\varrho = 1$ may lead to singularity). $R/R_0(D_e)$, $\varrho(D_e)$ and $\omega\tau_0(D_e)$ are fitted by:

$$\frac{R_b}{R_{b0}}(D_e) = \frac{R_b}{R_{b0}}(0) + a_1 D_e + a_2 D_e^2 \ ; \varrho(D_e) = \varrho(0) + a_1' D_e + a_2' D_e^2 \ ;$$

$$\omega\tau_0(D_e) = \omega\tau_0(0) + a_1'' \sin D_e \, e^{-a_2'' D_e} \quad (12)$$

with the fitted parameters equal to: $\frac{R_b}{R_{b0}}(0) = 1.009$; $a_1 = .0135$; $a_2 = -.0007$; $\varrho(0) = 0.088$; $a_1' = 0.0315$; $a_2' = -0.002$; and the time shift values with: $\omega\tau_0(0) = 0.211$; $a_1'' = 0.723$; $a_2'' = 0.316$ (with $R^2 = 0.98$) accounts for the inside bubble pressure response to an external pressure solicitation. These models will be used in the followings to estimate the effects of ultrasound on various parameters of a flotation cell.

4. Impact of ultrasound on governing equations for particle flotation

Modeling of flotation columns is a major step for optimizing recovery ([31], [32], [33]). Micro- and macro-scale models have been often used to predict the flotation cell kinetics. Macro-scale systems such as kinetic models have received most attention in the literature [34]. The column is modeled as a mass transfer unit using a kinetic model that is then used to predict the large-scale performance flotation system, provided the kinetics and hydrodynamics of the cell are well defined [24]. These models can then be used for developing control strategies in industrial applications.

4.1. Effect of ultrasound on the flotation rate constant

Flotation performance depends on the ability of bubbles to collect suspended particles and carry them to the surface, where a layer of froth can be removed over a launder [23] (Figure 2). The basic collection mechanism consists of collision and attachment of particles on the interfacial gas bubble surface. Gas (usually air or nitrogen) is injected at the base of the column to transport the floatable solids to the overflow. The volumetric gas flow-rate per unit cross-section, J_g, (also referred as superficial air rate [26]), is commonly used to describe bubble column operations. Gas dispersion parameters comprising bubble size (d_b), gas hold-up (ε_b), and derived parameter such as the bubble surface area flux ($S_b = 6J_g/d_b$) also control flotation performances [33]. Assuming the overall (k) and column (k_c) flotation rate constants being equal ($k_c = k$), they can be expressed as a function of the collection efficiency (E_k), bubble size (d_b) and superficial gas velocity (J_g) [24]:

$$k_c = \frac{3}{2}\frac{J_g E_k}{d_b} = \frac{S_b E_k}{4} = F S_b \propto d_b^n \ ; \quad n \in]-1,-1[\ \{\text{or } n \in]-3,-1.5[\text{ according to [35]}\} \quad (13)$$

where F is the floatability factor, which includes the contribution of particle size and hydrophobicity. When pilot and industrial scale experiments are not available at similar gas rates and bubble sizes, it is suggested [24] using a proportional fractal relationship between the bubble size (d_b) and the rate constant (k_c) established by [25] and [35]. The collection efficiency (E_k) can be estimated using the collision-particle probability models developed at the micro-scale [25].

Under *ultrasound*, the bubbles in flotation cell oscillate changing radius size and shapes according to the driving frequency and amplitude of the ultrasound as shown in Eq. (11). From Eq. (13), the column rate constant fluctuates against time t under the ultrasonic oscillation according to:

$$\frac{k_c(\omega t)}{k_{c0}} = \left(\frac{d_b(\omega t)}{d_{b0}}\right)^n = f^n(\omega t)\ ; \quad n \in]-1,-1[\ \{\text{or } n \in]-3,-1.5[\text{ according to [35]}\} \tag{14}$$

where k_{c_0} is the column rate constant rate without ultrasonic treatment, the other parameters being kept the same, $f(\omega\, t)$ is the function describing the radius-time bubble oscillations under ultrasonic solicitation. Eq. (14) shows that the rate constant k_c increases when the bubble radius d_b decreases under ultrasound compression. Given the fact that the oscillation time scale is very small compared to the flotation time-scale, we calculate the average value for $k_c(\omega\, t)/k_c$ over six periods for evaluating the overall recovery R_c (see next paragraph).

4.2. *Effect of ultrasound on the column recovery*

Assuming a non-ideal plug flow reactor with axial dispersion for the column so that the flotation process obeys first order kinetics, the final collection zone overall recovery R_c is predicted by [36], [24]:

$$R_c = 1 - \frac{4a \exp\left(\frac{1}{2}P_e\right)}{(1+a)^2 \exp(aP_e/2) - (1-a)^2 \exp(-aP_e/2)}\ ; \quad a = \left(1 + 4\,\frac{k\,\tau}{P_e}\right)^{1/2} \tag{15}$$

where $P_e = v_p L/D$ is the dimensionless Péclet axial dispersion number with v_p the superficial velocity of solid particles, L the collection zone length of the column, D the dispersion coefficient of the solid phase in the column, k_c, k the overall and collection zone first order constant rate, respectively. Using an axial dispersion model, the dispersion coefficient D of the solid phase in large columns can be predicted as a function of the column diameter d_c, superficial gas rate J_g and the % of solid in the pulp S according to the following semi-empirical model [24] $D = \alpha_1 d_c J_g e^{-\beta_1 S}$, with $\alpha_1 = 6.7\ 10^{-2}$ and $\beta_1 = 0.025$. A simplified expression for R_c is proposed by [25] as:

$$\frac{R_c}{R_\infty} = 1 - \frac{4}{3}\left[\frac{100}{9(10k_c\tau+12)} - \frac{12}{(k_c\tau+12)^2} - \frac{10}{9(k_c\tau+12)}\right] \tag{16}$$

where τ the mean residence time determined as the ratio between effective volume of the column (i.e. total volume - gas + froth volume) and volumetric feed flow rate, R_∞ is the recovery at $t = \infty$ (about $R_\infty = 90\%$), which depends on the ore mineralogy and flotation chemistry. The final overall recovery R_{fc} depends on the collection zone recovery R_c and on the froth zone recovery R_f [33], [37]:

$$R_{fc} = \frac{R_c\, R_f}{R_c(R_f - 1) + 1} = \frac{1}{1 + R_f^{-1}(R_c^{-1} - 1)}\ ; \quad R_f = \alpha \exp\left(-\beta\,\frac{L_f^{1.16}}{J_g^{.79}}\right)\ ; \quad \alpha = 85\%\ ; \quad \beta = 5.2\ 10^{-3} \tag{17}$$

where L_f is the froth depth, and J_g the superficial gas velocity. Finally, the overall recovery R_c depends on two dimensionless parameters, the collection first order *rate constant* k and the *Péclet number* P_e.

At the micro-scale, bubble radius changes impact the constant rate k_c through the collection efficiency E_k terms, and thus the overall recovery R_c, assuming the other parameters be kept constant (i.e. J_g, D, v_b, R_∞). As k_c depends as an inverse power function of the bubble radius $k_c \propto d^{-n}$, $n \in [-2,-1]$ (see Eq. (14)), when the bubble radius d_b decreases under ultrasound compression, the constant rate k_c increases, thus the overall recovery R_c improves (Figure 4a) , and vice versa when k_c decreases under bubble radius expand, the overall recovery R_c worsens.

4.3. *Effect of ultrasound on particle-bubble capture probability*

The probability P of a particle being collected by a gas bubble during flotation is given by Eq. (4). A generalized expression for the particle-bubble collision and attachment probabilities P_c and P_a, respectively, is proposed by [22], [38]:

$$P_c = A\left(\frac{d_p}{d_b}\right)^2\ ; \quad P_a = \sin^2\left[2 \arctan \exp\left\{\frac{-\left(45 + 8R_{eb}^{.72}\right) v_b t_i}{15 d_b\left(\frac{d_b}{d_p} + 1\right)}\right\}\right]\ ; \quad t_i = \frac{75}{\theta}\, d_p^{.6} \tag{18}$$

where d_p, d_b are the particle and bubble diameters, and A varies with the flow regime (i.e. Reynolds number), v_b the bubble raise velocity, t_i (in *s*) the induction time which is a function of the particle size

d_b (in m) and contact angle θ; $R_{eb} = v_b\, d_b\, \rho_f / \eta$ is the dimensionless bubble Reynolds number which combines the effect of bubble rise velocity v_b, mean bubble diameter d_b and density ρ_f and dynamic

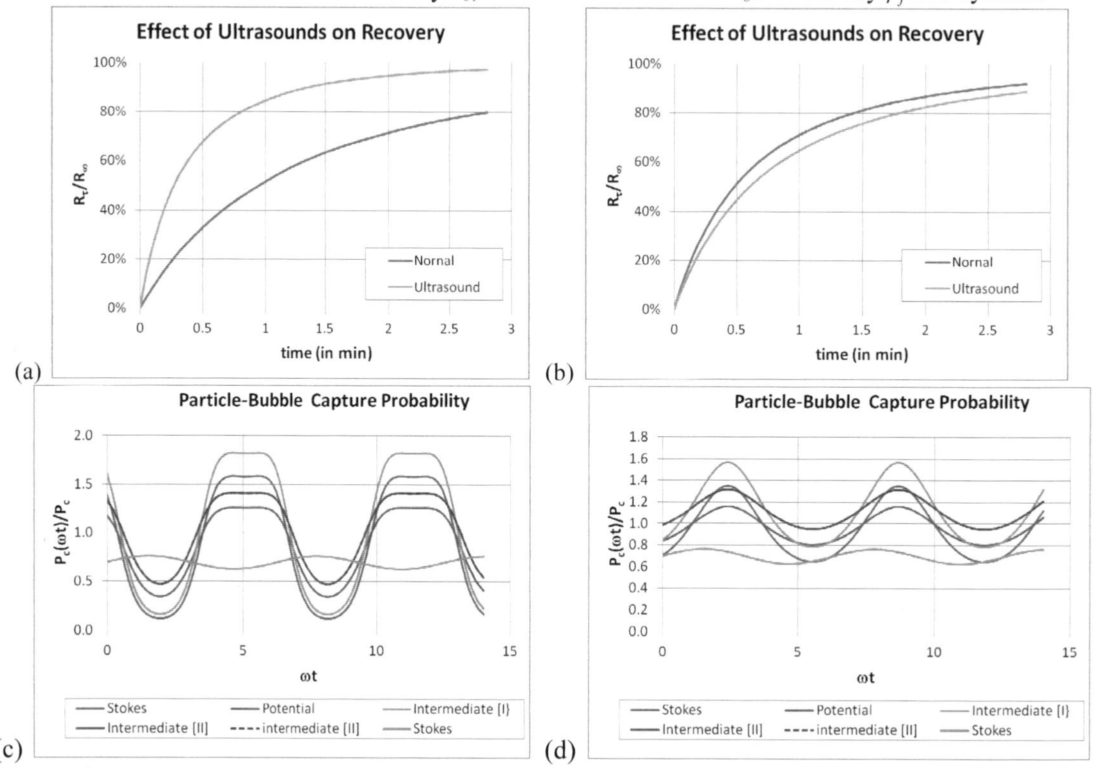

Figure 4. (a) Overall recovery R_c calculated using model (I) [15] improves by more than 20% when applying ultrasound. Similar results are reported for chalcopyrite by [37]. Experimental conditions: $P_e = 1.66$; $k_c = 1.25\text{s}^{-1}$; $n = -2$; $R/R_0 = 1$; $\varrho = 0.623$; $\omega\tau_0 = -0.374$; (b) Overall recovery R_c calculated using model (II) [16] improves by less than 5% when applying ultrasound. Experimental conditions: idem as in (a) except for $D_e = 4$; $R/R_0 = 1.052$; $\varrho = 0.182$; $\omega\tau_0 = -0.746$; (c) and (d) Calculated single bubble collision probability for various hydrodynamic models.

viscosity η of fluid around of the bubble; v_p is the particle velocity. R_{eb} values are used to describe the flow condition from *Newtonian flow*: ($R_{eb} \leq 1$), *intermediate I*: ($1 < R_{eb} < 20$), *intermediate II*: ($20 < R_{eb} < 400$), and turbulent: ($R_{eb} > 400$). The flow regime dependent coefficient A varies as: Newtonian flow [39] $A = 2/3$; intermediate I [38]: $A = 2/3\,(1+.178\, R_{eb})$; intermediate II [40]: $A = 2/3\,(1+.1875R_{eb}/(1+.249R_{eb}^{.56}))$.

The variation in bubble radius under ultrasound impacts the collision probability as $P_c(\omega t) / P_{c_0} = f^{-n}(\omega t)$ assuming a constant Reynolds number (Figure 4c and d). The attachment probability $P_a(\omega t) / P_{a_0}$ is a more complex function of ωt which can be calculated numerically. Globally, ultrasound has less impact on the collision probability than on the attachment probability; the relationship between the recovery impact and the collision and attachment probability is complex and not direct. An average increase in capture probability not necessary leads to an increase in recovery performance.

4.4. *Effect of ultrasound on collision angle*

The collision angle θ_c (Figure 3) depends weakly on particle size d_p, but strongly on the particle density and the dimensionless bubble Reynolds number R_{eb} and may be predicted by [21]:

$$\theta_c = \arccos(\delta)\,;\; x = \frac{3}{2} + \frac{0.91R_{eb}}{3.236+R_{eb}^{0.694}}\,;\; y = \frac{0.375R_{eb}}{1+0.217R_{eb}^{0.518}}\,;\; c = \frac{v_b}{v_p}\left(\frac{d_b}{d_p}\right)^2;\; \zeta = \frac{x+c}{3y}\,;\; \delta = \sqrt{\zeta^2+\frac{1}{3}} - \zeta \quad (19)$$

A simplified expression for θ_c which depends only on the hydrodynamic conditions (i.e. R_{eb}) is [32]:

$$Newtonian: \theta_c = 85.0 - 2.50 \log R_{e_b};\ Intermediate\ I: \theta_c = 85.0 - 12.49 \log R_{e_b};\ Intermediate\ II: \theta_c = 78.1 - 7.37 \log R_{e_b} \quad (20)$$

Ultra sound oscillation impacts the parameter c through the velocity and diameter of bubbles, and thus may increase the collision angle θ_c.

These above results demonstrate that improvement on overall flotation recovery R_c under ultrasound is a very complex process. It may be due to: (i) an *increase in particle-bubble collision* or *attachment probabilities* by reduction of the bubble radius, but it is not always the case; (ii) the overall recovery improves under ultrasound because the radius oscillation not only impacts the capture probability but also indirectly the constant rate and collision angle. Effects can be compensated or amplified.

5. Conclusions

This theoretical work shows that ultrasound effect on flotation can be predicted both at micro and macro scale, the major impact being the time-variation of bubbles volumes under pressure oscillation. The most direct consequences would be that the bubble ability for capturing particles increases when bubble size diminishes, by an increase in collision and attachment probability. This work also demonstrates a strongly dependency on the flow regime and on the bubble oscillation. If overall flotation recovery is chosen as criteria for measuring the impact of ultrasound, it has been shown that: (i) the recovery efficiency improve due to a reduction of the bubble radius increasing the particle-bubble collision and attachment probabilities by a factor ranging between 0.8 and 2 on average depending on the model used (or both depending on the flow regimes or/and on the model used to describe the radius oscillation); (ii) the improvement in overall flotation recovery under ultrasound is a very complex process involving not only the collision and attachment probabilities, but also the variation in bubble radius and flotation rate constant, effects being exaggerated or compensated depending on the flow regimes or/and on the model used to describe the radius oscillation; (iii) model I give similar results while model II seems to under estimate the bubble radius oscillation.

Several assumptions have been proposed during this theoretical work which might be confirmed or not by experimental work or direct measurement of the bubble oscillation to valid the different suggested models [41]. Non Newtonian regimes such as intermediate turbulent flows with intermediate Reynolds numbers seem to improve the overall flotation recovery. This point must be investigated more in future experimental works. In some flow regimes, it has been found numerically that ultrasound may decrease the overall flotation recovery, a point not fully understood yet but with important consequences in flotation cell design and applications. Finally, in certain conditions bubble may enter in resonance for a given frequency, and this implication has been not investigated here but needs some further theoretical works.

6. Acknowledgements

The work was supported by the Russian Foundation for Basic Research (grant 17-41-590974).
The authors would like to express their thanks for the support to Lorraine-Russia Arcus Project.

References

[1] Villar del R, Bouchard J, Desbiens A and Poulin E 2015. *Recent Developments in Flotation Column Instrumentation and Control: An Update*. Proceedings of the World Congress on Mechanical, Chemical, and Material Engineering (MCM 2015), Barcelona, Spain, July 20-21, Paper No. 127, 1-9.

[2] Mosai R and Dawe R A 2003. Gas attachment of oil droplets for gas flotation for oily wastewater cleanup. *Separation and Purification Technology*, **33**(3), 303-314.

[3] Beychok M R 1967. *Aqueous Wastes from Petroleum and Petrochemical Plants* (1st Ed.). John Wiley & Sons Ltd. LCCN 67019834. 370p.

[4] Packham R F and Richards W N 1972. *Water clarification by flotation - 1*. Technical paper TP. 87, the water research Association, Medmenham, Marlow, Buckinghamshire, SL7 2HD, England, 38p.

[5] Clingan B V and McGregor D R 1987. *Column flotation experience at Magma Copper Company, with*

related experience of other mineral processors, 116th Annual AIME-SME Meeting. Denver.

[6] Manlapig E V, Green C, Parkinson J W and Murphy A S 2001. *The technology and economic incentives for recovering coal from tailings impoundments*. SME Annual Meeting, Denver, Colorado, 26–28 February, Preprint 01-70.

[7] Kursun H and Ulusoy U 2015. Zinc Recovery from a Lead-Zinc-Copper Ore by Ultrasonically Assisted Column Flotation. *Particulate Science and Technology*, **33**(4), 349-356.

[8] Ozkan S G 2012. Effects of simultaneous ultrasonic treatment on flotation of hard coal slimes. *Fuel* **93**, 576-580.

[9] Royer J J, Monnin N, Pailot-Bonnetat N, Filippov L O, Filippova I V, Lyubimova T 2017. *Thermodynamics of ultra-sonic cavitation bubbles in flotation ore processes*. Part II, NAMES, (*this volume*)

[10] Brennen C E 1995. *Cavitation bubble dynamics*. Oxford University Press. 254p.

[11] Ozonek J 2012. *Application of hydrodynamic cavitation in environmental engineering*. CRC Press, Taylor & Francis Group, London, 126p.

[12] Sobhy A 2013. *Cavitation nano bubble enhanced flotation process for more efficient coal recovery*. PhD dissertation, Mining Engineering, University of Kentucky, UKnowledge, Lexington, US, 169p.

[13] Sobhy A and Tao D 2013. Nano bubble column flotation of fine coal particles and associated fundamentals. *Int. Journal of Mineral processing*, **124**, 109-116.

[14] Aks S and Vezzetti D 1980. Ultrasonic scattering theory I: Scattering by single objects. *Ultrasonic Imaging*, **2**, 85-101.

[15] Kim K and Kwak H Y 2007. Predictions of bubble behavior in sulfuric acid solutions by a set of solutions of Navier-Stokes equations. *Chemical Engineering Science*, **62**, 2880-2889.

[16] Foteinopoulou K and Laso M 2010. Numerical simulation of bubble dynamics in a Phan-Thien-Tanner liquid: non-linear shape and size oscillatory response under periodic pressure. *Ultrasonics*, **50**, 758-776.

[17] Deng C and Lizzi F 2002. A review of physical phenomena associated with ultrasonic contrast agents and illustrative clinical applications. *Ultrasound Med. Biol.*, **28**, 277-286.

[18] Spencer S, Bruniges R, Roberts G, Sharp V, Catanzano A, Bruckard W, Davey K and Zhang W 2012. An acoustic technique for measurement of bubble solids mass loading: (b) Monitoring of Jameson cell flotation performance by passive acoustic emissions. *Minerals Engineering*, **36-38**, 21-30.

[19] Qi B and Aldrich C 2002. Effect of ultrasonic treatment on zinc removal from hydroxide precipitates by dissolved air flotation. *Minerals Engineering*, **15**, 1105-1111.

[20] Yoon R and Luttrell G 1989. The effect of bubble size on fine particle flotation. *Min. Process. Extractive Metall. Rev.*, **5**, 101-122.

[21] Nguyen A V, Ralston J and Schulze H J 1998. On modelling of bubble-particle attachment probability in flotation, *Int. J. Miner. Process*, **53**, 225-249.

[22] Dai Z, Fornasiero D and Ralston J 2000. Particle-bubble collision models - a review. *Advances in Colloid and Interface Science*, **85**(2-3), 231-256.

[23] Shahbazi B, Rezai B and Koleini S M J 2010. Bubble-particle collision and attachment probability on fine particles flotation. *Chemical Engineering and Processing*, **49**(6), 622-627.

[24] O'Connor C and Mills P 1995. The influence of bubble size on scale-up of column flotation cells. *Minerals Engineering*, **8**(10), 1185-1195.

[25] Massinaei M, Kolahdoozan M, Noaparast M, Oliazadeh M, Yianatos J, Shamsadini R and Yarahmadi M 2003. Hydrodynamic and kinetic characterization of industrial columns in rougher circuit. *Minerals Engineering*, **22**,357-365.

[26] Yianatos J B 1989. *Column flotation modeling and technology*. In SAIMM Int. Coll.: Developments in Froth Flotation, Cape Town, South Africa, Vol. 2, 1-31.

[27] Yianatos J, Finch J and Laplante A 1987. The cleaning action in column flotation froths. *Trans. Inst. Min. Metall (Section C)*, **96**, 199-205.

[28] Koh P and Schwarz M 2006. CFD Modelling of bubble-particle attachments in flotation cells. *Minerals Engineering*, **19**(6), 619-626.

[29] Zhou Z A, Zhenghe X, Finch J A, Masliyah J H and Chow R S 2009. On the role of cavitation in particle collection in flotation - A critical review. II. *Minerals Engineering*, **22**, 419-433.

[30] Filippov L O, Matinin A S, Samyguin V D and Filippova I V 2013. - Effect of ultrasound on flotation kinetics in the reactor separator. Journal of Physics Conf. Ser., **416** (1). Doi:10.1088/1742-6596/416/1/012016.

[31] Derjaguin B and Dukhin S 1961. Theory of flotation of small and medium size particles. *Trans. Inst. Min. Metall.*, **70**, 221-246.

[32] Schulze H 1989. Hydrodynamics of bubble-mineral particle collisions. Min. Process. *Extractive Metall. Rev.*, **5**, 43-76.
[33] Finch J A and Dobby G S 1991. Column Flotation: a selected review. Part I. *Int. Journal of Mineral Processing*, **33**(1-4), 343-354
[34] Polat M and Chander s 2000. First-order flotation kinetics models and methods for estimation of the true distribution of flotation rate constants. *Int. J. Miner. Process*, **58**, 145-166.
[35] Yoon R 1993. Microbubble flotation. *Minerals Engineering*, **6**, 619-630.
[36] Dobby G and Finch J 1986. A model of particle sliding time for flotation size bubbles. *J. Colloid Interface Sci.*, **109**(2), 493-498.
[37] Cilek E C and Ozgen S 2010. Improvement of the Flotation Selectivity in a Mechanical Flotation Cell by Ultrasound. *Separation Science and Technology*, **45**, 572-579.
[38] Yoon R 2000. The role of hydrodynamic and surface forces in bubble particle interaction. *Int. Journal of Mineral Processing*, **58**(1-4), 129-143.
[39] Gaudin A 1932. *Flotation*. McGraw Hill Book Co. Inc., New York.
[40] Weber M and Paddock D 1983. Interceptional and gravitational collision efficiencies for single collectors at intermediate Reynolds numbers. *J. Colloid Interface Sci.*, **94**(2), 328–335.
[41] Lyubimov D V, Klimenko L S, Lyubimova T P, Filippov L O (2016). - The interaction of a rising bubble and a particle in oscillating fluid, *Journal of Fluid Mechanics*, **807**, 205-220.

IOP Conf. Series: Journal of Physics: Conf. Series **879** (2017) 012024 doi:10.1088/1742-6596/879/1/012024

Thermodynamics of ultra-sonic cavitation bubbles in flotation ore processes.

Royer[1] J J, Monnin[1] N, Pailot-Bonnetat[1] N, Filippov[1,2] L O, Filippova[1,2] I V, Lyubimova[2] T

[1]Université de Lorraine, GeoRessources, ENSG, 2 rue Doyen Marcel Roubault, BP10162, F-54505, Vandœuvre-lès-Nancy, France.
[2]National University of Science and technology "MISIS", 117049 Moscou, Russia
[3]Perm State University, Perm, Russia.

Email: jean.jacques.royer@gmail.com

Abstract. Ultra-sonic enhanced flotation ore process is a more efficient technique for ore recovery than classical flotation method. A classical simplified analytical Navier-Stokes model is used to predict the effect of the ultrasonic waves on the cavitations bubble behaviour. Then, a thermodynamics approach estimates the temperature and pressure inside a bubble, and investigates the energy exchanges between flotation liquid and gas bubbles. Several gas models (including ideal gas, Soave-Redlich-Kwong, and Peng-Robinson) assuming polytropic transformations (from isothermal to adiabatic) are used to predict the evolution of the internal pressure and temperature inside the bubble during the ultrasonic treatment, together with the energy and heat exchanges between the gas and the surrounding fluid. Numerical simulation illustrates the suggest theory. If the theory is verified experimentally, it predicts an increase of the temperature and pressure inside the bubbles. Preliminary ultrasonic flotation results performed on a potash ore seem to confirm the theory.

1. Introduction

Froth flotation is a widely commonly used effective process in mineral processing ([1], [2], [3]), wastewater cleaning [4], coal industry [5], and hydrocarbon pollution remediation [1]. The ability of air bubbles to selectively adhere to specific mineral surfaces and separating hydrophobic from hydrophilic materials ([1], [6]) makes froth flotation an effective technique widely used in the industry. Recent works have investigated froth flotation enhanced by ultrasonic stimulation ([3], [7]). In the previous paper [8], we have investigated the ultra sonic stimulation on flotation both at micro and macro scales showing that the overall flotation recovery increases due to an increase in bubble ability for capturing particles, especially because of higher particle-bubble collision and particle-bubble attachment probabilities.

In this work, we focused on the thermo-dynamical aspects induced by ultra-sonic treatment at the bubble scale assuming the same model for ultra sonic wave as in [8] i.e. the theoretical oscillating bubble radius time dependent curves is modeled by a trigonometric polynomial previously ([9], [10]). A thermodynamics approach is suggested to model the energy exchanges between the liquid and the gas bubble submitted to an ultrasonic treatment; the temperature and pressure behavior of a bubble is then calculated together with the internal energy, total work, and heat exchange assuming various gas models (including polytropic process, Soave-Redlich-Kwong, Peng-Robinson, and ideal gas). It is shown that the application of ultrasonic external fields increases the temperature and pressure of the gas trapped in bubbles.

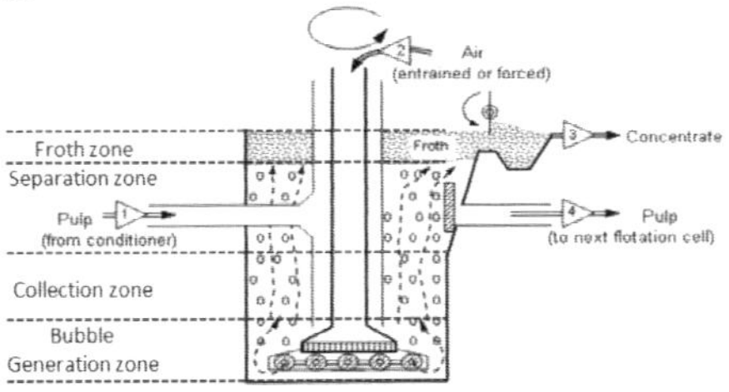

Figure 1. (a) Schematic cross-section with the different zones: 1 - the pulp enters the cell from a conditioner, and flows to the bottom of the cell; 2 - small air bubbles are passed down as a vertical impeller; 3 - collection of the mineral concentrate froth from the top of the cell; 4 - pulp flows to another cell. (Modified after http://www.wikiwand.com/en/Froth_flotation).

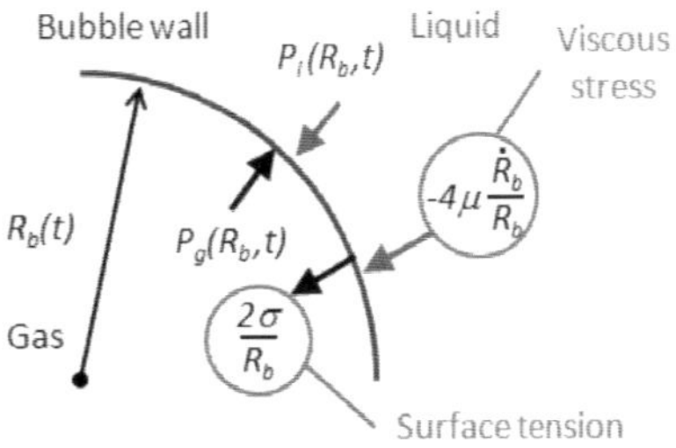

Figure 2. Schematic view of bubble-particle interaction and forces acting on the bubble surface including the gas surface tension, the liquid viscous stress, the liquid pressure P_l, the gas pressure P_g, t is time, R_b the bubble radius, $\dot{\boldsymbol{R}}_b$ its derivate against time, μ the cinematic liquid viscosity, and σ surface tension (after [11])

2. Governing equations for bubble evolution under ultrasonic waves

Hydrodynamics of single bubble immersed in a liquid subjected to ultrasonic waves can be described by the Navier-Stokes equations applied to the gas inside the bubble, initially at rest, and to the incompressible liquid adjacent to the bubble wall. The Navier-Stokes equations can be solved analytically in spherical coordinates [9] or numerically using a radial formulation [10].

2.1. Solving Navier-Stokes equations

A bubble at rest with mean initial radius R_0, at pressure P_0, and temperature T_0, is immersed in an unbounded incompressible liquid submitted to an external time-dependent pressure at its boundary $P_\infty(t)=P_0+P(t)$, where t the time, and $P(t)$ the time-dependent pressure time (including negative and positive fluctuations). Far from the bubble the temperature of the liquid is assumed to be constant at T_∞. It is also assumed that the pressure and the temperature inside the bubble are always uniform, and therefore is time-dependant. The viscous Newtonian fluid is assumed incompressible with constant density ρ. The bubble has a spherical shape preserved during all the process, with radius $R(t)$ oscillating through time under ultrasound. Neglecting gravitational forces, the Navier-Stokes equations (or mass and momentum conservation equations) describe the flow field surrounding the bubble [10]. The first one is the continuity of the liquid:

$$\left(\frac{\partial\rho}{\partial t}\right) + \boldsymbol{\nabla}.\left(\rho\,\mathbf{u}(r,t)\right) = 0 \tag{1}$$

with $\boldsymbol{u}$ the liquid velocity, r the radial distance to the bubble center, and t the time. As the bubble is assumed to be spherical, Eq.(1) can be simplified into $\partial\,(r^2\,u(r,t)) = 0$ using spherical coordinates, with solution for the radial velocity:

$$u(r,t) = \frac{1}{r^2}F(t) \tag{2}$$

where $F(t)$ is a function of time determined by the boundary conditions. Assuming no mass transfer across the bubble boundary, the wall velocity in a local coordinate system attached to the bubble is equal to the change rate of its radius $u(R_b,t) = \dot{R}_b$ (where $\dot{R}_b$ is the derivation of R_b against time t), thus $F(t) = R_b^2(t)\dot{R}_b$, and consequently Eq.(2) becomes:

$$u(r,t) = \left(\frac{R_b(t)}{r}\right)^2\dot{R}_b \tag{3}$$

Neglecting the dynamic viscosity of the liquid, and substituting the radial component of the velocity, the radial component of the Navier-Stokes equation applicable to the liquid can be written as:

$$\frac{1}{\rho}\frac{\partial P_l}{\partial r} + \frac{\partial u}{\partial t} + u(r,t)\frac{\partial u}{\partial r} = 0 \qquad -\frac{1}{\rho}\frac{\partial P_l}{\partial r} = \frac{R_b^2\,\ddot{R}_b + 2\,R_b\,\dot{R}_b^2}{r^2} - \frac{2\,R_b^4\,\dot{R}_b^2}{r^5}$$

$$\Rightarrow -\frac{1}{\rho}\int_{P_l(R_b,t)}^{P_\infty(t)} dP = \int_{R_b}^{\infty}\left(\frac{R_b^2\,\ddot{R}_b + 2\,R_b\,\dot{R}_b^2}{r^2} - \frac{2\,R_b^4\,\dot{R}_b^2}{r^5}\right)dr \Rightarrow \frac{P_l(R_b,t) - P_\infty}{\rho} = R_b\ddot{R}_b + \frac{3}{2}\,\dot{R}_b^2 \tag{4}$$

IOP Conf. Series: Journal of Physics: Conf. Series **879** (2017) 012024 doi:10.1088/1742-6596/879/1/012024

where $P_l(R_b,t)$ is the liquid pressure on the bubble wall which is equal to the sum of all forces acting on the wall bubble (Figure 2): (i) the *gas pressure* in the bubble P_g, (ii) the *viscous stress* σ_{rr} of the liquid; and (iii) the *superficial tension* τ. It comes $P_l = \sigma_{rr} + \tau + P_g$. The radial component of the *viscous stress* σ_{rr} acting on the bubble wall depends on the liquid cinematic viscosity μ [12]:

$$\sigma_{rr}(r,t) = 2\mu\,\frac{\partial u}{\partial r} = -\,4\mu\frac{\dot{R}_b}{R_b(t)} \tag{5}$$

The *surface tension* τ acting in the bubble is given by the Laplace's formula $\tau(t) = -\frac{2\sigma}{R_b(t)}$ where σ is the surface tension. A process for an ideal gas is *polytropic* and obeys the relation $PV^m = B$ if the ratio of energy transfers as heat δQ to energy transfer as work δW remains constant ($\delta Q/\delta W = cst = K$) at each infinitesimal step, m is the *polytropic coefficient* which depends on the nature of the process (and of the gas). For ideal gas, specific values of m vary between *isothermal* ($m = 1$) to *isentropic* (adiabatic or reversible) ($m = \gamma = 1.4$ for air), γ being the *adiabatic coefficient*. Assuming an ideal gas polytropic transformation for the gas trapped in the bubble, the gas pressure is equal to:

$$P_g(t) = P_{g_0}\left(\frac{R_{b_0}}{R_b(t)}\right)^{3m} \tag{6}$$

with P_{g_0} the initial gas pressure in the bubble at rest. Assuming that at $t = 0$, $P_l(R_b,0) = P_0$ and $\dot{R}_b = 0$, the initial pressure in the bubble becomes $P_{g_0} = \frac{2\sigma}{R_{b_0}} + P_0$. Then, from Eqs. (5), and (6), the total pressure at the bubble boundary P_l can be written as:

$$P_l(R_b,t) = \left(\frac{2\sigma}{R_{b_0}} + P_0\right)\left(\frac{R_{b_0}}{R_b(t)}\right)^{3m} - 4\mu\frac{\dot{R}_b}{R_b(t)} - \frac{2\sigma}{R_b(t)} \tag{7}$$

Other gas models (including ideal gas, Soave-Redlich-Kwong, and Peng-Robinson) can be used to predict the evolution of the internal bubble pressure (see §2.3.).

2.2. *Effect of ultra sound on the bubble pressure boundary*

From $P_\infty(t) = P_0 + P(t)$, and equating Eqs.(4) and (6), the time-dependent pressure term $P(t)$ is equal to:

$$P(t) = \left(\frac{2\sigma}{R_{b_0}} + P_0\right)\left(\frac{R_{b_0}}{R_b(t)}\right)^{3m} - \frac{2\sigma}{R_b(t)} - \rho\left[R_b\ddot{R}_b + \frac{4\mu}{\rho}\frac{\dot{R}_b}{R_b} + \frac{3}{2}\dot{R}_b^2\right] - P_0 \tag{8}$$

The general model can be used to describe the time evolution of the bubble radius $R_b(t)$ given by $R_b(t) = R_{b_0}f(t)$ with $f(\omega t) = 1 + \varrho\cos(\omega(t+\tau_0) + \upsilon\pi/2)\{1 - \mu'[1 - \exp\{\sin(\omega(t+\tau_0))\}]\}$ with $\varrho < 1$, where the coefficients (R/R_0, ϱ, $\omega\tau_0$, $v = \mu' = 1$) are constant or function of the Deborah number D_e ($R/R_0(D_e)$, $\varrho(D_e)$, $\omega\tau_0(D_e)$, $v = \mu' = 0$) [7], where $D_e = \lambda\omega$ (with λ the polymer relaxation time) [10]. Thus, Eq.(8) can be rewritten as:

$$P(t) = \left(\frac{2\sigma f^{-1}}{R_{b_0}} + P_0\right)[f^{-3m+1} - 1] + P_0 f^{-3m}[1 - f] - \rho R_{b_0}^2\left[f\,\ddot{f} + \frac{4\mu}{\rho R_{b_0}^2}\dot{f}f^{-1} + \frac{3}{2}\dot{f}^2\right] \tag{9}$$

Eq.(9) gives explicitly the time-dependent pressure term $P(t)$ as a function of the ultra sonic wave function f.

2.3. *Pressure inside the bubble*

Eq.(9) gives the pressure of the liquid at the bubble boundary assuming a polytropic process for the ideal gas trapped in the bubble. Several other gas models (including Soave-Redlich-Kwong, and Peng-Robinson) can be used to predict the evolution of the internal bubble pressure.

Polytropic ideal gas process: (P): $P_g(t)\,V_b(t)^m = P_{g0}\,V_{b0}^{\,m} = B$ where B is a constant, as $V_b(t) = 4/3\pi\,R^3_b(t)$, reporting all constant terms in B', it comes: $P_g(t)\,R_b(t)^{3m} = P_{g0}\,R_{b0}^{\,3m} = B'$ (with $B' = 3B/4\pi$), then using the general model for describing the time-dependent bubble radius submitted to ultrasound, the gas pressure inside the bubble is given by: $P_g(t) = \left(\frac{2\sigma}{R_{b_0}} + P_0\right)f^{-3m}$.

Soave-Redlich-Kwong gas model (SRK): $P_g(t) = \frac{\mathcal{R}T_b(t)}{V_m - b} - \frac{a\alpha}{V_m(V_m + b)}$ with $a = \frac{0.427\mathcal{R}^2T_c^2}{P_c}$; $b = \frac{0.08664\mathcal{R}T_c}{P_c}$ where T_c and P_c are the critical temperature and pressure for the gas, respectively (for air $V_m = 22.4\,\mathrm{l.mol^{-1}}$), and $\Re = 8.314\ \mathrm{J.(K{\cdot}mol)^{-1}} \approx 2\ \mathrm{cal.(K{\cdot}mol)^{-1}}$ is the ideal gas constant.

Normally, $\alpha = (1 + (0.48508 + 1.55171\omega - 0.15613\omega^2)(1 - T_r^{0.5}))^2$ with ω the *acentric factor*, and T_r the dimensionless temperature $T_r(t) = T_b(t)/T_c$. Assuming no change of state in the bubble, the acentric factor can be neglected, so a simplified expression for $\alpha = (1 + 0.48508(1 - T_r^{0.5}))^2$ can be used.

Peng-Robinson gas model (PR) is similar to the SRK model $P_g(t) = \frac{\mathcal{R}T_b(t)}{V_m - b} - \frac{a'\alpha}{V_m(V_m + 2b') - b'^2}$ with $a' = \frac{0.457235\mathcal{R}^2T_c^2}{P_c}$; $b' = \frac{0.077796\mathcal{R}T_c}{P_c}$; $\alpha = (1 + \kappa(1 - T_r^{0.5}))^2$ with $\kappa = 0.37464 + 1.54226\omega - 0.26992\omega^2$, but neglecting the acentric factor gives: $\alpha = (1 + 0.37464(1 - T_r^{0.5}))^2$. The temperature $T_b(t)$ can be calculated using the polytropic model.

Generalized SRK-PR model: The SRK and PR model can be rewritten in a single equation: $P_g(t) = \frac{\mathcal{R}T_b(t)}{V_m - b} - K$ where $K = K_{SRK} = \frac{a\alpha}{V_m(V_m + b)}$, and $K = K_{PR} = \frac{a'\alpha}{V_m(V_m + 2b') - b'^2}$, for the SRK and PR models.

2.4. Temperature inside the bubble

A *polytropic* transformation for an *ideal* gas model is assumed for estimating the uniform temperature inside a homogeneous gas bubble.

Polytropic process: $T_b(t)\, V_b(t)^{m-1} = T_{b0}\, V_{b0}^{m-1} = C$ (from reporting the ideal gas equation $PV = n\mathfrak{R}T$ in the definition $PV^m = B$), where the constant $C = B/n\mathfrak{R}$; as for pressure, it comes: $T_b(t) = T_{b0} f^{3(1-m)}$.

Ideal gas model (I) : $P_g(t)\, V_b(t) = n\mathfrak{R}\, T_b(t)$ and $P_{g0}\, V_{b0} = n\mathfrak{R}\, T_{b0}$ where n is the amount of gas (in moles), $\mathfrak{R} = 8.314$ J.(K·mol)$^{-1}$ being the ideal gas constant (equal to the product of the Boltzmann constant and the Avogadro constant). When ultra sounds are applied, it comes: $T_b(t) = T_{b_0} f^{3(1-m)} \left(\frac{2\sigma}{R_{b_0}P_0} + 1\right)$ with a pressure equal to $P_g(t) = P_{g_0} f^{-3m} \left(\frac{2\sigma}{R_{b_0}P_0} + 1\right)$.

2.5. Temperature in the liquid around the bubble

The fluid temperature T_l around the bubble is given by the heat equation in spherical coordinates:

$$\frac{\partial T_l}{\partial t} = \frac{\alpha_l}{r^2}\frac{\partial}{\partial r}\left(r^2 \frac{\partial T_l}{\partial r}\right) - u(r,t)\frac{\partial T_l}{\partial r} \tag{10}$$

where α_l is the liquid thermal diffusivity, u(r,t) the fluid radial velocity around the bubble, and T_l the liquid temperature. Due to the nonlinearities, this equation has no exact analytic solutions. However, a relevant approximation of the solution valid when the thickness of the thermal boundary layer surrounding the bubble is small compared to its radius is given by [13]:

$$T_\infty(t) - T_B(t) = \sqrt{\frac{\alpha_l}{\pi}} \int_0^t \frac{R_b^2(x)\left(\frac{\partial T}{\partial r}\right)_{r=R_b(x)}}{\sqrt{\int_x^t R_b(y)^4 dy}}\, dx \tag{11}$$

An approximation of the analytical solution can be found if the bubble surface is assumed to be maintained at constant temperature. Assuming that the spherical bubble with a constant average radius R_{b0} and a constant average temperature $\tilde{T}_B$ at the bubble surface, is placed at rest in a stagnant fluid at constant temperature $\tilde{T}_f$ (far away from the bubble), the average temperature $\tilde{T}(r)$ in the surrounding fluid is an inverse distance function of the distance r to the bubble center given by [14] $\frac{\tilde{T}(r) - \tilde{T}_f}{\tilde{T}_B - \tilde{T}_f} = \frac{R_{b0}}{r}$. Notice that the temperature profile is not dependent on the fluid thermal conductivity. The decrease in temperature is very fast locally, for small bubbles (around *100μm*), at a distance $r = 10R$ (*1mm*), the liquid temperature increases only of about *3°C* for a temperature difference of $\Delta T = \tilde{T}_B - \tilde{T}_f \approx 30°C$.

3. Energy balance and thermodynamic considerations

The energy balance in a bubble evolving under an external field such as ultrasonic treatment can be now calculated assuming a polytropic transformation and considering different gas models (Ideal, SRK, and PR). The first law of thermodynamics gives the total variation of the internal energy ΔU_{tot} of a system as the sum of the total work W_{tot} and the total heat Q_{tot} supplied to or escaping from the system through its boundaries: $\Delta U_{tot} = W_{tot} + Q_{tot}$.

Table 1. Synthesis of theoretical formula and results from various numerical experiments

Parameters	State equations		
	Ideal gas	Soave-Redlich-Kwong	Peng-Robinson
$T_b(t)$	$T_b(t) = T_{b_0} f^{3(1-m)}\left(\frac{2\sigma}{R_{b_0}P_0}+1\right)$	$T_b(t) = T_{b0} f^{3(1-m)}$	
$\tilde{T}_b(in°C)$	$88.02°C$	$61.62°C$	
$\tilde{T}_b(in\ \mathrm{K})$	$361.17K$	$334.77K$	
$P_g(t)$	$P_g(t) = \frac{n\mathcal{R}T_b(t)}{V_b(t)}$	$P_g(t) = \frac{\mathcal{R}T_b(t)}{V_m - b} - \frac{a\alpha}{V_m(V_m+b)}$	$P_g(t) = \frac{\mathcal{R}T_b(t)}{V_m - b} - \frac{a'\alpha}{V_m(V_m+2b') - b'^2}$
$\tilde{P}_g(in\ MPa)$	0.158	0.124	
$W_{tot}(t)$	$W_{tot}(t) = W_0\left[f^{3(1-m)} - 1\right]$	$W_{tot}(t) = W_0\left[f^{3(2-m)} - f_0^{3(2-m)}\right]$; $W_0 = \frac{(m-1)n\mathcal{R}\,T_{b_0}}{m\,(m-2)}$	
$\tilde{W}_{tot}(in\ 10^{-8}\,\mathrm{J})$	0.087	62	
$Q_{tot}(t)$	$Q_{tot}(t) = Q_0\left[f^{3(1-m)} - 1\right]$	$Q_{tot}(t) = 4\pi Q_0 t\, f\left(f^{3(1-m)} - 1\right)$; $Q_0 = k\,R_{b_0}T_{b_0}$	
$\tilde{Q}_{tot}(in\ 10^{-8}\,\mathrm{J})$	0.13	14.2	
$\Delta U_{tot}(t)$	$\Delta U_{tot}(t) = (W_0 + Q_0)\left[f^{3(1-m)} - 1\right]$	$\Delta U_{tot}(t) = W_{tot}(t) + Q_{tot}(t)$	
$\Delta U_{tot}(in\ 10^{-8}\,\mathrm{J})$	0.217	76.2	

3.1. Polytropic models (P)

Internal energy of a bubble: for an ideal gas $dU = nc_V\,dT_g = \frac{n\mathcal{R}\,dT_g}{\gamma-1} = \frac{n\mathcal{R}\,T_g}{\gamma-1}(1-m)\frac{dV_b}{V_b}$ where n is the number of moles of gas, c_V is the molar constant volume (c_P pressure) heat capacity of the gas, and $\gamma = c_P/c_V$ the heat capacity ratio (γ=5/3 = 1.67 for monatomic gas with three degree of freedom, and 7/5 = 1.4 for diatomic gas with five degree of freedom, γ = 1.4 for air). After some arithmetic, it comes: $\Delta U_{tot} = \frac{P_{g_2}V_2 - P_{g_1}V_1}{\gamma-1} = U_0\left[f^{3(1-m)} - 1\right]$, with $U_0 = \frac{V_0P_0}{\gamma-1}\left(\frac{2\sigma}{R_{b_0}P_0}+1\right)$. If the transformation is isothermal ($m=1$), and the total internal energy variation is null $\Delta U = 0$ (adiabatic transformation).

Total work: Assuming a polytropic process so $P_g(t)\,V_b(t)^m = P_{g0}\,V_{b0}^{\ m}$, the total work done by the system when the volume bubble changes from V_1 to V_2, is given by: $W_{tot} = -B\int_1^2 P_g(t)dV = -B\int_1^2 \frac{dV}{V^m(t)}$. For integration, one must separate the *isothermal* ($m=1$) to the *non-isothermal* ($m\neq1$) cases; it gives after some arithmetic: $m = 1$, $W_{tot} = B\ln\left(\frac{V_1}{V_2}\right)$; $m\neq1$ $W_{tot} = \frac{P_{g_2}V_2 - P_{g_1}V_1}{m-1}$. The total work can be rewritten in terms of temperature as: $W_{tot} = \frac{n\,\mathcal{R}\,(T_2-T_1)}{m-1}$. Finally, reporting in the total work expression the bubble pressure $P_g(t)$ for an ideal gas polytropic process, and $V_b(t) = 4/3\pi R^3_b(t)$, it comes: $m\neq1$, $W_{tot} = W_0\left[f^{3(1-m)} - 1\right]$ where $W_0 = \frac{V_0P_0}{m-1}\left(\frac{2\sigma}{R_{b_0}P_0}+1\right)$ with V_0, P_0, T_0 are the volume, pressure, and temperature of the bubble at rest; for the isothermal case ($m=1$), $W_{tot} = -\,W_0'\ln(f)$ with $W_0' = 3n\mathcal{R}\,T_0$.

Heat supplied by the liquid Q_{tot} can be evaluated applying the first law of thermodynamics, for $m\neq1$ $Q_{tot} = \Delta U_{tot} - W_{tot} = (U_0 - W_0)\left[f^{3(1-m)} - 1\right] = Q_0\left[f^{3(1-m)} - 1\right]$, with $Q_0 = V_0P_0\left(\frac{2\sigma}{R_{b_0}P_0}+1\right)\left(\frac{1}{\gamma-1} - \frac{1}{m-1}\right)$. For an isothermal transformation ($m = 1$) $Q_{tot} = -\,W_{tot} = W_0'\ln(f)$; when $m = \gamma = 1.4$, the transformation is adiabatic as the total heat supplied to the system is null ($Q_{tot} = 0$), $\Delta U_{tot} = W_{tot}$.

3.2. Generalized SRK-PR models with m≠1 *(isothermal case excluded)*

Total work: assuming a polytropic transformation, the gas pressure P_g and the bubble volume V_b are related by $P_g(t)\,V_b(t)^m = P_{g0}\,V_{b0}^{\ m} = C$; reporting the pressure given by the generalized SRK-PR gas state equation, it comes: $\left(\frac{\mathcal{R}T_b(t)}{V_m - b} - K\right)V_b^m(t) = C$. By differentiation $dV_b = -\frac{\mathcal{R}V_b(t)dT_b}{m\left(\mathcal{R}T_b(t) - K(V_m - b)\right)}$.

(a) 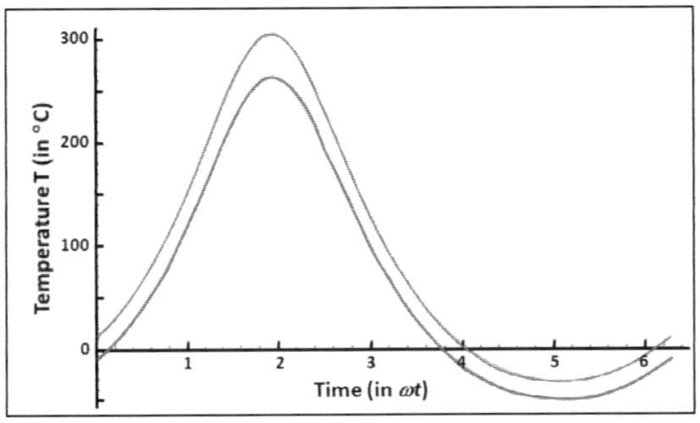

(b) 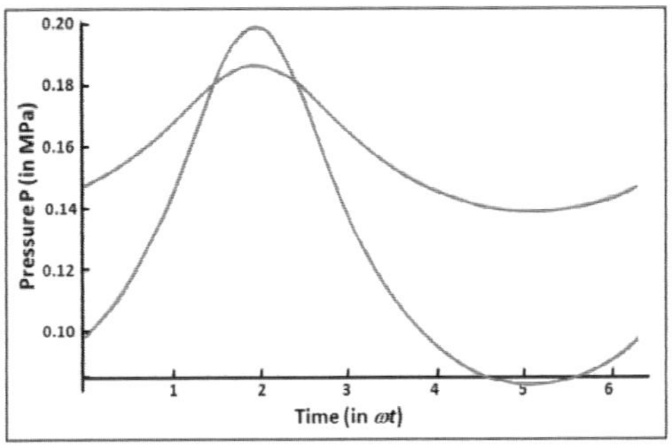

Figure 3. (a) Calculated temperature inside a bubble for ideal (red), and polytropic (blue) gas model; (b) calculated pressure inside a bubble for ideal (blue), and SRK-PR (red) gas model. Average increase in temperature is about 43°C, and pressure variation about 0.022 *MPa*.

So the total work is given by $W_{tot} = -\int_1^2 P_g(t)dV_b = \int_1^2 \left(\frac{\mathcal{R}V_b}{m(V_m-b)}\right) dT_b$, but as bubbles are spherical, $V_m - b \approx 4/3\pi R^3_{b0}$ so $\frac{V_b}{(V_m-b)} \approx \left(\frac{R_b(t)}{R_{b_0}}\right)^3 = f^3$; however, $dT_b = 3(1-m)T_{b_0}f^{2-3m}df$ so by integration, the total work is equal to: $W_{tot} = W_0\left[f^{3(2-m)} - f_0^{3(2-m)}\right]$ with $W_0 = \frac{(m-1)\mathcal{R}\, T_{b_0}}{m\,(m-2)}$ (valid for a molar volume, $n=1$), where f_0 is the f function at initial time $t=0$.

3.3. Heat transfer

Two models, *simple conductive* and *convective with a mobile bubble*, are investigated.

Simple conduction: the bubble is assumed to be motionless in the fluid with uniform surface temperature. The Fourier's law in spherical coordinates gives the heat flux Φ at the surface of the bubble: $\Phi = \left(\frac{k}{R_b(t)}\right)\left(T_b(t) - T_{b_0}\right)$ where k is the thermal conductivity (for water at $T_\infty = T_{b0} = T_0 = 298.15$K (25°C), k = 0.6071 $W.m^{-1}K^{-1}$). The total heat transfer through the bubble surface S is equal to: $Q_{tot}(t) = S\Delta t\Phi = 4\pi R_b^2(t)\Delta t\left(\frac{k}{R_b(t)}\right)\left(T_b(t) - T_{b_0}\right)$. Starting from $t=0$ to $\Delta t = t$ after some arithmetic, and observing $\frac{T_b(t)}{T_{b_0}} = f^{3(1-m)}$, it comes: $Q_{tot}(t) = 4\pi Q_0 t\, f\left(f^{3(1-m)} - 1\right)$ with $Q_0 = k\, R_{b_0} T_{b_0}$.

Convective model with a mobile bubble: The motion of the bubble trough the flotation column creates a convective component for the heat transfer with water. Considering the bubble surface as isothermal, and applying the analytic method for laminar free convective heat transfer from isothermal spheres developed by [9], the heat flux at the surface bubble is: $\Phi = k\left(\frac{1}{R_b(t)} + \frac{1}{\sqrt{\alpha\pi t}}\right)\left(T_b(t) - T_{b_0}\right)$ where k is the thermal conductivity of the water, $\alpha = \frac{k}{\rho c_P}$ the thermal diffusivity, c_P the constant pressure heat capacity, $\rho = \frac{\rho_0}{1+\beta(T_b(t)-T_{b_0})}$ the temperature dependent water density with at $T_0 = 298.15K$ ρ_0=997.05 $kg.m^{-3}$, and β the water volumetric temperature expansion coefficient $\beta = 2.10^{-4}K^{-1}$. After some arithmetic, it comes: $\Phi = k\, R_{b_0}^{-1} f\, T_{b_0}\left(f^{3(1-m)} - 1\right)\Phi_c$ with $\Phi_c = f^{-1} + \frac{R_{b_0} f\sqrt{c_P \rho_0}}{\sqrt{k\pi t\left(1+\beta T_{b_0}\left(f^{3(1-m)}-1\right)\right)}}$, being the convective term. Thus, the total heat transfer is given by $Q_{tot}(t) = 4\pi Q_0 t\, f\left(f^{3(1-m)} - 1\right)\Phi_c$, this expression is similar to the conductive case, excepted that it is multiplied by the convective term Φ_c.

4. Numerical simulation

Numerical simulations were made using the following parameters ([9], [7]): P_0=0.142MPa, σ=0.0728$N.m^{-1}$, R_{b0}=500-700μm, $\omega\tau_0$=-0.374, R/R_0=1, ω=28.5kHz, m=1.2, μ=0, ν=1, μ=0.025$N.s.m^{-2}$, ρ=1,800$kg.m^{-3}$, T_0=298.15K, n=1mol, $\mathfrak{R}$=8.314$J.mol^{-1}.K^{-1}$, and γ=1.4. Results are summarized Table 1.

IOP Conf. Series: Journal of Physics: Conf. Series **879** (2017) 012024 doi:10.1088/1742-6596/879/1/012024

4.1. Average temperature inside the bubble

Temperature inside a single bubble submitted to ultrasonic waves was calculated over a period against time (ωt), and fluctuates as a sinusoid (Figure 3a). Due to the compression of the gas, instantaneous temperature can reach punctually values as high as *300℃*, but also decrease under *0℃* during decompression. The average value of the temperature $\widetilde{T_b} = \frac{1}{2\pi}\int_0^{2\pi} T_b(\omega t)d(\omega t)$ was calculated by integration over one period of time. For the polytropic gas model, $\tilde{T}_b^P = 334.77K = 61.62°C$, a value corresponding approximately to one third less than for the ideal gas model ($\tilde{T}_b^I \approx 88.02$°C). Keeping the key parameters in the range $0 < \varrho < 1$ and m=1.2, prevents singularities for the bubble behavior and shows that the instantaneous temperature of the bubble is always inferior to 600K (not thousand of degrees as reported in [15]).

4.2. Average pressure inside the bubble

Pressure in the bubble fluctuates also as a sinusoid (Figure 3b), over a range of about *0.10MPa* given the boundaries conditions used for the reported numerical experience. The SRK and PR gas models give identical results with bigger amplitude variations compared to the ideal gas model. The average value of the bubble pressure $\widetilde{P_g} = \frac{1}{2\pi}\int_0^{2\pi} P_g(\omega t)d(\omega t)$ calculated by integration over one period of time is $\tilde{P}_g^{SRK} = 0.124MPa$ for the Peng-Robinson gas model and $\widetilde{P}_g^I = 0.158MPa$ for ideal gas model.

4.3. Average internal work and head transfer between the bubble and the liquid

The *average work* done over a period by a bubble of radius equal to *500-700µm* has been estimated at $\widetilde{W}_{tot}^{Ig} = 4.95 - 13.58\times10^{-5}J$ for the ideal gas model, and at $\widetilde{W}_{tot}^{SRK/PR} = 3.64-10.00\times10^{-3}J$ for the Peng-Robinson gas model. The huge difference between the two models is in part explained by the higher power term affecting *f* in the Peng-Robinson model. The conduction model gives an average heat transfer of $\widetilde{Q}_{tot}^{Ig} = -8.08 - 22.18\times10^{-3}J$.

5. Thermodynamics variation against *m* and *ϱ*

5.1. Effect of polytropic coefficient m on P, T, ΔU, W_{tot}, Q_{tot},

P and T: Bubble average temperature and pressure increase when the process moves from isothermal to adiabatic (isentropic or reversible) both for the ideal and SRK gas model. Numerical results show unrealistic temperature and pressure for values greater than $m \approx 1.21$ (Figure 4a), indicating that heat and work flows in opposite directions ($\delta Q/\delta W > 0$), and the compression/decompression of bubbles would be closer to an isothermal than adiabatic process. The isothermal process ($m=1$) can be excluded, as heat is transferred through the bubble boundary from gas to the water given the thermal conductivity of water, so *m* would be in the range $1<m<1.21$. High values near the adiabatic conditions ($\Delta U \approx 0$) indicate that energy does not escape but accumulates in the bubble, increasing dramatically both the temperature and the pressure.

5.2. Effect of ultra sonic wave amplitude ϱ

Assuming a range $0 < \varrho < 1$ for the amplitude coefficient of the ultra sonic wave [7], and a polytropic process for an ideal or SRK gas, the temperature and pressure increase when the amplitude increases. For $0 < \varrho < 0.5$, the temperature inside the bubble increases smoothly with ϱ up to about 100°C, to reach a sharp increase up to 300°C for $\varrho \approx 0.8$, as more energy is carried by the ultrasonic wave . The over-pressure inside bubble also increases with ϱ from approximately *0.1* to *0.22MPa*.

5.3. Effect of polytropic coefficient m on thermodynamics

Internal energy ΔU_I (ideal gas model), total work W_{tot} (SRK gas model), and heat exchanges Q_{tot} (convective exchanges) have been calculated at the bubble scale over one period of time (Figure 5). As for average inside bubble temperature and pressure, they all reach high values nearby adiabatic conditions for ($m> 1.21$).

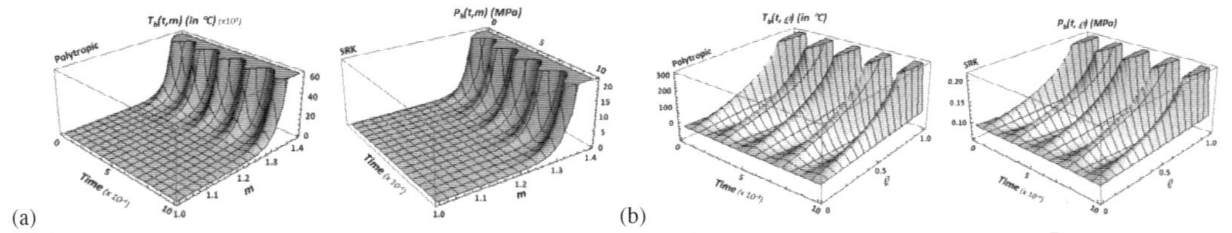

(a) Temperature Pressure (b) Temperature Pressure

Figure 4. Calculated temperature and pressure inside a bubble for polytropic and SRK models varying (a) the polytropic coefficient *m* and (b) the amplitude ϱ of the ultra sonic wave.

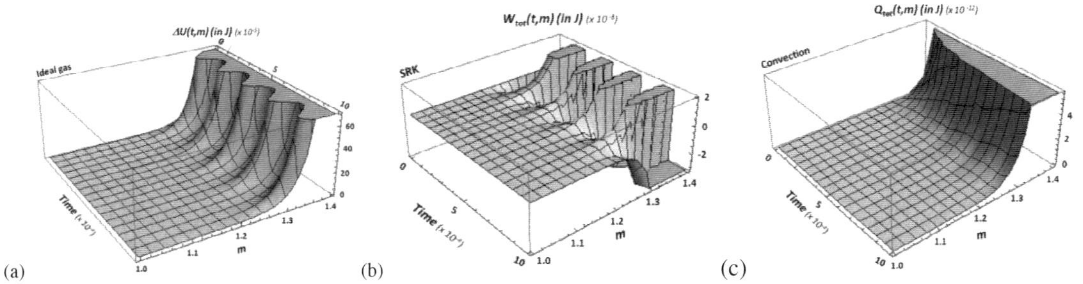

(a) Variation of bubble internal energy ΔU (b) Total work W_{tot} (c) Total convective heat exchange Q_{tot}

Figure 5. Calculated energy balance for polytropic and SRK models varying the polytropic coefficient *m* *(for* $R_0=13\mu m$*)*.

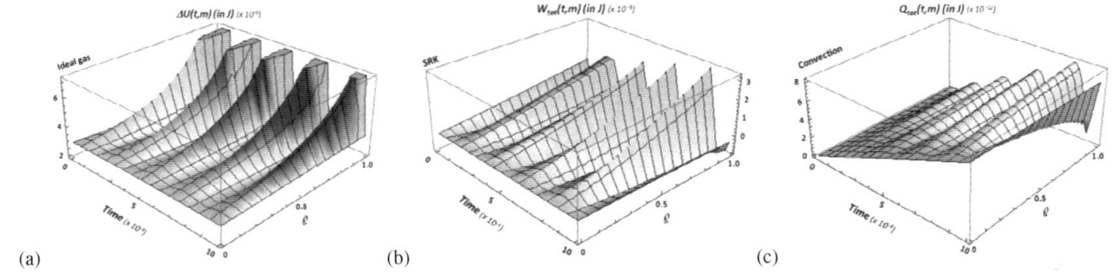

(a) Variation of bubble internal energy ΔU (b) Total work W_{tot} (c) Total convective heat exchange Q_{tot}

Figure 6. Calculated energy balance for polytropic and SRK models varying amplitudes ϱ of ultra sonic wave *(for* $R_0=13\mu m$*)*.

5.4. Effect of ultra sonic wave amplitude ϱ on thermodynamics

The same energy balance was performed when varying the ultrasonic amplitude ϱ (Figure 6). As for average inside bubble temperature and pressure, the internal energy and total work increase as amplitude increases. For the ideal gas model, the total work and total heat transfer are similar, but differ when applying a convective model of heat exchange (and SRK gas model). In this last case, the total heat seems to increase against time, indicating that the cooling of the bubble by the surrounding liquid is too slow, leading to an overheating of the gas trapped in the bubble.

6. Conclusions

This theoretical work demonstrates that the ultrasound effect on thermodynamics variables describing the temperature, pressure, internal energy, total work, and heat exchanges can be predicted at the bubble scale. The most direct impact is that under ultra sonic stimulation, the work required to modify the bubble volume is converted into heat that increases the average temperature of the gas trapped in the bubble, up to about *60°C*, or more, depending on the ultrasonic wave amplitude, the nature of the gas and the process involved (isothermal or adiabatic). This heat is transferred to the surrounding liquid by conduction (or convection) through the bubble surface, maintaining average bubble temperature equilibrium by cooling. However, given the micro bubble size, the warming of the surrounding liquid is limited at about 2°C, despite than nearby the bubble boundary, higher temperatures can be observed, implying in some case, the vaporization of the liquid, and thus the growing of the bubble. Inside bubble pressure increases by about 0.14MPa. Several assumptions have done during this theoretical work which may to be confirmed or not by experimental work, or by direct measurements of the bubble characteristics. The isothermal case is a very interesting case, because it induces a null internal energy, meaning that all the work done by the pressure forces on the bubble is integrally converted into heat supplied to the liquid around the bubble, without any lost due to the bubble heating/cooling, limiting the overall energy consumption due to ultrasonic stimulation. This favorable case should be searched when running a flotation process. Some authors suggest [12] the gas contained in the bubble does not correspond to a unique gas (air), but rather to a mixing of air and vapor created by boiling of the liquid surrounding the bubble. In this last case, it is required to take into account the energy used to vaporize the liquid in the energy balance, which is according to [12] higher than the energy used for heating/cooling the bubble, a point not investigated in this work.

7. Acknowledgements

The authors would like to express their thanks for the support to the Lorraine-Russia Arcus Project. The work was supported by the Russian Foundation for Basic Research (grant 17-41-590974).

References

[1] Clingan B V and McGregor D R 1987. *Column flotation experience at Magma Copper Company, with related experience of other mineral processors*, 116th Annual AIME-SME Meeting. Denver.
[2] Villar del R, Bouchard J, Desbiens A and Poulin E 2015. *Recent Developments in Flotation Column Instrumentation and Control: An Update*. Proceedings of the World Congress on Mechanical, Chemical, and Material Engineering (MCM 2015), Barcelona, Spain, July 20-21, Paper No. 127, 1-9.
[3] Kursun H and Ulusoy U 2015. Zinc Recovery from a Lead-Zinc-Copper Ore by Ultrasonically Assisted Column Flotation. *Particulate Science and Technology*, **33**(4), 349-356.
[4] Mosai R and Dawe R A 2003. Gas attachment of oil droplets for gas flotation for oily wastewater cleanup. *Separation and Purification Technology*, **33**(3), 303-314.
[5] Sobhy A 2013. *Cavitation nanobubble enhnaced flotation process for more efficient coal recovery*. PhD dissertation, Mining Engineering, University of Kentucky, UKnowledge, Lexington, US, 169p.
[6] Manlapig E V, Green C, Parkinson J W and Murphy A S 2001. *The technology and economic incentives for recovering coal from tailings impoundments*. SME Annual Meeting, Denver, Colorado, 26–28 February, Preprint 01-70.
[7] Royer J J, Filippov L O and Filippova I. V. 2012. *Improving fine mineral ore flotation using ultrasound*. Internal report, GeoRessources, Nancy, France, 22p.
[8] Filippov L O, Royer J J and Filippova I V 2016. *Improving efficiency in flotation ore recovery using ultra-sonic cavitation bubbles. Part I.* (this volume), 9p.
[9] Kim K and Kwak H Y 2007. Predictions of bubble behavior in sulfuric acid solutions by a set of solutions of Navier-Stokes equations. *Chemical Engineering Science*, **62**, 2880-2889.
[10] Foteinopoulou K and Laso M 2010. Numerical simulation of bubble dynamics in a Phan-Thien-Tanner liquid: non-linear shape and size oscillatory response under periodic pressure. *Ultrasonics*, **50**, 758-776.
[11] Nguyen A V, Ralston J and Schulze H J 1998. On modelling of bubble-particle attachment probability in flotation, *Int. J. Miner. Process*, **53**, 225-249.
[12] Brennen C E 1995. *Cavitation bubble dynamics*. Oxford University Press. 254p.
[13] Plesset M S and Zwick S A 1952. A Nonsteady heat diffusion problem with spherical symmetry. *Journal of applied physics*, **23**, 95-98.
[14] Carslaw H S and Jaeger J C 1959. *Conduction of heat in solids*. Oxford Univ. Press, 517p.
[15] Qi B and Aldrich C 2002. Effect of ultrasonic treatment on zinc removal from hydroxyde precipitates by dissolved air flotation. *Minerals Engineering*, **15**, 1105-1111.
[16] Beychok M R 1967. *Aqueous Wastes from Petroleum and Petrochemical Plants* (1st ed.). John Wiley & Sons Ltd. LCCN 67019834. 370p.

Dynamics of an ensemble of spherical particles under translational vibrations of linear polarisation

T P Lyubimova[1,2], Y M Maslova[1], O O Fattalov[1,2]

[1]Institute of Continuous Media Mechanics UB RAS, Perm, Russia
[2]Perm State University, Perm, Russia

lubimova@psu.ru

Abstract. This work studies the dynamics of a set of rigid spherical particles in an oscillating viscous liquid under translational vibrations. Experimental data on the influence of vibration parameters and fluid properties on the formation and characteristics of structures are obtained. In the frequency range from 5 Hz to 27 Hz, particles align to form linear structures oriented perpendicular to the direction of the vibrations. For frequencies greater than 27 Hz, the particles are collected on the walls of the cavity. Distance between the linear chains increases with the vibration amplitude according to a close-to-linear law, but decreases inversely as a function of the vibration frequency.

1. Introduction

Studies of vibration effect on multiphase systems are very important for numerous applications, such as extraction and processing of minerals, chemical technology, metallurgy, construction materials industry [1], building [2], and medicine for treatment of some neural and muscular diseases [3]. The use of vibrations has literally revolutionized many industries, providing large technical and economic effects.

Investigations in the field of vibrational hydrodynamics are also of great importance for theoretical reasons. The fundamental questions of the vibration effect on liquids, gases and multiphase media are described in the books and papers by many authors [4-7].

Bodies suspended in a liquid and subject to vibrations can exhibit unusual behavior. According to the Archimedes' principle, it is well known that a body immersed in a liquid at rest sinks if its density is greater than the density of the liquid, and floats upward in the opposite case. However, under the action of vibrations, the behavior of solids can be changed. Chelomei [8] observed that dense balls immersed in a vertically vibrating cylinder filled with a liquid can rise up and stabilize above a layer of less dense liquid.

The behavior of a body immersed in a liquid contained in a recipient submitted to high-frequency oscillations was analyzed theoretically by [9]. The averaged equations were derived and it was shown that the vibrations change the buoyancy conditions of the solid due to arising of average vibrational force of attraction to the nearest wall. The existence of average attraction force was confirmed experimentally in [10].

The theoretical interaction of two particles in an oscillating liquid was further investigated by [11]. It was shown that an average attractive force appears between particles when the vibrations are perpendicular to the line connecting the particle centers, and an average repulsive force for the vibrations propagating directly along this line.

IOP Conf. Series: Journal of Physics: Conf. Series **879** (2017) 012025 doi:10.1088/1742-6596/879/1/012025

The interactions of a small number of spherical particles immersed in an oscillating fluid, were studied by Wunenburger et al [12]. They conducted a series of experiments in which small metallic spherical particles were placed in a shallow cavity submitted to horizontal sinusoidal vibrations. The observations have shown that the spherical particles form equidistant chains oriented perpendicular to the direction of propagation of the vibrations.

Voth [13] observed clusters and hexagonal lattices formed by spherical particles in a vertically oscillating cavity filled with a mixture of water and glycerol. Thomas and Gollub [14] studied the fluctuations of these clusters and concluded that the forces arising between the particles are not just the forces of pair interaction. To explain these interactions, various mechanisms have been proposed, including the effect of stationary flow.

The results of experimental and numerical study of the interaction between two metallic spherical particles placed in a liquid-filled container performing horizontal oscillations are presented in [15]. It is found that there is an equilibrium distance between the particles at which a transition occurs from the attractive force at large distances between the surfaces of the spheres, to the repulsive force at small distances.

The interaction of two cylindrical solids with parallel axes in a pulsating flow of viscous fluid directed perpendicular to the plane passing through the axes of the cylinders was investigated in the absence of gravity [16]. It was found that at large distances, the interaction forces tend to bring the solids closer when the viscosity effects are small. With an increase in the relative role of viscosity, i.e. when the distance between the solids becomes smaller or the frequency decreases, the decelerating effect of the friction decreases the velocity drop and reduces the attraction force. At distances less than a certain critical value, the decelerating effect of the viscosity becomes so strong that the interaction force changes of sign: the attraction effect switches into a *repulsive effect.* This critical distance is of the order of the Stokes length; i.e. it increases with increasing viscosity and/or with the decrease in frequency. An analytical expression for the force acting on the rigid particle in a non-uniform pulsational flow was obtained in [17].

The present paper investigates experimentally the behavior of a system of rigid spherical particles in a container filled with a viscous incompressible fluid and subject to translational harmonic vibrations of high frequency and small amplitude.

2. Experimental settings and investigation method

A multiphase system constituted by rigid particles suspended in a viscous liquid is considered. The container with the liquid is subject to translational vibrations of linear polarization. It is known that under the action of a vibrational force, two spherical particles immersed in an inviscid or low-viscous liquid are attracted if the vibrations are perpendicular to the line connecting the centers of the two particles, and repelled if the vibrations are parallel to the line connecting the particle centers [1]. As the viscosity increases, the sign of the interaction force between the two particles changes to the opposite. In the case of large number of particles, the particles form specific patterns under the action of vibrational forces [2].

Experiments were carried out using an electrodynamic vibrator V650 based on a LDS vibrator. The vibrations were controlled manually using a vibrometer and a generator. Steel balls of the diameter $d = 4.5$ *mm* were placed in a rectangular cavity of size $110 * 200 * 15$ *mm* made of Plexiglas. A digital camera fixed on a tripod above the experimental setup schematically presented in Figure 1, allows tracing the dynamics of the patterns in the cavity.

The vibration frequency in the cavity was set by the generator and the vibration amplitude was varied with an amplifier. The vibration amplitude was changed by steps of 10% of the maximum power of the amplifier, which was 100%. When the required frequency and amplitude of the signal were set, the vibrator was turned on simultaneously to the ventilation unit which is necessary for cooling the vibrator. Under the action of vibrations, the spherical particles randomly located in the cavity at the initial time started to form linear clusters, and after some time when the system reached a stationary

IOP Conf. Series: Journal of Physics: Conf. Series **879** (2017) 012025 doi:10.1088/1742-6596/879/1/012025

mode, periodicity could be detected between clusters. At this time, the digital camera shot the cavity, and the amplitude of the vibrations was measured with the help of the vibrometer.

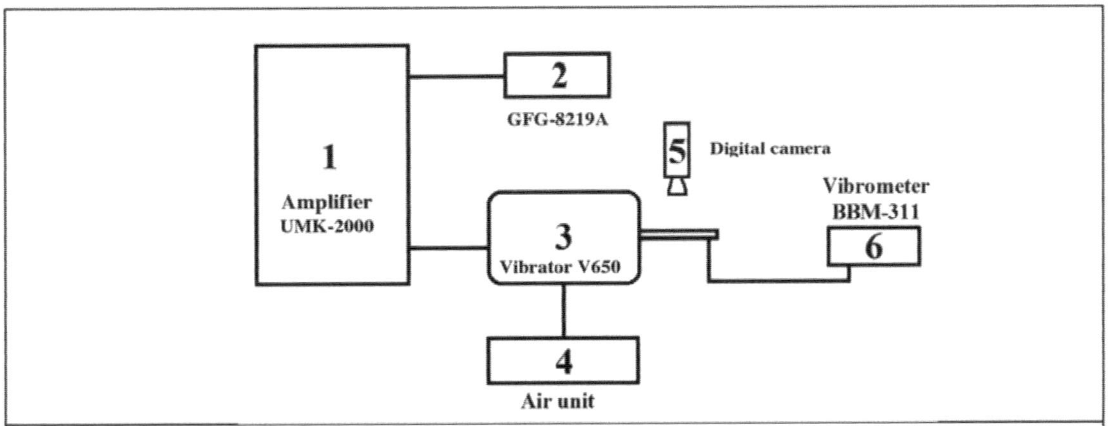

Figure 1. Schematic experimental setup. 1 - amplifier UMK-2000, 2 - signal generator of special form GFG-8219A, 3 - electrodynamic vibrator V650, 4 - air handling unit, 5 - digital camera, 6 - portable vibrometer BBM-311

3. Experimental results

The present study included two series of experiments. In the first series, the range of vibration parameters during which the periodic structures of particles were formed in the liquid, was investigated. It was found that the formation of structures consisting of linear chains occurs in the frequency range from 5 Hz to 27 Hz. At frequencies greater than 27 Hz, the particles were collected at one of the walls of the cavity. This quasi-equilibrium state of the systems is similar to quasi-equilibrium state of gas-liquid system in a rectangular container subjected to high frequency horizontal vibrations discovered in [18] where this phenomenon was called "orienting effect of vibrations": interface is oriented perpendicular to the vibration axis. As it was shown in [18] this quasi-equilibrium state corresponds to the minimum of average Lagrangian of the pulsational motion.

Figure 2 shows the experimental photos in a quasi-stationary states of the system formed by steel balls immersed in a water-glycerol solution in the container oscillating with amplitude $A = 0.94$ mm. The system parameters are: number of particles $N = 150$, particle radius $r_0 = 2.25$ mm, liquid density ρ = $1.2 \cdot 10^3$kg/m^3, particle density ρ_0 = $7.8 \cdot 10^3$kg/m^3, kinematic viscosity of the liquid $\nu = 8.7 \cdot 10^{-6}$ m^2/s, amplitude of container vibrations $A = 0.94$ mm. It is found that during the first rather short stage, the action of vibrations leads to the formation of clusters consisting of several particles. Further evolution of the system consists in the formation of linear chains oriented perpendicular to the direction of vibrations. Why-> develop this point in the discussion

In our opinion, this behavior could be explained on the basis of the analysis of the interaction force between the particles in viscous oscillating fluid at different distances between the particle centers and different mutual orientations of the vibration axis with respect to the line connecting the particle centers (see Introduction Section).

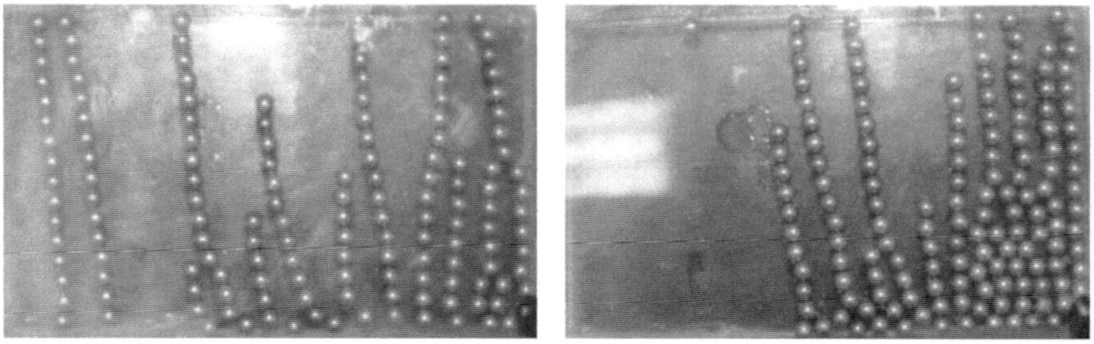

Figure 2. Experimental photos of structures in a quasi-stationary state of the system, formed by steel balls in a water-glicerol solution at the vibration frequencies 24.6 Hz (left photograph) and 28 Hz (right photograph).

The second series of experiments was carried out with 50% water-glycerol solution at a density of $\rho = 1.126 \cdot 10^3$ kg . m^{-3} and a kinematic viscosity $\nu = 3.2 \cdot 10^{-6}$m^2 . c^{-1}, in which 250 steel spherical particles of the radius $r_0 = 2.25$ mm and density $\rho_0 =$ $7.8 \cdot 10^3$ kg . m^{-3} were immersed. Figure 3 shows the dependences of spatial period of the formed quasi-stationary structures (the distance between the chains) T on the vibration amplitude for two fixed values of the vibration frequency obtained in these experiments. As one can see in Figure 3a., at fixed frequency of vibrations, the spatial period of the structures increases quasi linearly against the vibration amplitude. Comparison of the curves obtained for different frequencies shows that the spatial period of the structures decreases with increasing vibration frequency (see, Figure 3b).

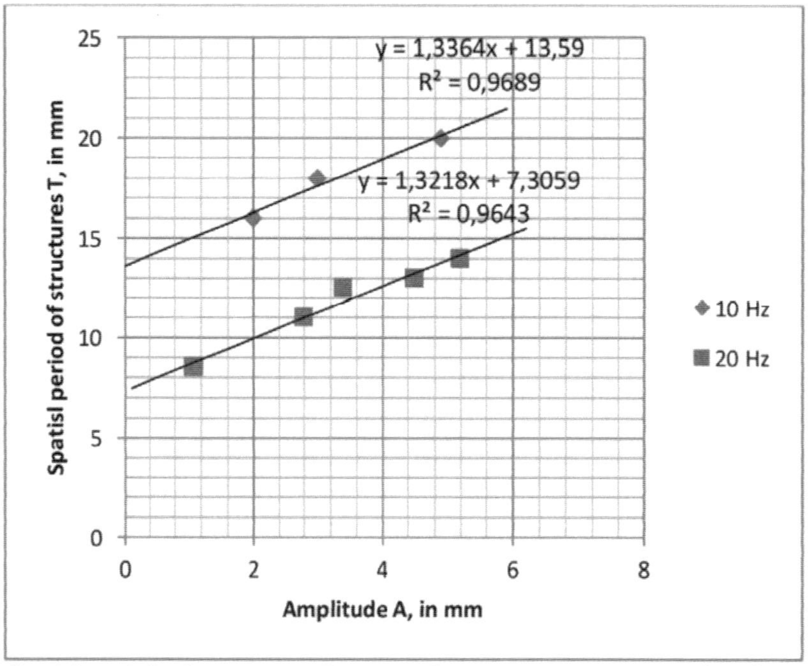

Figure 3a. Spatial period of structures (T *in mm*) between structures as a function of vibration amplitude (A *in mm*) at frequencies 10 Hz and 20 Hz

IOP Conf. Series: Journal of Physics: Conf. Series **879** (2017) 012025 doi:10.1088/1742-6596/879/1/012025

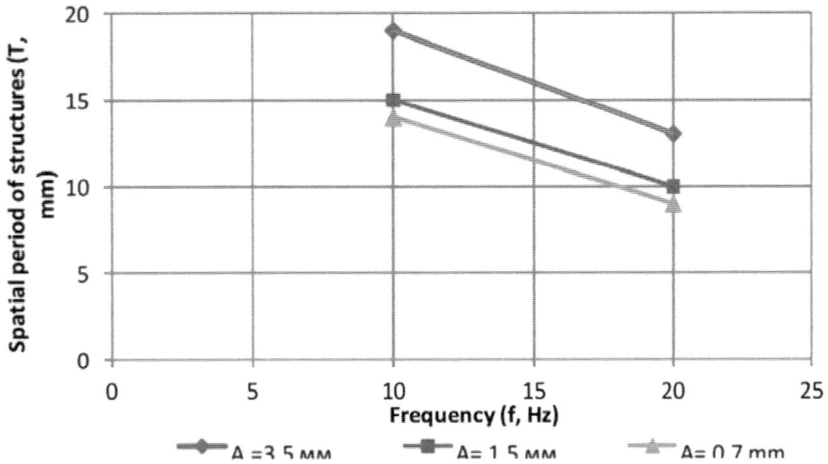

Figure 3b. Dependence of the spatial period of structures on the frequency of vibrations at three fixed values of the amplitude.

4. Conclusions

The dynamics of rigid particles in oscillating viscous liquid have been investigated experimentally. Two series of experiments were carried out. In the first series, the range of vibration parameters was studied, where periodic structures of particles are formed in the liquid. It was found that particles aligned to form linear structures oriented perpendicular to the direction of the vibrations are observed in the frequency range from 5 Hz to 27 Hz. At frequencies greater than 27 Hz, the particles are collected at one of the walls of the cavity.

The second series of experiments were devoted to the investigation of the dependence of the spatial period of the formed structures (the distance between the chains) on the vibration parameters. It is found that the spatial period of the structures increases with the vibration amplitude growth according to a close-to-linear law. Comparison of the curves obtained at different frequencies shows that the spatial period of the structures decreases with the increase in the vibration frequency.

5. Acknowledgments

The work was supported by the Russian Science Foundation (grant 14-21-00090).

6. References

[1] Blekhman II, 1988. *What can vibration do? About "vibrational mechanics" and vibration technology*. Nauka, Moscow, 208 p.
[2] *Vibrations in technology*: A handbook in 6 vol. 1978-1981 M: Mechanical Engineering.
[3] Kreimer A Y, 2002. *Vibration as a therapeutic factor*. Tomsk: Publishing house of Tomsk University, 259 p.
[4] Lamb G, 1947. *Hydrodynamics*. M L Gostekhizdat, 928 p.
[5] Betchelor J, 1973. *Introduction to the dynamics of liquids*. M: The World, 760 p.
[6] Loitsyansky L G, 1973. *Mechanics of fluid and gas*. M.: Science, 848 p.
[7] Landau L D, Lifshitz E M, 1986. *Theoretical Physics*: Textbook. In 10 vol. T. 6. Hydrodynamics. 3rd ed., Revised. M .: Science. Ch. Ed. Fiz-mat. Lit., 736 p.
[8] Chelomei V N, 1983. Paradoxes in mechanics caused by vibrations, *Dan SSSR*. **270**(1). 62-67.
[9] Lyubimov, D V, Lyubimova, T P, Cherepanov, A A, 1987. On a motion of solid body in a vibrating fluid. In: Zhukhovitsky, E.M. (ed.) Convective Flows, pp. 61-70.

[10] Hassan S., Lyubimova T P, Lyubimov D V, Kawaji M, 2006. Effects of vibrations on particle motion near a wall: Existence of attraction force. Int. J. Multiphase Flow, Vol.32, Issue 9, pp.1037-1054

[11] Lyubimov D V, Cherepanov A A, Lyubimova T P, 1992. The motion of a solid body in a liquid under the influence of a vibrational field. Rev. Proc. of the First Int. Symp. on Hydromechanics and Heat / Mass Transfer in Microgravity, Gordon and Breach, 247-251.

[12] Wunenburger R, Carrier V, and Garrabos Y, 2002. Periodic order induced by horizontal vibrations in a two-dimensional assembly of heavy beads in water. *Phys. Fluids* **14**(7), 2350-59.

[13] Voth G A, Bigger B, Buckley M R, Losert W, Brenner M P, Stone H A, Gollub J P, 2002. Experimental Measurements of Stretching Fields in Fluid Mixing, Phys. Rev. Lett. **88**, 254501-4..

[14] Thomas C C, Gollub, J P, 2004. Structures and chaotic fluctuations of granular clusters in a vibrated fluid layer. *Phys. Rev. E* **70**(6), 061305.

[15] Klotsa D, Swift M R, Bowley R M, King P J, 2009. Chain formation in spheres. *Phys. Rev.* E.. **79**(2), 021302.

[16] Lyubimova T P, Lyubimov D V, Shardin M, 2011. Interaction of rigid particles in a pulsational flow. *Microgravity Science and Technology*, **23**(3), 305-309.

[17] Zaichkin E V, Lyubimov D V, 2001. The behavior of a body suspended in a liquid in the field of torsional vibrations. Vibration effects in hydrodynamics. *Permian*, **2**, 97-109.

[18] Lyubimov D V, Cherepanov A A, Lyubimova T P, Roux B. 1997. Orienting effect of vibrations on the interphase. C.R.A.S., t.325, Serie II b, 391-396

IOP Conf. Series: Journal of Physics: Conf. Series **879** (2017) 012026 doi:10.1088/1742-6596/879/1/012026

Experimental study of formation and dynamics of cavitation bubbles and acoustic flows in NaCl, KCl water solutions

K A Rybkin[1], Yu K Bratukhin[1], T P Lyubimova[1,2], O Fatallov[1,2], L O Filippov[3,4]

[1]Perm State University, Perm, Russia
[2]Institute of Continuous Media Mechanics, Perm, Russia
[3]Universite de Lorraine, Nancy, France
[4]National University of science and technology, Moscow, Russia

lubimova@psu.ru

Abstract. The acoustic flows and the phenomena associated with them arising under the action of ultrasound of different power on distilled water and aqueous solutions of a mixture of NaCl and KCl salts of various concentrations are studied experimentally. It is found that in the distilled water, under the action of ultrasound, the appearance of inertial and non-inertial cavitation bubbles takes place, then the formation of stable clusters, the distance between which depends on the power of the ultrasound source is observed. Experiments show that an increase in the mass concentration of salts in water leads to the decrease in the average diameter of the arising inertial cavitation bubbles and to the gradual decrease in their number, up to an almost complete disappearance at nearly 13% of the concentration of the salt mixture in the water.

1. Introduction

Solution of the problems associated with the practical application of ultrasound is impossible without knowledge of the acoustic field characteristics, i.e. the spatial distribution of sound pressure or the intensity of the sound wave for the case under consideration, and also without the analysis of the cavitation processes occurring in liquids under the action of ultrasound. The use of ultrasonic sources is associated with certain problems. One of them is the complexity of creating a uniform acoustic field, since even the flat radiator used in the setup [1] gives strongly non-uniform field. Since the source itself can have physical irregularities on the surface, the field observed when it is switched on looks extremely inhomogeneous, which significantly complicates the experimental investigation and interpretation of the results obtained.

Separate attention should be paid to the effect of ultrasonic radiation on solid inclusions in the liquid [2], since the distribution of cavitation bubbles on a solid sample is still insufficiently studied. It is known, for example, that the influence of ultrasound on liquid-solid systems is largely related to the chemical effects of ultrasound, which are mainly associated with the collapse of cavitation bubbles [3]. Each collapse of the bubble potentially leads to a local increase in pressure up to 500 atm and a temperature of about 5000 K, and also causes shock waves, microcurrents and surface erosion of solid inclusions [4]. These and other effects, occurring when ultrasound is applied to liquid media, attracted attention of specialists in the field of ore enrichment. As a result of the studies, some positive effects of ultrasonic on ore flotation were revealed [5].

The possibilities of increasing the efficiency of flotation, using ultrasound, on one hand, and use as a mother liquor of seawater on the other hand, are of interest at present due to the acute shortage of fresh water in certain regions [6-9]. When using solutions of KCl and NaCl salts as test liquids, one can expect a significant change in both the behavior of the whole system under study, [10,11] and the formation of air bubbles during flotation [12].

This work is devoted to an experimental study of the formation and dynamics of cavitation bubbles produced by an ultrasonic radiator in saturated solutions of KCl and NaCl. Works in this direction are not known to the authors of this article. The study is motivated by the importance of studying the effect of ultrasound on the process of flotation of potassium chloride in saturated brines of a mixture of potassium and sodium salts.

2. Experimental setup and experimental procedure

The experiment was carried out in a cell 110 * 116 * 160 mm^3 in the form of a parallelepiped (Fig. 1). The cell was made of plexiglass 3 mm thick. The central part of the cell was penetrated from the side by a laser sheet created by a cylindrical lens and a green DPSS laser KLM-532. To record acoustic currents, a high-speed camera was used. The shooting speed varied from 10 to 100 frames per second. As test liquids, saturated solutions of a mixture of chemically pure KCl and NaCl salts were used, in a weight ratio of 6/4. For comparison, experiments were performed with degassed distilled water.

As a source of ultrasound, a metal disk with a diameter of 88 mm was used, placed so that the center of the radiator coincided with the center of the bottom of the cell. He connected to an ultrasonic oscillator with a frequency of 40 kHz and a maximum power of 100 watts. It was possible to vary the power (60%, 80% and 100% of the maximum).

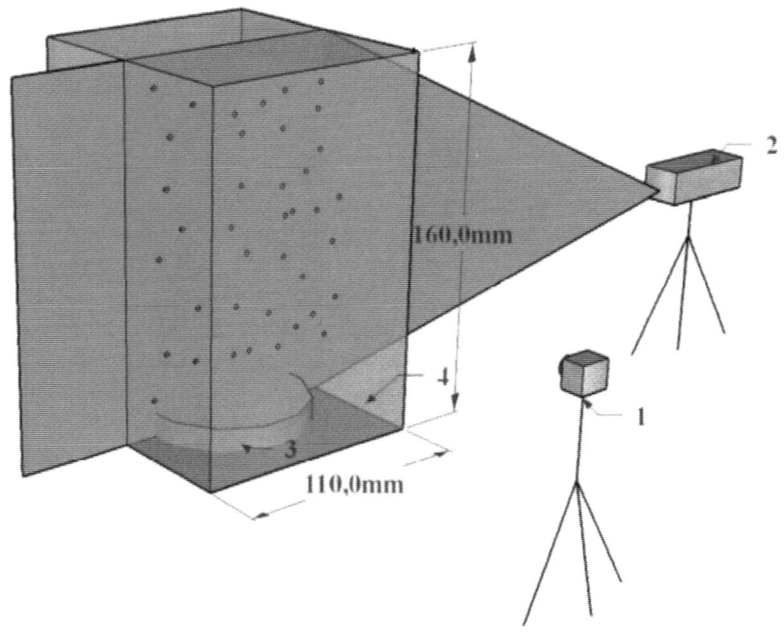

Figure 1. Experimental setup: 1) high-speed camera, 2) laser, 3) ultrasound source, 4) plexiglass cuvette

It is known [12] that the ultrasonic field of a plane acoustic radiator is concentrated in a cylindrical volume of diameter D and height Z_0 (Fig. 2). Starting from the distance $Z_0 = D / (\Delta\ \lambda)$, the field widens conically. The interval from the emitter to Z_0 is called the near zone or the Fresnel zone. The region where $Z > Z_0$ is called the far zone, or the Fraunhofer zone. In this zone, the pressure amplitude decreases proportionally to the distance from the radiator.

For experiments with degassed distilled water at a frequency of *40 kHz*, the speed of sound propagation is $c = 1500\ m / s$, wavelength $\lambda = 37.5\ mm$. The width of the Fresnel zone under these conditions *is* $Z_0 - 2.34\ cm$, the Fraunhofer zone begins at a distance of about 2 cm from the source, after which the ultrasonic field begins to expand conically. The sine of the angle α between the direction of propagation of the ultrasonic wave and the beam generator is related to the geometric parameters of the setup by the formula $\sin\alpha = A^{\lambda} / D$ where, for a circular plate $A = 1.22$, the calculations in this case are given $\sin\alpha = 0.18$, whence $\alpha = 10$.

For experiments with saturated NaCl and KCl salt solutions at a frequency of 40 kHz, the acoustic wave velocity is $c \approx 1.87 * 10^3\ m / s$, the wavelength $\lambda = 46.7\ mm$, and the Fresnel zone is limited by the distance from the surface of the ultrasonic radiator $Z_0 = 1.46\ cm$. The angle of deflection of the propagation of an ultrasonic wave from the beam generatrix in salt solutions is $\alpha = 15$.

In the near zone, for $Z < Z0$, the intensity of ultrasonic oscillations can have several maxima. In the far zone, for $Z > Z0$, the wave intensity has one maximum and decreases with distance from the beam axis.

The cavitation bubbles that arise in the liquid also affect the propagation of an acoustic wave [13].

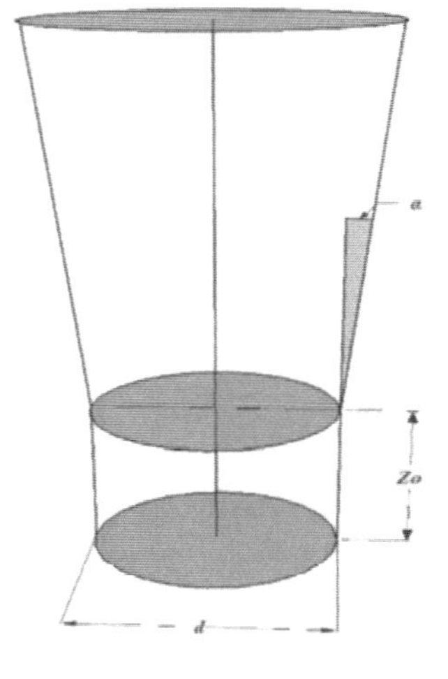

Figure 2. Idealized field of a plane acoustic radiator

As measurements have shown, the temperature of the liquid changes under the influence of ultrasonic vibrations (see Fig. 3, which shows the experimental temperature change curve obtained with a copper-constantan thermocouple). This leads to the fact that the process of bubble induction and the type of acoustic flow vary with time.

Quite sharp temperature fluctuations against the background of its growth under the action of the general heating of the liquid, observed in the experiment after switching on the ultrasonic radiator, are related, apparently, to the collapse of cavitation bubbles that occurs near the junction of the thermocouple [3].

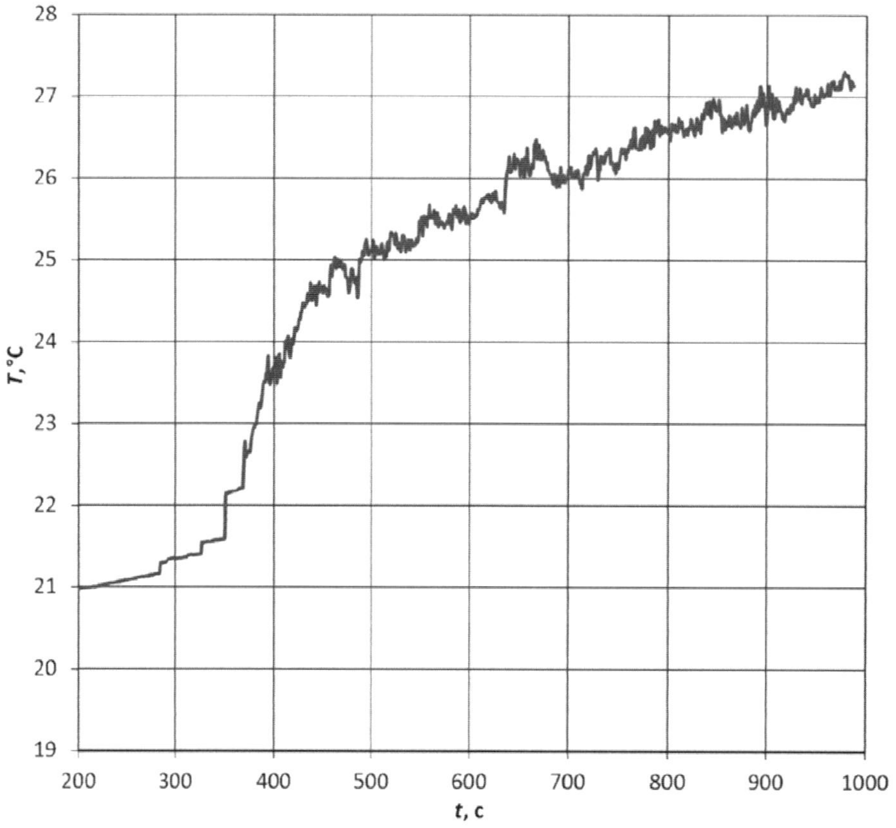

Figure 3. Change in the temperature of degassed distilled water under the action of ultrasound at a frequency of 40 KHz at 100% of the emitter power, at a point 5 cm from the bottom

3. Results of experiments

3.1. Experiments with distilled water

Experiments with distilled water, as in [14-15], revealed the presence of two types of ultrasonic cavitation. The first of these is inertial cavitation, which is characterized by rapid growth of the bubble in the negative pressure phase. At the onset of the half-cycle of compression, the inertial bubbles collapse sharply, and local heating and hydrodynamic perturbations in the form of microscopic waves, cumulative jets and microflows of liquid can occur. From inertial bubbles, comet-shaped streamers are formed (see Fig. 4a).

a) 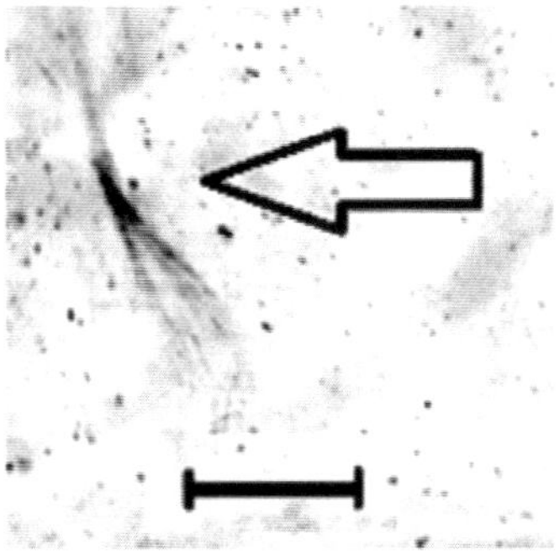

b) 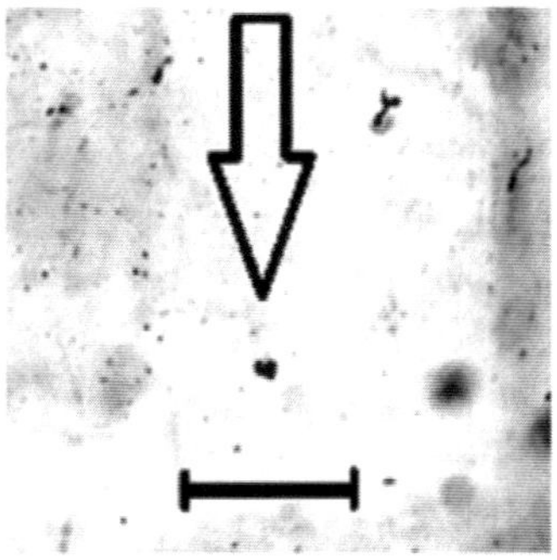

Figure 4. a) Streamer from inertial cavitation bubbles; b) non-inertial bubble in distilled water. Photos were taken at 60% of the power of the ultrasonic radiator

The second type of ultrasonic cavitation is a non-inertial (or stationary) cavitation, in which bubbles grow due to rectification diffusion or are compressed to a resonance size as a result of gas dissolution, and then oscillate for a long time (Figure 4b).

Figure 4b shows that non-inertial bubbles have a much larger size than inertial bubbles. However, the difference is not only in size. The lifetime of inertial bubbles is less than one tenth of a second, i.e. of the order of several periods of oscillation of the acoustic wave (this time was determined with the help of a high-speed camera, with a sampling rate of 100 frames per second). For non-inertial bubbles, the lifetime is much longer than the period of oscillation of the acoustic wave. According to the results of observations, for ultrasound frequency of 40 kHz, the lifetime of non-inertial bubbles, in some experiments, was more than three minutes. Being in the zone of low pressure, the bubble performs translational oscillations with an unchanged amplitude in one plane. For fixed parameters, non-inertial bubbles arise at the same points in space, called the antinodes of a standing wave. Then, translational oscillations of the bubble and deformation of its surface occur, while the bubble is able to capture another non-inertial bubble popping up next to it (Fig. 5a-d). Such absorption can occur many times, until the diameter of the bubble becomes critical, and the Archimedes force does not exceed the value of the Bjerkness force [16-17], due to the pressure gradient that "draws" the bubble into the antinode of the standing wave and, thereby, holds it in some area of space.

Gas bubbles in a liquid are moved from the equilibrium state by changing the external pressure and perform periodic radial oscillations. Two bubble pulsating in phase are attracted to each other by the Bjerknes force, but near the contact, attraction is counteracted by repulsion, which arises due to nonlinear interaction of the Bjerknes force and the viscous force. For oscillations of a single bubble as a result of nonlinear interaction of the radial and deformational modes at their resonance, the energy of the radial oscillations transfers to the energy of deformational oscillations with a large amplitude increase. These effects provide a qualitative explanation of experiments on the coalescence and fragmentation of pulsating bubbles in a liquid.

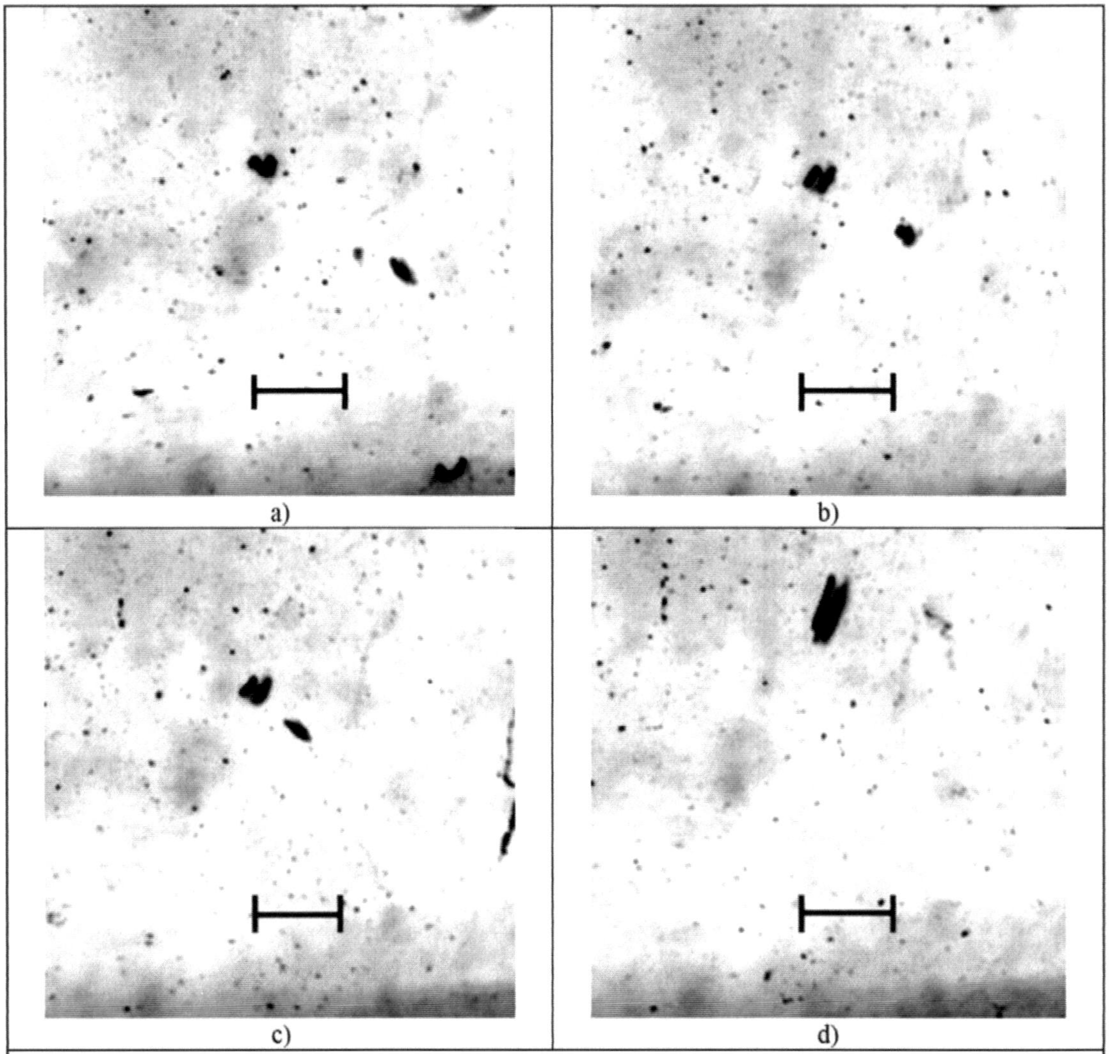

Figure 5. Dynamics and fusion of two non-inertial bubbles and the beginning of the emergence of a bubble of critical diameter. The length of the risks is 1 cm

In the experiments, vertical cluster formation from inertial bubbles was also observed; moreover, the distance between adjacent bubbles in the cluster, at a given power, remained unchanged in the future; The system entered the stationary mode. In Fig. 6 a-c, three photographs are shown resulting from addition of 50 frames with a shooting frequency of 100 frames per second for different powers of the ultrasonic radiator at a fixed frequency of US 40 kHz for a) 60% b) 80% c) 100% of the maximum power. The areas of dark spots are the areas of drift of inertial cavitation bubbles drawn into the antinodes of a standing wave. With increasing emitter power P, the distance between cavitation bubbles l in a vertical cluster increases linearly (Fig. 7).

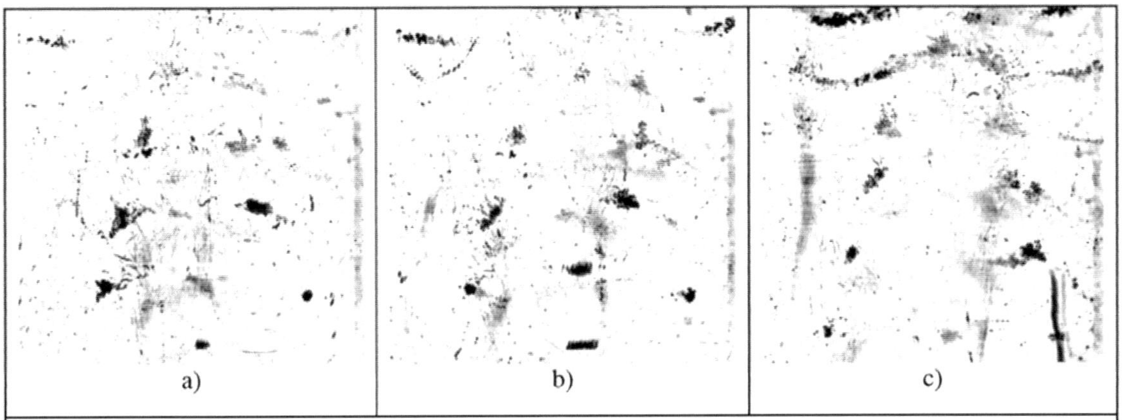

Figure 6. Distilled water, dark spots drift area of inertial cavitation bubbles

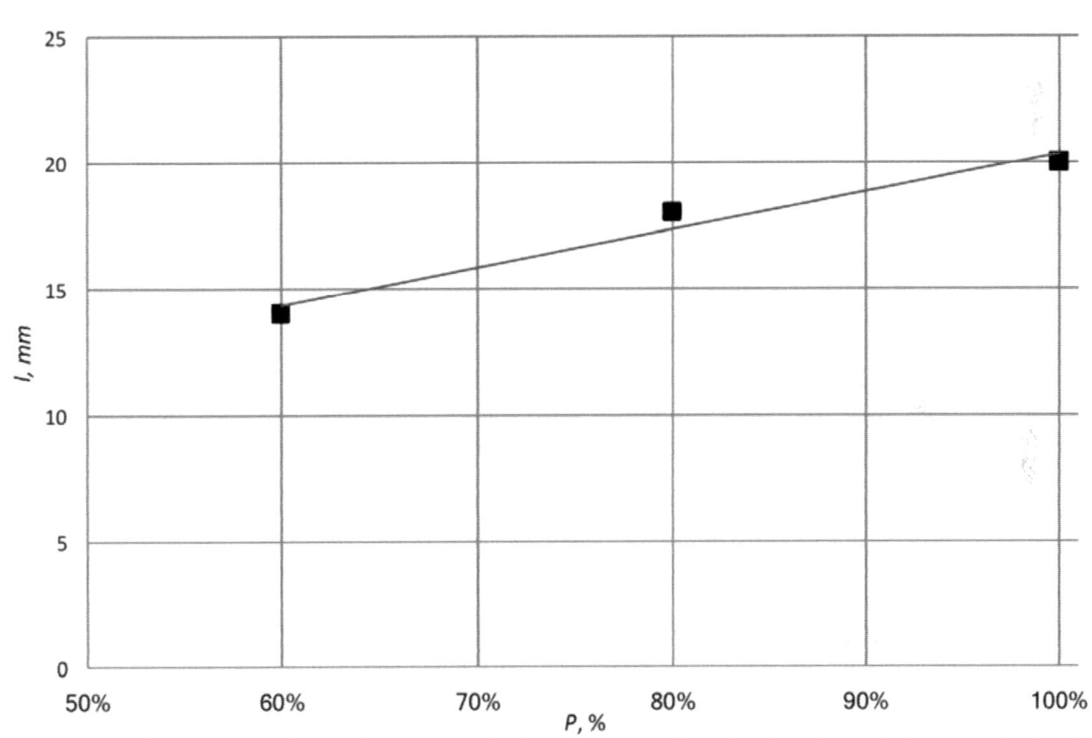

Figure 7. Dependence of the distance between non-inertial bubbles in a vertical cluster on the power of the radiator

3.2. Experiments with aqueous solutions of NaCl and KCl salts

Experiments with saturated solutions of NaCl and KCl salts showed that in this case, at any given power of ultrasound, only inertial cavitation bubbles arise. In Fig. 8 shows the result of combining 300 frames made at a rate of 10 frames per second for a saturated solution of NaCl and KCl salts in a ratio of 4: 6, respectively. As noted above, the wavelength in a saturated salt solution at an ultrasound frequency of 40 KHz is $\lambda = 46.7\ mm$. The distance between the singular points that are local sources of bubble streamers located above the ultrasonic radiation source in the experiment was $\lambda / 2$ and did not depend on the power of the ultrasonic radiation.

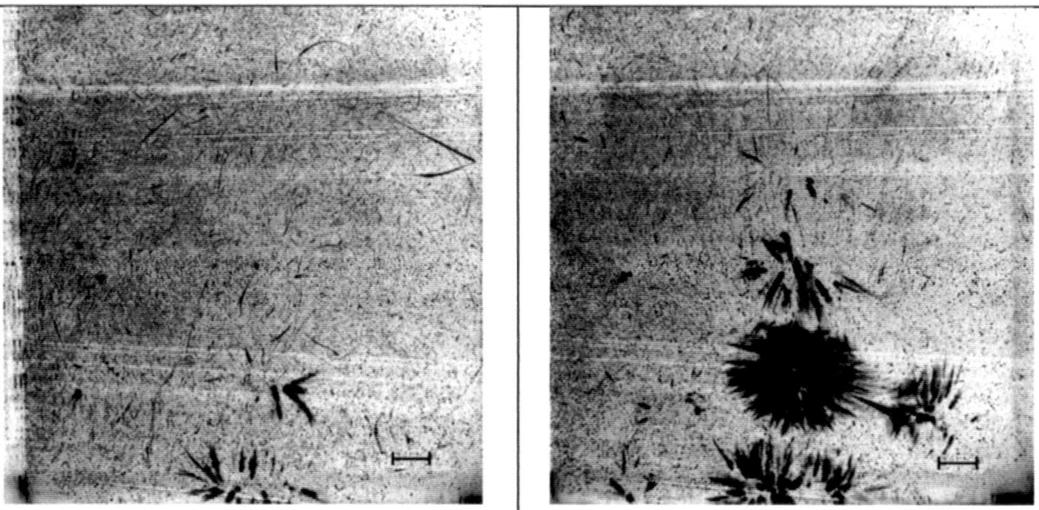

a) b)

Figure 8. The emergence of inertial cavitation bubbles in a saturated solution of NaCl and KCl salts. a) power US 60%, frequency 40 KHz, b) power US 100%, frequency 40 KHz

From the photographs obtained during the experiments it can be concluded that as the intensity of the streamers' appearance increases, the type of flow that appears in the cell changes, apparently streamers that break the uniformity of the antinodes, thereby knocking out the levitating bubble from the stable zone to the high-pressure zone, after which the bubble begins to float.

A series of experiments was also conducted to study the formation and dynamics of cavitation bubbles in solutions with different concentrations of NaCl and KCl. These experiments have shown that at low salt concentrations (below 2%), there are simultaneously two types of cavitation, with the threshold of inertial cavitation higher than in distilled water. Below this threshold, during the whole series of experiments, comet-shaped streamers do not appear. With the increase in the concentration from 2% to 13%, the frequency of the appearance of streamers increases, and the average diameter of the levitating bubbles decreases. At a salt concentration of 13% virtually all levitating bubbles disappear.

In [7], devoted to the flotation in sea water, the size of the air bubbles injected by the impeller in various salts including the solutions of NaCl and KCl was measured. It was found that the bubble diameter decreases with the increase of salt concentration and the dependence of the bubble diameter on the concentration is qualitatively similar for different salts. The origin of the bubbles in [7] is different from our case: in [7] this is the injection of the air into the aqueous solution of salt under pressure, and in our case cavitation bubbles arise under the action of ultrasound, however in both cases the growth of the salt concentration leads to the decrease in the bubble size, which can have a beneficial effect on the flotation process.

Figure 9 shows the dependences of the normalized bubble diameter $D = d / d_0$ (curve 1) and of the normalized number of bubbles $U = N / N_0$ (curve 2) obtained in the experiments, on the mass concentration of the salt in the solution, n, %, where d is the bubble diameter in the salt solution, d_0 is the diameter of the freely levitating non-inertial cavitation bubble in the distilled water, N is the number of levitating bubbles in the salt solution, and N_0 is the number of non-inertial bubbles of the diameter greater than $d = 1\ mm$ in distilled water.

Calculation of the relative diameter and number of cavitation bubbles was made on the basis of processing of photographs obtained by irradiating the cuvette with a vertical laser sheet Fig. 1. The photos obtained in this way were downloaded into the software package Wolfram Mathematica where by means of the built-in functions their sequential processing was carried out.

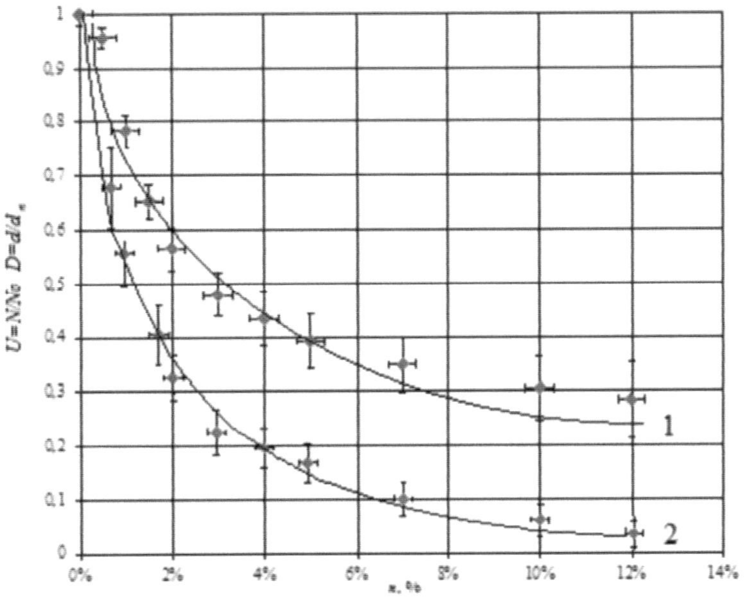

Figure 9. Dependences of the normalized diameter of the bubble (upper curve) and of the normalized number of inertial bubbles (lower curve) on the salt concentration

4. Conclusions

The appearance and dynamics of cavitation bubbles in water and salt solutions have been studied. It was found that a significant change in the physicochemical properties of water upon its saturation with salts of NaCl and KCl leads to a qualitative change in the regime of induction of cavitation bubbles. In distilled water, both inertial and non-inertial cavitation bubbles are observed, and in saturated salt solutions only non-inertial cavitation bubbles whose zones of occurrence are strictly fixed, and their location does not depend on the power of the ultrasonic source.

There is also the appearance of clusters consisting of cavitation bubbles located in the antinodes of a standing wave in which, under certain conditions, stable levitating vesicles arise.

The dependence of the distance between drift regions in a vertical cluster of non-inertial bubbles in distilled water on the emitter power at a fixed emitter frequency of *40 kHz* is determined.

Acknowledgements

The work was supported by the Russian Foundation for Basic Research (grant N 17-41-590974).

References

[1] Tran B C, Seo J, Hall T L, Fowlkes J B & Cain C A 2003. Microbubble-enhanced cavitation for noninvasive ultrasound surgery. *Ultrason. Ferroelectr. Freq. Control,* **50**(10), 1296–1304.

[2] Kauer M et al 2017. Visualization and optimization of cavitation activity at a solid surface in high frequency ultrasound fields. *Ultrasonics Sonochemistry,* **34,** 474–483.
[3] Leighton T 2012. The Acoustic Bubble. *Academic Press.*
[4] Suslick K S 1990. Applications of Ultrasound to Materials Chemistry. *Sonochemistry Science,* 247, 1439–1445.
[5] Filippov L O, Matinin A S, Samiguin V D & Filippova I V 2013. Effect of ultrasound on flotation kinetics in the reactor-separator. *Journal of Physics: Conference Series,* 416.
[6] Castro S, Miranda C, Toledo P & Laskowski J 2013. Effect of frothers on bubble coalescence and foaming in electrolyte solutions and seawater. *International Journal of Mineral Processing,* **124**, 8–14.
[7] Jeldres I, Forbes L & Cisternas L 2016. Effect of seawater on sulfide ore flotation: A review. *Mineral Processing and Extractive Metallurgy,* **37**(6), 369–384.
[8] Wei T, Peng Y & Vink S 2016. The joint action of saline water and flotation reagents in stabilizing froth in coal flotation, *International Journal of Mineral Processing,* **148,** 15-22.
[9] Jeldres R, Arancibia-Bravo M, Reyes A et al 2017. The impact of seawater with calcium and magnesium removal for the flotation of copper-molybdenum sulphide ores. *Minerals Engineering* **109.**
[10] Brennen E 1977. Cavitation and bubble dynamics. *Annual Review of Fluid Mechanics,* **9.**
[11] Lauterborn W and Ohl C 1998. The Peculiar Dynamics of Cavitation Bubbles. *A lied Scientific Research,* **58**(1–4), 63–76.
[12] Sprawls P 1989. *Ultrasound Production and Interactions,* Web site.
[13] Margulis M A 2000. Sonoluminescence. *Physics-Uspekhi,* **43** (3), p 259.
[14] Zijlstra A, Fernandez Rivas D, Gardeniers H, Versluis M, & Lohse D 2015. Enhancing acoustic cavitation using artificial crevice bubbles. *Ultrasonics,* **56**, 512–523.
[15] Ashokkumar M 2011. The characterization of acoustic cavitation bubbles. *Ultrasonics Sonochemistry,* **18**(4), 864–872.
[16] Matula T, Cordry S, Roy R. & Crum L 1997. Bjerknes force and bubble levitation under single-bubble sonoluminescence conditions. *J. Acoust. Soc. Am,* **102**, 1522-1527.
[17] Matula T 2003. Bubble levitation and translation under single bubble sonoluminescence conditions. *The Journal of the Acoustical Society of America,* **114(**2), 775-781.